Contemporary Topics in

POLYMER SCIENCE

Volume 5

Contemporary Topics in
POLYMER SCIENCE

Volume 5

Edited by

E. J. Vandenberg

Hercules Incorporated
Wilmington, Delaware
Currently with Arizona State University
Tempe, Arizona

PLENUM PRESS · NEW YORK AND LONDON

The Library of Congress has cataloged this work as follows:

Contemporary topics in polymer science. v. 1–
New York, Plenum Press, c1977–

    v. ill. 24 cm.

    Vols. before v. 2 cataloged separately in L.C.
    Vols. 1–   comprise the proceedings of the American Chemical Society, Division of Polymer Chemistry's meetings.
    Key title: Contemporary topics in polymer science, ISSN 0160-6727.

    1. Polymers and polymerization—Collected works.
QD380.C63   547′.84′08   79-641014
                                                       MARC-S

Library of Congress   78[8311]

ISBN-13: 978-1-4612-9706-2   e-ISBN-13: 978-1-4613-2759-2
DOI: 10.1007/978-1-4613-2759-2

Proceedings of the Eleventh Biennial Polymer Symposium of the Division of Polymer Chemistry on High Performance Polymers, held November 20–24, 1982, at the Cerromar Beach Hotel, Dorado Beach, Puerto Rico

© 1984 Plenum Press, New York

Softcover reprint of the hardcover 1st edition 1984

A Division of Plenum Publishing Corporation
233 Spring Street, New York, N.Y. 10013

PREFACE

The Eleventh Biennial Polymer Symposium of the
Division of Polymer Chemistry, Incorporated of the
American Chemical Society was held November 20-24, 1982
at the Cerromar Beach Hotel, Dorado Beach, Puerto
Rico. The theme of the meeting was "High Performance
Polymers."

On this occasion Professor Herman F. Mark received
the Fourth Division of Polymer Chemistry Award for his
outstanding achievements and his unique missionary role
in the development of Polymer Chemistry. Professor
Mark was the premier organizer of many important firsts
in polymer chemistry, to name just a few - the first
polymer journal, the pre-eminent Journal of Polymer
Science; the first U.S. academic center of Polymer
Science at the Brooklyn Polytechnic Institute which led
to a long procession of eminent polymer scientists; the
"High Polymer" Monograph series. In the Division of
Polymer Chemistry, he was the first secretary-treasurer
and chairman in 1955.. A detailed biography follows
along with Professor Mark's reminiscences on the Early
Days of Polymer Science, the topic of his Award lecture.

It was indeed a pleasure and ultimate honor to be
the Chairman and organizer of the technical program of
this Symposium. The fourteen invited lectures are
given herein. I have tried and believe succeeded in
presenting important current research by leading
workers on High Performance Polymers. To recognize the
importance of nature's high performance polymers, I
have presented these first, including the fundamental
role of carbohydrate polymers (Marchessault), and the
rapidly developing field of controlling natures methods
via gene synthesis (Caruthers) towards a new era of
bio-synthetic methods which have as yet unpredictable
effects on life and polymer science. A paper on
Polymeric Monolayers (Ringsdorf and Dorn) strives to

bridge the gap between nature's methods and more
conventional polymer synthetic methods. Three papers
highlight our recent achievement of ultra high strength
and ultra high modulus fibers from organic polymers, as
first recognized as theoretically possible by Professor
Mark many years ago. The areas covered are polymeric
carbon (Fitzer), polyolefins (Ward) and polyesters
(Jackson). Two papers highlight high temperature
capability combined with high strength as shown for
some interesting new aromatic polymers (Stille) and,
for ultimate high temperature performance, one of the
recent all inorganic polymers (Dhingra). Another paper
(Relles) covers a recent significant development on
polyetherimides which provide premium properties with a
new material which can be fabricated by conventional
methods. The last five papers present important
developments relating to polymer applications – two on
the important developments on polymers with metal-like
conductivity (Wegner and Baughman), one on electronic
applications for high temperature polymers (Economy);
one on a novel approach to high temperature, ionomeric
membranes (Eisenberg, Harris, Yeager et al) and finally
a paper on the difficult problems relating to polymer
flammability (Pearce).

The program also included twenty poster
contributions which will be published elsewhere by the
individual authors. These papers were very important
in increasing the scope and interactions at the meeting
and I am very appreciative of the effort put forth by
these contributors.

Many contributed to the success of this Symposium.
I cite especially: all the invited lecturers,
Professor J. C. Salamone and S. C. Israel for the early
support and site selection efforts, Ms A. B. Salamone
for publicity, Dr. B. M. Culbertson for handling the
poster sessions, Dr. G. A. Stahl for the many
arrangement details on site, Ms. Newman for a myriad of
travel arrangements, and organizational details for the
Symposium, the Executive Committee of the Polymer
Division for their understanding support and help;
Hercules Incorporated, and especially Ms. Janis
Phillips, for providing time, facilities etc for
organizing this symposium, and the session leaders

(Vogl, Bikales, Wooten, Culbertson, Pearce and Hwa) for
their excellent on site handling of the various
sessions and the excellent discussions periods.   I also
thank the Continental Packaging Company, Dow Chemical
Company and E. I. du Pont de Nemours and Company for
their financial support of the Award Reception in honor
of Professor Mark.   The continued support of the Dow
Chemical Company for the Division of Polymer Chemistry
Award is also gratefully recognized.   The editor also
acknowledges partial financial support for the foreign
speakers by the Donors of The Petroleum Research Fund
administered by the American Chemical Society.

E. J. Vandenberg
Wilmington, Delaware

CONTENTS

Professor Herman F. Mark

Winner of the 1982

Division of Polymer Chemistry Award

American Chemical Society

DIVISION OF POLYMER CHEMISTRY AWARD TO HERMAN F. MARK

INTRODUCTION OF HERMAN F. MARK

Eli M. Pearce

Polytechnic Institute of New York
333 Jay Street
Brooklyn, New York 11201

Members of the Polymer Division, Guests and friends of Herman Mark -

I am privileged to introduce Herman F. Mark this evening as the Polymer Division Award winner. Previous winners of this award were Dr. Paul J. Flory of Stanford, Dr. Carl S. Marvel of the University of Arizona and the late Dr. Maurice Huggins.

It would be impossible to comment in detail about such a full and rich career. If I were to do so, the occasion would be similar to that where the chairman of a meeting was upset about a speaker who had talked much beyond his time -- and was continuing to do so. He hurriedly rushed to the dais with the gavel and accidently hit a member sitting at the dais on the head. The individual who had been hit, as he was sliding underneath the table was heard to say "Hit me again, I can still hear him".

Instead, I would like to comment very briefly on a few aspects of Herman Mark's career: -- his world travels, the title of Geheimrat bestowed by his colleagues, and just a few of his many accomplishments.

1

In regard to his travels -- Professor Mark has
traveled many times around the world, and continues to
lecture, teach and consult everywhere. His travels are
legion -- and I will not dwell on this further.

You know much about his science and his visionary
ideas but probably much less about the man. He is
known affectionately by his colleagues as Geheimrat --
a funny title chosen by his colleagues precisely
because it suggests the sort of man he isn't --
Geheimrat being a Prussian title formerly bestowed by
German Kaisers on professors who were prominent in some
field of research, and to Europeans a standard
caricature or stereotype of a stuffy antique, with a
bristling mustache, an overstuffed belly and somewhat
aloof and stiff with dignity.

Geheimrat is a man who has inspired all who have
known him to continue to achieve and advance.
Traditionally, this would start in graduate school by
setting ones future goals -- by the titles that he
would bestow on you -- the graduate student was always
called "doctor", the postdoctoral fellow as called
"professor", the professor was many times called
"chairman", and so forth.

Herman F. Mark is acknowledged throughout the
worldwide scientific community as "the Father of
Polymer Science" having excelled not only as a
scientist, but as a communicator of science. His
intelligence, warmth, sensitivity and wit have enabled
him to be a magnificent teacher of others, as well as a
scientist without comparison in his discipline. As a
polymer scientist, he has accomplished the following:

- The first interpretation of the crystal structure of
  stretched rubber;
- The first X-ray analysis of silk and noting the
  possible existence of a helicoidal conformation of
  protein molecules.
- Conformations of protein molecules.

- The foundation for the chair structure of cellulose.
- Key contributions to the kinetic theory of rubber elasticity.
- The first application of chromatographic techniques on higher polymers.
- Key contributions to the general theory of the degradation of long chain molecules.
- First experimental observation of second-order transition phenomena.
- Theoretical and experimental basis for vinyl polymerization in solution, suspension and emulsion.
- First use of electron microscope to measure the molecular weight of polymers;

and many more contributions difficult to summarize. Attributions made during this symposium have emphasized contributions that he has made which foresaw future developments in polymer science.

In addition, Professor Mark has founded the Journal of Polymer Science, published and edited many volumes in the field, and has been professionally active. He is acknowledged throughout the scientific community as the Father of Polymer Science. He received the National Medal of Science in 1980, over 30 medals and awards from all over the world, and almost 20 honorary degrees. He has published over 500 original and review articles and about 20 books. Most recently, he received the Polymer Division Education Award in 1982. It is most appropriate that Professor Mark be honored with the Polymer Division Award, especially since his is a founder of the Division.

I introduce you to the Geheimrat.

THE EARLY DAYS OF POLYMER SCIENCE

Herman F. Mark

Polytechnic Institute of New York
333 Jay Street
Brooklyn, New York   11201

SYNOPSIS

A short review is given of the events which occurred between 1926 and 1947 and led ultimately to a consolidation of Staudinger's view concerning the existence of macromolecules as he had postulated them.

Three conferences may be taken as characteristic for the early development of polymer science: The meeting of the Deutsche Naturforschergesellschaft in Duesseldorf in 1926 where Staudinger successfully upheld his concept of the existence of macromolecules against powerful opposition, the Faraday Society Meeting in Cambridge in 1935 where Carothers presented his classification of addition and condensation polymers, and the First Polymer Conference of the International Union of Pure and Applied Chemistry in Liege in 1947 where polymer science established itself as an accepted and vigorous member of chemical disciplines.

INTRODUCTION

Since the beginning of his existence, man has
strongly relied on the use of natural organic polymers for food, clothing, and shelter. When he
ate meat, bread, fruit, or vegetables and drank
milk, he was feeding on proteins, starch, cellulose
and related polymeric materials; when he put on
clothing made of fur, leather, wool, flax, and cotton
he used the same natural polymers; and when he protected himself against wind and weather in tents
and huts, he constructed these primitive buildings
of wood, bamboo, leaves, leather, and fabrics,
which again all belong to the large family of organic polymers. In addition to the above-mentioned
types, there is rubber, many other gums and resins
and bark.

Even later, when higher levels of civilization
were reached, organic polymers were essential ne-
cessities in peace and war. All books in the famous
library of ancient Alexandria consisted either of
cellulose (paper) or protein (parchment), and
they consist of these materials in all libraries
of the world up to the present day. All transportation on land and sea throughout many centuries
operated on wooden chariots and ships which were
put in motion with the aid of ropes and sails
made entirely of such cellulosics as flax, hemp,
or cotton. The music of all string instruments
is produced by the vibrations of wooden, resin-
treated boards; and all famous paintings together
with many of the most valuable statues consist of
cellulose, lignin, and paints which are held together by polymerized terpenes on such materials
as paper, canvas, or wood. Bow and arrow are
cellulose, lignin, resin, and proteins; catapults
and siege towers were made of wood and moved with
ropes and - until about 100 years ago - all sea
battles were fought with wooden ships which were
maneuvered with the aid of cellulosic sails and
ropes.

While, in this way, natural organic polymers
literally dominated the existence and welfare of
all nations, little was known about their composition and nothing at all on their structure. In
each sector - food, clothing, transportation, com-

munication, housing, and art - highly sophisticated
craftsmanship developed which was sparked by human
intuition, creativity, zeal, and patience and led
to accomplishments which will ever deserve the highest admiration of generations to come.

But even when the chemistry of organic compounds
became a respectable scientific discipline in the
early decades of the last century, the all-impor-
tant helpers of mankind - proteins, cellulosics,
starch, and wood - were not in the mainstream of
organic chemical research.

Why?

Because somehow they did not seem attractive at
that time for a truly scientific study since they
did not respond to the then existing methods for
isolation, purification, and analysis. The experimental backbone of organic chemistry in those days
was dissolution, fractional precipitation, and
crystallization or distillation; it worked and still
works with all ordinary organic compounds such as
sugar, glycerol, fatty acids, alcohol, and gasoline
but fails with cellulose, starch, wool, and silk.
These materials cannot be crystallized from solution and cannot be distilled without decomposition.

This fundamental and embarrassing difference
between the natural organic materials and the ordinary organic chemicals warned the chemists of
the last century that there might be some essential
and basic difference between these two classes of
substances and that one would have to develop spe-.
cial, new, and improved experimental methods to
force the second class into the realm of truly
scientific studies.

The breakthrough came in the early decades of
this century, mainly through the adoption of physical methods, such as improved optical devices like
the ultramicroscope, and the application of the
ultracentrifuge, of new viscometers, diffusion cells
and, most of all, through the systematic use of x-
ray and electron diffraction to fibers, membranes,
and tissues. A decade of intense research on cellu-

lose, proteins, rubber, and starch wound up with
the following fundamental results:

1.   All investigated materials consist of <u>very
large molecules</u>. Whereas the molecular weights of
ordinary organic substances such as alcohol, soap,
gasoline, or sugar range from about 50 to 500, the
molecular weights of the natural organic building
materials range from 50,000 to several millions, a
fact which earned for them, through Staudinger, the
name "giant molecules" or "macromolecules."

2.   Most of them have the shape of <u>long flexible
chains</u> which are formed by the multifold repetition.
of a base unit. One often refers to this unit as
a "monomer" (monos is the Greek word for "one" and
meros is the Greek word for "part") and to the
macromolecule itself as "polymer) (polys is the
Greek word for "many").

3.   If a sample consisting of regularly built,
flexible chains is exposed to mechanical deformation, the individual macromolecules are <u>oriented</u>
and show a tendency to form thin, elongated bundles of high internal regularity which are usually
referred to as "crystalline domains" or, simply,
as crystallites. Depending upon the nature of the
material and the severity of the treatment, a different percentage of the material undergoes crystallization whereas the rest remains in the "amorphous" or "disordered" state so that any given sample - fiber, film, rod, or disk - consists of two
phases: amorphous and crystalline.

It was found that the crystalline domains contribute to strength, rigidity, high melting characteristics, and resistance against dissolution,
whereas the amorphous areas impart softness, elasticity, absorptivity, and permeability.

As soon as the study of natural polymers had
started to establish these ground rules, chemists
were strongly tempted to synthesize equivalent
systems from simple, available and inexpensive raw
materials. The years from 1920 to 1940 brought ever-
increasing successful efforts to:

1. Provide for more and cheaper basic building units by the synthesis of new monomers.

2. Work out efficient equations to describe quantitatively the mechanism of polymerization and polycondensation.

3. Establish quantitatively the molecular weight and molecular microstructure to arrive at polymer characterization.

4. Explore the influence of the structural details on the different ultimate properties - molecular engineering.

Learning originally from nature and following up on the established principles, scientists and engineers succeeded in producing a wide variety of polymeric materials which outdo their original native examples in many ways and in most cases, are much more accessible and less expensive.All this gave a tremendous lift to the important industries of man-made fibers, films, plastics, rubbers, coatings, and adhesives and made everybody's life richer, safer and more comfortable. Statistics show that in 1980 about 10 million tons of man-made fibers were produced and sold in the world with a total value of more than 30 billion dollars. During the same year about 40 million tons of synthetic plastics were produced which represent a total value of about 30 billion dollars. Figures of the same order of magnitude hold for synthetic rubbers, coatings, and packaging materials. As a consequence synthetic organic polymers have become a significant factor in the economy of all industrialized countries in the world.

THREE MILESTONES

For three decades from the 1920's to the 1960's enormous progress was made in understanding the structure of native polymers and synthesizing artificial counterparts. Instead of giving a detailed chronological account of these years events, I shall describe three representative meetings which took place during this period, because they have become classical milestones of modern chemical history.

   The _first_ took place in Duesseldorf where the
Gesellschaft Deutscher Naturforscher und Ärzte met
in 1926, the second was a meeting of the Faraday
Society in Cambridge in 1935 and the third occur-
ed in 1948 in Liege, Belgium and was the Inaugural
Session of the High Polymer Commission of IUPAC
(The International Union of Pure and Applied Chemistry). I had the privilege to present lectures at
all three conferences and was chairman of the last.

Duesseldorf Meeting, 1926

   The meeting in Duesseldorf was arranged in order
to confront Staudinger, the protagonist of the
macromolecular concept, with several other dis-
tinquished scientists who were still reluctant to
admit the existence of giant organic molecules and
adhered to the idea that many natural substances -
cellulose, silk, rubber - consist of small units
which are held together by exceptionally strong
intermolecular forces like those in Werner's complex compounds or in hydrogen-bonded systems. In
his lecture, Staudinger told an audience of several
hundred classical organic chemists that he had established the existence of molecules which are a thousand times larger than those which they themselves
were studying in their laboratories. One of them
commended: "We are shocked like zoologists would
be if they were told that somewhere in Africa an
elephant was found who was 1500 feet long and 300
feet high."

      To the classical arguments in favor of tightly
knit systems of small units, which were presented
by prominent scientists such as Pringsheim, Berg-
mann, and Waldschmidt-Leitz, a new one had recently
been added; namely, the fact that the crystallographic basic cells of cellulose, silk, and rubber
are so small that they only can contain particles
with molecular weights less than one thousand.

      According to the then existing classical crys-
tallographic teaching, a molecule cannot be larger
than the basic œll. This is where my lecture was
supposed to contribute to the discussion by explaining that, under certain conditions, a chemical molecule could well be larger than the basic crystallo-

graphic cell and could even be as large as the en-
tire crystallite.

The presentations were followed by rather ani-
mated discussions which proved that the differences
in opinion were not yet eliminated. At the end
Willstaetter, who presided, summarized his position
by saying: "Such enormous organic molecules are
not to my personal liking but it appears that we
all shall have to become acquainted with them."
In fact, the "theory of the small units" lost more
and more ground and in science and engineering the
macromolecular character of the principal natural
polymers became an accepted fact and immediately
encouraged the synthesis and study of any material
which could readily be prepared from available monomers which would undergo polymerization in one way
or another. Thus began the first phase of polymer
chemistry as a rapidly expanding empirical scouting
for new polymers and their principal properties with
the intention of arriving at a general survey of the
width and depth of this new brand of organic chemistry.

Cambridge Meeting, 1935

In 1935 the Faraday Society arranged a meeting
on "Phenomena of Polymerization and Polycondensation"
in Cambridge, England. It was an international conference, the large size and high level of which
proved the enormous progress which the young branch
of polymer science had made during the last decade.
There was no question anymore about the existence
of macromolecules, for the contributions dealt with
the mechanism of polymerization reactions, the
determination of molecular weights, the properties
of specific polymers as a result of their molecular
structure, and potential areas of application. For
the first phase the words "observation" and "preparation" were valid; for this second phase the
words were "measurement" and "quantitive characterization." In addition to the old guard - Staudinger,
Meyer, Rideal and myself - a new upcoming generation of excellent scientists presented their work -
Melville, Schulz, Houwink, Astbury and others. The
outstanding figure of this meeting was undoubtedly
Wallace H. Carothers who had come from Wilmington

to give an account of the momentous studies which
he and his associates had carried out during the
last decade.

At the end of the symposium, everybody was con-
vinced that polymer chemistry had grown into a
full-scale science with unexpected new vistas for
intensification of understanding and expansion of
application. Universities and industrial organizations started to compete with each other to enter
the field at any point from fundamental aspect to
practical evaluation. This second, rapidly ascen-
ding phase of polymer science was first slowed down
and later vigorously accelerated by World War II.
The United States and Canada in particular were
literally running away in basic and applied areas.
Essential fundamental details in polymerization
mechanisms including suspension and emulsion polymerization were clarified, and the groundwork for
x-ray examination and IR spectroscopy of polymeric
systems was firmly established.

After the war the first close contact in science
and engineering was made by visiting professors
from the United States to Europe. Such positions
were occupied among others by Alfred, Doty, Mesro-
bian, Tobolsky, Overberger and myself, who gave
several series of lectures in Western Europe right
after the war. The interest was understandably
very pronounced, and after contact with the presi-
dent and the secretary general of IUPAC in Paris
it was decided to organize in the summer of 1947
an International Polymer Symposium in Liege and
to initiate a Commission on High Polymers within
the Division of Physical Chemistry. There did not
yet, at that time exist scientific contact with
Germany, Italy, Japan and the USSR with its satel-
lites, but all other countries where work on polymers was done sent strong representative delega-
tions.

Liege Symposium, 1947

At the conference the state of the art was reviewed in a series of comprehensive papers, and
the main lines for further progress were in the
center of the discussion. Much time was spent to
report progress made in England and America to the

scientists on the continent. This included synthesis and technology of such polymers as polyethylene, nylon and polyester, new methods for structure determination such as x-ray Fourier analysis
and polarized infrared; and the beginning of spin
resonance and the elaborate use of light-scattering
techniques. Present were most scientists who, within the next decade, emerged as leaders of the various branches of polymer science. From England
came, among others, Astbury, Bawn, Bernal, Evans,
Melville, Rideal, and Thompson. Italy contributed
Nasini and Natta; France sent Champetier, Chapiro,
Magat and Sadron. They had the opportunity to
meet Hermans, Houwink, and Staverman from the
Netherlands; Errera and Smets from Belgium; Claesson
and Ranby from Sweden; Ant-Wuorinen from Finland;
W. Kuhn, K.H. Meyer and Signer from Switzerland.

This conference initiated a vigorous and systematic cooperation of all Western polymer scientists which soon resulted in the essential clarifi-
cation of all phases of condensation and addition
polymerization including copolymerization, graft,
and block polymerization, in an impressive build-
up of elastomer and plastics technology, in an almost complete domination of synthetic polymers in
the coatings and packaging field. As more polymers with more complicated structures became available the methods for their quantitative characterization had to be sharpened. All scattering techniques - light, x-rays, electrons, and neutrons -
received a thorough overhauling; resonance processes - nuclear and electronic magnetic movements -
were built up to admirable precision and, most of
all, the study of the chemical structure as well
as the state or order (regularity, crystallinity)
of polymers by vibrational (infrared and Raman)
spectroscopy was intensified and expanded. Today
this technique can be called the most informative
and powerful tool for the characterization of even
very complex polymeric systems.

CONCLUSIONS

In fact, all these physical methods became more
and more necessary because of the enormous number
of new polymers, of the increasing complexity of

their structure and of the wide variety of their
application. In order to design a macromolecule
for a special use, many detailed structural conditions have to be fulfilled which require very pre-
cise methods to establish their existence and to
control their durability in use.

One of these methods is the vibrational spectroscopy which has its roots in the late 1920's and
early 1930's. One of them was the fundamental
understanding of molecular vibrations on the basis
of quantum mechanics; it was first put in evidence
by the absorption of infrared radiation and later
also found in the modulations of scattered visible
light in the Raman effect. The two methods comple-
ment each other dramatically because of their different response to the selection rules which control the transition probabilities between different
vibrational states of a molecular framework. The
other root was a gradual and substantial improvement of the experimental techniques, such as stronger and more uniform sources of the primary radiations, higher resolution in the spectroscopic part
of the equipment, and, perhaps most of all, more
sensitive and reliable receivers.

Another important new and powerful method which
greatly advanced our knowledge of the molecular
structure of organic systems including polymers
are the magnetic resonance methods: electron spin
and nuclear spin which provide information on the
exact relative position of certain atoms - C,N.H -
to each other.

And, of course, diffraction experiments - electrons, x-rays and neutrons - some of which already
have been important in the earliest days of polymer science are now reaching a new and highly sophisticated level and are an indispensible help
for the microcharacterization of polymeric systems.

Thus: the almost limitless number of new experimental techniques for the preparation of new polymeric systems and the availability of powerful tools
for their characterization open up a bright future
for our field of interest: Further growth in the
synthesis of new useful materials and ever better
understanding of their structure and behavior.

CARBOHYDRATE POLYMERS:

NATURE'S HIGH PERFORMANCE MATERIALS

R. H. Marchessault

Xerox Research Centre of Canada
2480 Dunwin Drive
Mississauga, Ontario, L5L 1J9

SYNOPSIS

The term carbohydrate polymers describes the ubiquity
in nature of molecular systems containing carbohydrates.
The property of conformational restriction in
polysaccharides make them candidates for being the initial
self-ordering molecules of prebiotic evolution. This same
property in the complex carbohydrate moiety of
glycoproteins is the basis of carbohydrate-mediated
information transfer through cell surface oligosaccharides
interacting with each other or with lectin-like proteins in
cell-cell recognition processes.

Exopolysaccharides in their capacity to induce
synthesis of antibodies (i.e., as immunogens) or their
reactivity with antibodies (i.e., as antigens) have been
on the ground floor of the development of molecular
biology. The oligosaccharide repeating unit is the
chemical expression of immunological character and their
availability for attachment to synthetic carriers is a
first step toward manmade vaccines. The exopolysaccharide
gums display ordering characteristics in solution which are
not matched by the plant polysaccharides or synthetic
polyelectrolytes.

Finally, the term carbohydrate polymers can be stretched to include natural polyalkanoates based on hydroxy acids of carbohydrate origin. The chemistry and properties of poly-β-hydroxybutyrate (PHB) make it a biomass transducer. A material which is intermediate between carbohydrates and fats, PHB is readily derived from microbial systems. Its potential as a natural thermoplastic, a biomedical implant and a source of chemicals from biomass is described.

INTRODUCTION

Carbohydrate polymers derive from nature's capacity for converting carbohydrate molecules into polyacetals. Figure 1 shows how chemical bonding between a pyranose sugar's C-1 and with one of several different hydroxyls can occur in four different ways. In addition, the configuration of the bond at C-1 may be axial or equatorial (α or β) leading to steric isomers. Thus 2 glucopyranose units can be acetal-linked in 8 different ways [1]. These types of macromolecules are usually referred to as polysaccharides. However, the past decade has seen a major development in our understanding of naturally occurring polysaccharides systems some of which involve carbohydrates covalently linked to other chemical components, e.g., protein, lipid, lignin etc., hence the term carbohydrate polymers better describes the ubiquity in nature of molecular systems containing carbohydrates.

In addition, the natural polysaccharides studied by chemists as the first models of synthetic polymers tended to be homopolysaccharides or cellulose derivatives. As chemists probed deeper into the fabric of biomaterials, the inherent complexity of natural systems became apparent. As the molecular architecture of branching and the copolymeric design of polysaccharides started to unravel, the name complex polysaccharide also came into use [1a].

The variety of polysaccharides produced in natural systems has never been estimated. For example, more than 25,000 algal species have been described but the glycans of only about 150 have been investigated. [2] The evidence that the D-sugars predominate in nature and that evolutionary selection has not produced an "infinite variety" by playing upon all possible molecular permutations is overwhelming. Some of the more commonly available polysaccharides are shown in Table I. Most of these materials are commercially available although not necessarily from the source indicated e.g. chitin for commerce is obtained from crustacean shells, not fungi. While all materials in Table I are linear, some are copolymers, including homoglucans with variable linkage types. Significantly different functional properties in cell walls are achieved by interrupting the $\beta(1\rightarrow4)$ linkage of cellulose with some $\beta(1\rightarrow3)$ linkages as in cereal glucans and lichenan[3,4].

The rich variety of materials which are being revealed as a result of biological studies falls into the general category of complex polysaccharides and relate to: peptidoglycans, glycoproteins, proteoglycans, lipopolysaccharides.

While structures are still being unraveled, the past decade has brought a much clearer understanding than was the case two decades ago. For example Figure 2 shows the difference between glycoproteins and proteoglycans. In both cases there is a covalent association with protein: in one case, a globular molecule depicted by an α-helix segment is shown with "antenna" oligosaccharides. In the other case, a proteoglycan molecule is depicted as a high molecular weight helical hyaluronic acid backbone (HA) with sidechains of "core protein" to which are attached short branches of sulfated polysaccharides. Glycoproteins and proteoglycans are two distinct fields of biochemistry the latter involving the so-called mucopolysaccharides of higher animals. We shall comment in this article only on the glycoproteins.

The recent realization that microbial polysaccharides are composed of regular repeating units involving from 1 to 7 carbohydrate units, and the fact that they are of crucial importance in the immune reaction has prompted widespread investigation. Because they surround the cell, these

Table 1 : Some Linear Carbohydrate Polymers

Source	Common Name	Sugars*	Chain**
Plants	Cellulose	Glc	$\beta(1\rightarrow4)$
	Cereal Glucans	Glc	$\beta(1\rightarrow3)$, $\beta(1\rightarrow4)$
	Amylose	Glc	$\alpha(1\rightarrow4)$
	Pectic Acid	Gal-A	$\alpha(1\rightarrow4)$
	Inulin	Fruct.	$\beta(2\rightarrow1)$
	Mannans	Man(Gal)	$\beta(1\rightarrow4)$
	Xylans	Xyl (Glc-A, Ara)	$\beta(1\rightarrow4)$
Algae	Carrageenan	Gal	$\beta(1\rightarrow3)$/ $\alpha(1\rightarrow4)$
	"Green Seaweed"	Xyl	$\beta(1\rightarrow3)$
	Laminaran	Glc	$\beta(1\rightarrow3)$
Fungi	Chitin	N-Acetyl Glc	$\beta(1\rightarrow4)$
	Pullulan	Glc	$\alpha(1\rightarrow4)$/ $\alpha(1\rightarrow6)$
	Pustulan	Glc	$\beta(1\rightarrow6)$
	Nigeran	Glc	$\alpha(1\rightarrow3)$/ $\alpha(1\rightarrow4)$
	Lichenan	Glc	$\beta(1\rightarrow3)$, $\beta(1\rightarrow4)$
	Scleroglucan	Glc (Glc.)	$\beta(1\rightarrow3)$
	Xanthan Gum	Glc, Man, Glc-A	$\beta(1\rightarrow4)$
	Curdlan	Glc	$\beta(1\rightarrow3)$
	Dextran	Glc	$\alpha(1\rightarrow6)$

*   Bracketed sugars are present as substituents on the
    main chain.

** A comma indicates a statistical copolysach.; the "slash"
    symbol indicates an alternating copolysach.

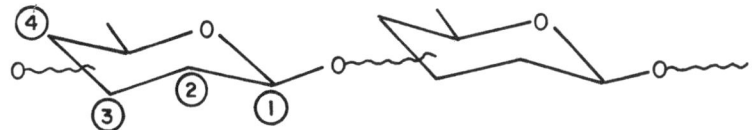

Fig. 1   Schematic of possible polysaccharide structures from
          pyranose sugars. The C-1 carbon is in the β configur-
          ation and the "chair" pyranose ring is $^4C_1$.

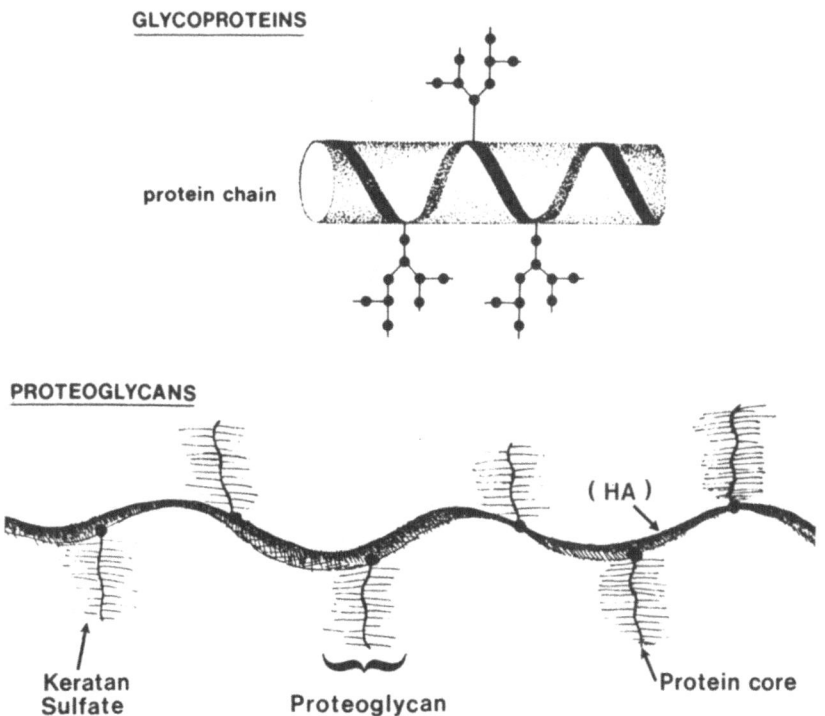

Fig. 2   Models depicting the difference between glyco-
          proteins and proteoglycans. (cf. text, p. 3)

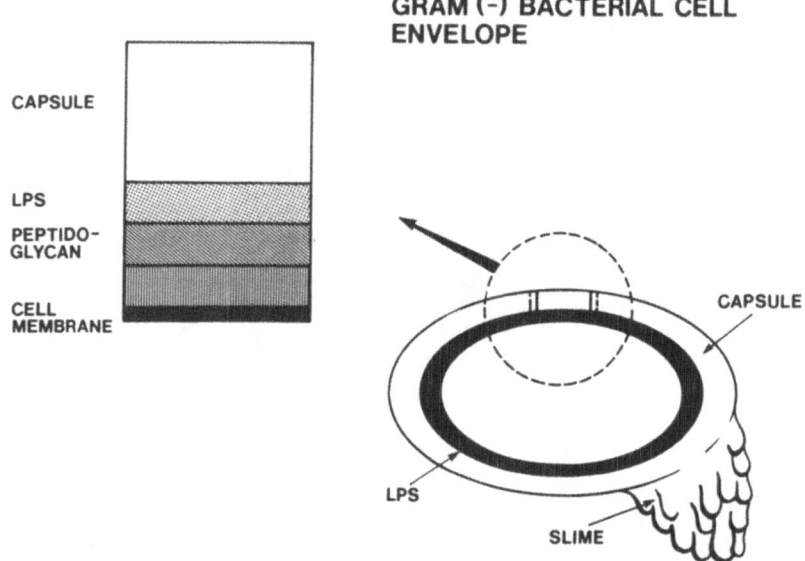

GRAM (-) BACTERIAL CELL ENVELOPE

CAPSULE

LPS

PEPTIDO-GLYCAN

CELL MEMBRANE

CAPSULE

LPS

SLIME

Fig. 3   Diagramatic representation of gram negative bacterial cell envelope.

polysaccharides are frequently referred to as capsular and since the capsule frequently sloughs off into the culture medium, the general term exopolysaccharide or slime is applied. Figure 3 is a general schematic of the components in the gram negative bacterial cell envelope. Typical examples would be polysaccharides from:

Pneumococcus[5]
Escherichia coli
Klebsiella[6]

Serological testing has revealed seventy-seven different types of Klebsiella and structures for the capsular polysaccharide have been proposed for 55. Some typical Klebsiella polysaccharide structures are listed in Table 2.

A new phase of carbohydrate chemistry and technology is emerging as the aftermath of the OPEC crisis and in step with the new biology. Renewable resources have appeal in their own right but coupled with the standards of high performance materials and the chemical industry's need for organic raw materials, the carbohydrates of biomass stand in a new light.[7] This is especially true when the emerging genetic engineering component is added to existing microbial technology.[8] Based on high performance criteria, the specialized properties of carbohydrate polymers, especially carbohydrate-mediated information transfer, are like "missing links" in the research objectives of synthetic polymer chemists. There is no better way to illustrate this than to examine new concepts in evolution involving polysaccharides. To do so, however, we should first look at the conformational and crystallization properties of cellulose which, in nature, literally lead to self-assembly. Then we will survey some unique chemical recognition properties of polysaccharides, followed by a look at microbial polysaccharides and finally some thoughts on chemicals from biomass.

CELLULOSE CONFIRMATION AND NASCENT CRYSTALLIZATION

The conformation accessible to a pair of contiguous
residues in the chain, (i.e., cellobiose) can be estimated
by constructing a steric map. Treating β-D-glucopyranose.
as a rigid body, the various conformations of a
disaccharide are generated by rotations ϕ and ψ around the
glycosidic bonds,,respectively, as shown in Figure 4 which
includes a β(1→6) linked disaccharide as well as the
classical phosphate diester link of the nucleotides. As
the number of flexible bonds between repeating units
increases, the number of conformations which are
energetically probable does also. Hence, the molecule on
the left in Figure 4 is the more conformationally
restricted molecule of the three depicted in Figure 4.
This statement is quantified for each molecule by means of
its energy surface. In the case of cellobiose, for each
pair of values of ϕ and ψ, the distance between the
nonbonded atoms are calculated and either a "contact
distance map" or "energy map" can be constructed.[9] The
former, shown in Figure 5, simply gives the boundary of the
conformational space, within which there are no steric
overlaps. The allowed distance criteria[9, 10] used for
such a map are well known. The normal limit distances are
those derived from observed interatomic non-bonded
distances in crystal structures of related small molecules.
The extreme limit distances allow for compression of the
atoms.

The contact distance map in Figure 5 shows that the
accessible conformations are very limited for the
cellobiose moiety. It comprises only 5% of the (ϕ, ψ)
surface, indicating that the freedom of rotation of the
residues is highly restricted. This fact means that the
cellulose molecule has a strong bias for stretches of
contiguous glucose units which are in a ribbon-like
arrangement corresponding to the family of ϕ, ψ inside the
boundary of the contact distance. This ribbon-like
conformation is also that in the cellulose crystallite
which in turn is alligned in the fiber axis direction of
natural fibers such as ramie, flax, wood etc. This

Table 2:   Some Klebsiella Polysaccharides*

K-TYPE	STRUCTURE	REF.
K2	$-\overset{3}{}$Glc $\overset{1\ 4}{\underset{\beta}{}}$ Man $\overset{1\ 4}{\underset{\beta}{}}$ Glc $\overset{1}{\underset{\alpha}{}}$ — with $\overset{3}{\underset{1}{}}\alpha$ GlcA (arrow)	(70)
K4	$-\overset{3}{}$Glc $\overset{1\ 2}{\underset{\alpha}{}}$ GlcA $\overset{1\ 3}{\underset{\alpha}{}}$ Man $\overset{1\ 3}{\underset{\alpha}{}}$ Glc $\overset{1}{\underset{\beta}{}}$	(71)
K8	$-\overset{3}{}$Glc $\overset{1\ 3}{\underset{\beta}{}}$ Gal $\overset{1\ 3}{\underset{\beta}{}}$ Gal $\overset{1}{\underset{\alpha}{}}$ — with $\overset{4}{\underset{1}{}}\alpha$ GlcA	(72)
K16	$-\overset{3}{}$Glc $\overset{1\ 4}{\underset{\alpha}{}}$ GlcA $\overset{1\ 4}{\underset{\beta}{}}$ Fuc $\overset{1}{\underset{\alpha}{}}$ — with $\overset{4}{\underset{1}{}}\beta$ Gal	(73)
K17	$-\overset{4}{}$GlcA $\overset{1\ 3}{\underset{\beta}{}}$ Rha $\overset{1\ 4}{\underset{\alpha}{}}$ Glc $\overset{1\ 2}{\underset{\alpha}{}}$ Rha $\overset{1}{\underset{\beta}{}}$ — with $\overset{3}{\underset{1}{}}\alpha$ Rha (arrow)	(74)

*   The large arrow indicates the bacteriophage attack site.
The writer acknowledges the help of: Angela V. Savage,
Ph.D. thesis, The Faculty of Graduate Studies, U. of
Brit. Columbia (1980) in preparing Table 2.

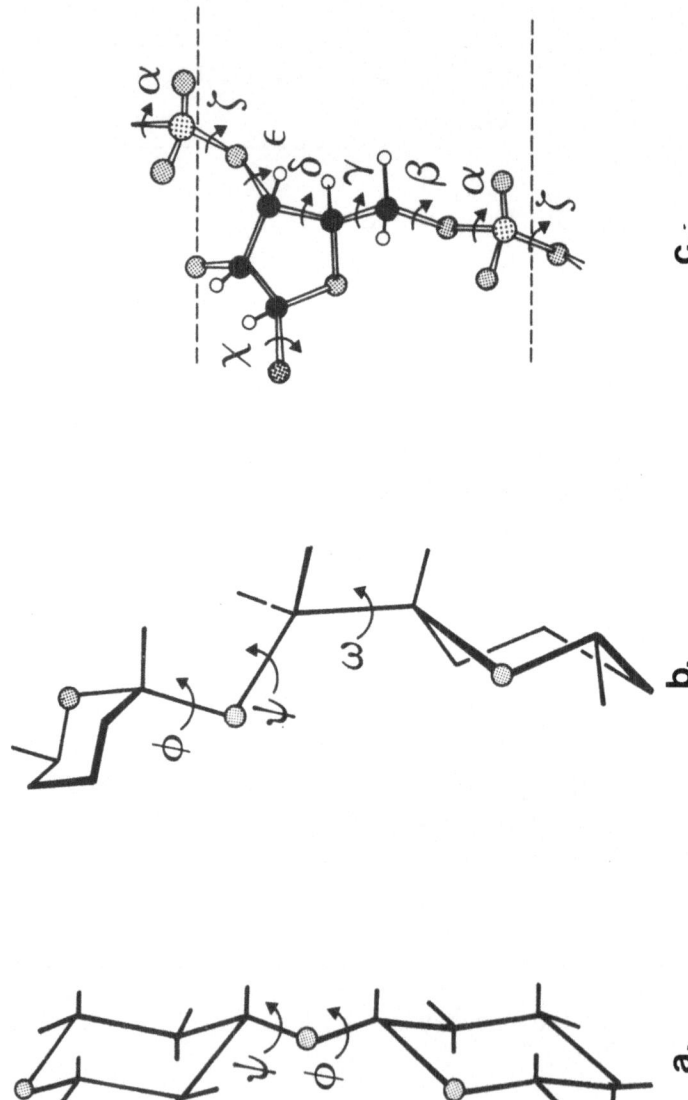

Fig. 4   From left to right are shown the molecular conformations for:
β(1→3) glucan, α(1→6) glucan and phosphodiester link between ribose
units with the labelled dihedral angles.

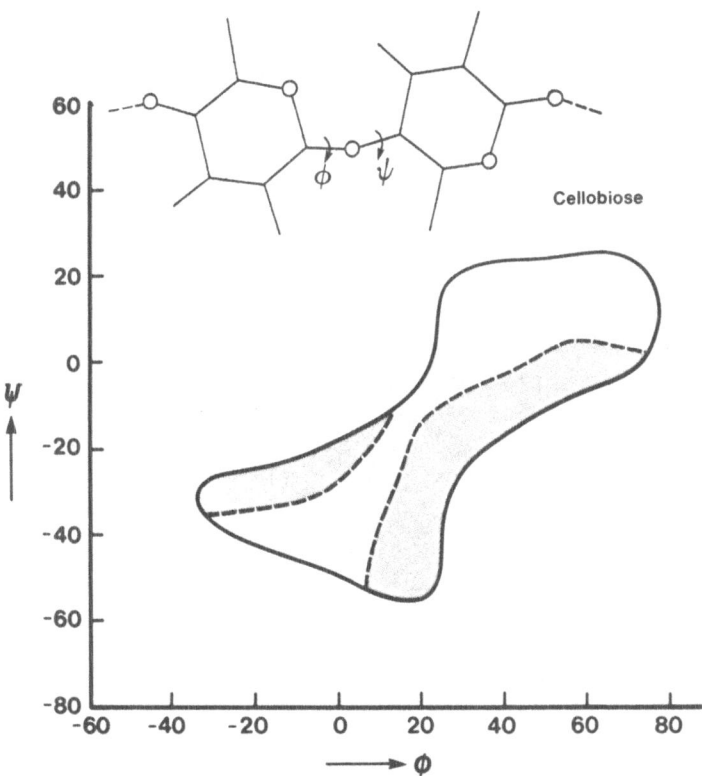

Fig. 5   Schematic of the Cellobiose unit showing the torsion
          angles φ and ψ. Also included is the contact dis-
          tance map for cellobiose constructed using "extreme
          limit" criteria. The shaded areas correspond to
          conformations where intramolecular hydrogen bonds
          are possible.

structural continuity, provided by the coincidence of the preferred conformation ribbon axis, the fiber axis and the plant stem is illustrated in Figure 6.

The remarkable fact, is that cellulose shows no variations in density along its crystalline microfibrils. This can be demonstrated by the technique of diffraction contrast electron microscopy.[11] This shows that microfibrils of the alga <u>Valonia ventricosa</u> have rigorously parallel edges 200 A across and a continuous crystalline structure over several micrometres (>10,000 A). Because of the beam sensitive nature of cellulose, the resolution depends on the chemical stability of the specimen to the electrons and it is in the order of 30 to 50 A. Imperfections in the crystalline order must be less than this size.

One can associate conformational rigidity with the high level of lattice perfection of native cellulose. However, regenerated cellulose does not have this same perfection. Only by evoking the phenomenon of simultaneous biosynthesis and crystallization[12,13,14] leading to a nascent morphology can we account for this remarkable fine structure: self-ordering at the earliest possible stage. Nor is the phenomenon limited to cellulose since other native polysaccharides, e.g. chitin[15] and paramylon[16] show nearly 100% microfibrillar crystalline order in the native state.

EVOLUTION AND CONFORMATIONAL PERSPECTIVES OF GLYCANS

Two aspects of evolution concern the theoretical biologist:

- Prebiotic evolution, i.e. how did the initial self-ordering take place?

- Structural evolution, i.e. the relation of morphological complexity to molecular structure.

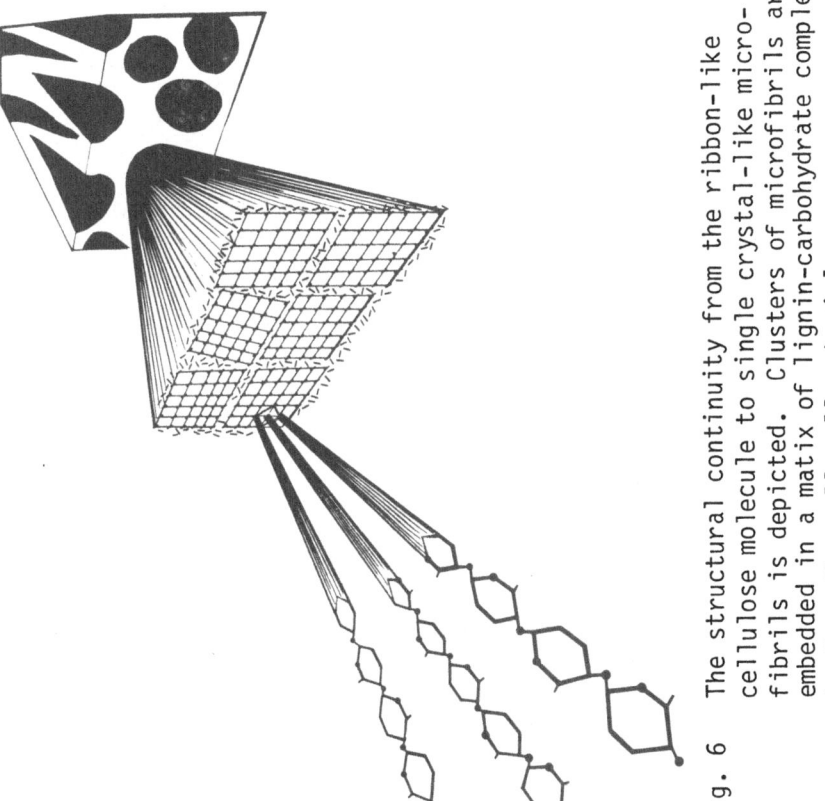

Fig. 6   The structural continuity from the ribbon-like cellulose molecule to single crystal-like micro-fibrils is depicted. Clusters of microfibrils are embedded in a matix of lignin-carbohydrate complex to form the cell wall material.

The first topic is usually treated in terms of the familiar nucleic acid-protein systems with which we are familiar today. However, a novel proposal for evolution from polysaccharides by A. W. Burgess[17] illustrates some high performance characteristics of glycans and relates to what has been said above concerning cellulose conformation.

If one compares the conformational energy surface of proteins and polysaccharides in terms of the usual dihedral angles ϕ, ψ which determine local conformations and flexibility, the results show that 15% of the conformations for the peptide bonds of the alanyl peptide are within 5 kcal/mole of the energy minimum. On the other hand, rotation around the $\beta(1 \rightarrow 4)$ glycosidic linkage for cellobiose shows that less than 5% of the possible conformations are allowed in this energy limit and when intramolecular hydrogen bonds are considered the allowed area is close to 1% (Fig. 5). In general, Burgess argues[17] that polysaccharides are far more rigid than polypeptides or nucleic acids.

The necessary property to allow prebiotic evolution is thought to be formation of a self-ordering system. A set of conditions for prebiotic evolution from polysaccharides might be:

- they form readily from simple sugars

- they are conformationally restricted

- the conformations contain considerable redundancy

- the repetitive formation of a given polysaccharide is autocatalysed.

There is strong experimental evidence in support of the first two points. The monosaccharides themselves being derived from such simple molecules as acetaldehyde, formaldehyde and water all of which are thought to have been present in the prebiotic earth.

The redundancy condition implies that changes in the sugar units or occurrence of side chains do not necessarily destroy the conformational preference of the molecule. Redundancy is to be associated with repetitive and similar three dimensional structures over the range of compositions.[17] There are many demonstrations of this in polysaccharides, e.g.:

- $\beta(1 \rightarrow 4)$ xylans maintain their threefold helical conformation in the crystal even with extensive substitution on the backbone[18,19]

- $\beta(1 \rightarrow 4)$ mannans maintain their 2_1 helix (ribbon-like) in spite of galactose side chains or changing from mannose to glucose in the backbone.[20,21]

The autocatalytic synthesis of repetitive polysaccharide structures is intimately linked with insolubility of the polysaccharide product compared to the repeating unit. The remarkable fibrillar or extended chain nature of some polysaccharides in their nascent state is attributed to conformational rigidity coupled with simultaneous polymerization and crystallization[12,13] which might have been the essential self-ordering phenomenon of early prebiotic synthesis.

If one examines the structural evolution in the glycans of the algae, support for these concepts is forthcoming. Since algae are living fossils (e.g. the blue-green alga are estimated to have appeared some 3 billion years ago) one may expect to see some of the conditions for prebiotic evolution demonstrated by examining their biotic evolution. For example, Painter[22] notes that the evolution of "morphological complexity has been accompanied by the replacement of irregular glycan structures by more regular ones" which contain a smaller number of different sugar residues. In the red and green algae one can trace the evolutionary pathway:

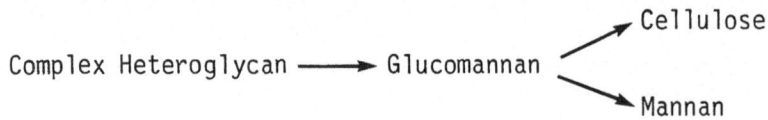

This evolutionary trend is an illlustration of both
the redundancy condition and the autocatalytic assembly
concept. The latter concept is associated with a matrix
role of the cellulose crystal lattice in relation to
biosynthesis. Cellulose and mannan in algae are more highly
crystalline and less soluble than glucomannan.[14] The
response to the environmental pressures has been the
development of the ultimate in polymer crystalline texture:
the cellulose microfibril.

Nevertheless, cellulose has been a slow developer and
scholars associate its first clear appearance with the
microscopic green seaweeds of the Ordovician period, i.e.
celllulose is probably no more than 500 million years old.
Prior to this, complex polysaccharides may have played a
role in the development of primitive cells walls. One such
is the glycolipid or lipopolysaccharide.

Current models for membrane structures usually are
based on the bipolar properties of phospholipids. However
lipopolysaccharides are prominent in the cell wall of gram
negative bacteria and evolutionists propose[17] that
primitive cells could have derived from vescicles based on
glycolipids. Recent studies on synthetic glycolipids has
brought to light their liquid crystalline properties which
result from hydrogen-bonded carbohydrate moieties and
intercalated alkyl chains.[23] Similarly, the liquid
crystalline properties of cellulose and cellulose
derivatives[24] are being discovered and this too
demonstrates rigidity and self-ordering properties of
importance in the context of fiber spinning[25].

CARBOHYDRATE-MEDIATED RECOGNITION

Glycoproteins are exceedingly important components of mammalian cell membranes and it is becoming accepted that their oligosaccharide moieties, complex carbohydrates, serve as probes wherewith the cell interacts with its environment. Conversely, the environment delivers signals to the interior of the cells through cell surface oligosaccharides. This carbohydrate-mediated recognition is intimately involved in cell-cell recognition. For example, the adherence of bacteria is a prerequisite for infection of cells[26].

This adherence is in numerous cases inhibited specifically by low concentrations of monosaccharides that mimic the glycoside component of cell surface glycoproteins.[27] The phenomenon is illustrated in Figure 7. Where the bacterial surface may be considered to be covered with "bacterial surface lectins" which have been saturated by an excess of "receptor analogues".

The term lectin itself refers to a naturally occuring, carbohydrate-binding protein of non-immune origin which agglutinates cells and/or precipitates complex carbohydrates.[28] Lectins represents a class of proteins which react specifically with free carbohydrates or carbohydrates found in polysaccharides leading inevitably to agglomeration if the receptor molecules are at a cell surface. Thus the inhibition of bacterial adherence is easily understood in terms of lectin-like receptors at the bacterial surface which are saturated by the excess of carbohydrates (cf. Fig. 7).

The chemistry of cell surface oligosaccharides is now well-understood.[29,30,31] They are isolated as multiantennary structures with the chemically defined polarity of a polysaccharide structure terminating with an amino acid, usually asparagine (cf. Fig. 8). Structural and conformational work is showing that these molecules maintain a stable three-dimensional structure in solution which makes them capable of interacting with complementary

BACTERIAL ADHERENCE

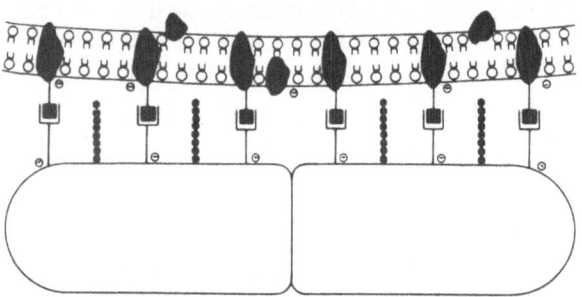

ADHERENCE INHIBITOR

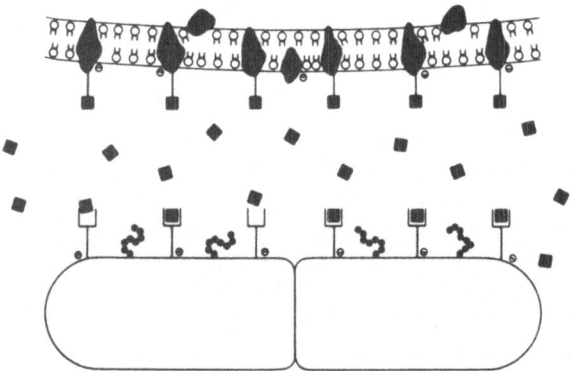

Fig. 7   Schematic showing adherence between "square"
          receptor units of glycoproteins in the membrane and
          complementary ligands at the surface of bacteria.
          An excess of "receptor analogues" inhibits[26]
          adherence.

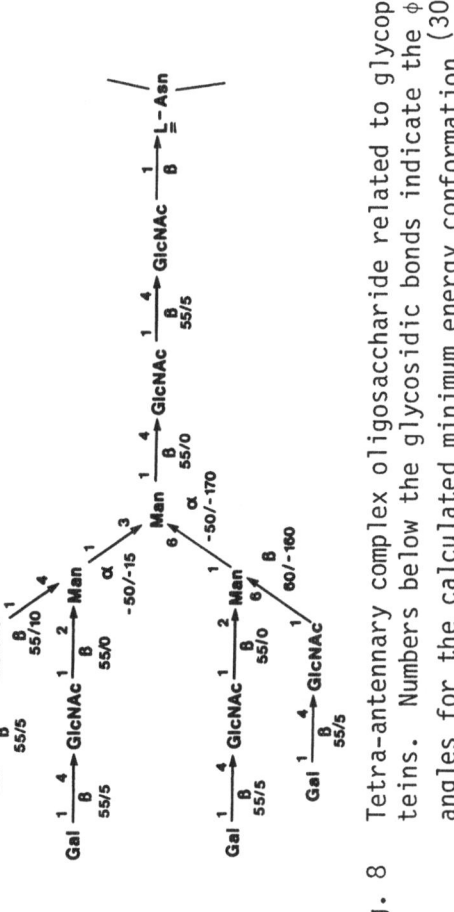

Fig. 8 Tetra-antennary complex oligosaccharide related to glycoproteins. Numbers below the glycosidic bonds indicate the ϕ, ψ angles for the calculated minimum energy conformation.(30)

molecules. By combining high resolution NMR, x-ray
diffraction and hard-sphere calculations, a full three-
dimensional structure for these molecules is being
derived[30,31] for isolated molecules of this type which
are too large for the direct crystallographic methods. At
the same time, difference-fourier data from protein
crystallography is providing three-dimensional information
on the shape of these molecules as they are complexed to
proteins.[32,32a]

Evidence is accumulating that relatively minor changes
in the primary monosaccharide sequence produces changes in
conformation which determines binding.[33,34] Fig. 9 is an
example of this where a biantennary glycopeptide with two
different conformations generated by rotating angle ω i.e.
between C-6 and C-5, at the Man $\alpha(1{\rightarrow}6)$ Man unit, are depicted
as CPK space-filling models[33]. Docking characteristics of
the two conformations with respect to a receptor molecule are
bound to differ and these differences are sensed by NMR
methods[34].

This interaction is already a form of information
transfer in which the information resides not in a nucleic
acid or protein molecules but rather in an oligosaccharide.
Similar oligosaccharides and similar conclusions are found
with the blood group substances, the critical antigens
which detemine blood types [35, 36]. The conclusion is that
the recognition is conformationally based and plays upon
the structural rigidity of these oligosaccharides.

This rapidly developing field offers tremendous scope
for preventing bacterial colonization and infections in
humans. The lectin/carbohydrate specific interactions are
potentially useful for a wide range of industrial
applications since the preparation of plant lectins are
readily scaled[28]. Biomedical diagnostics are but one
example of potential uses. At the same time elucidation of
these three dimensional oligosaccharide structures are
providing an understanding of the pathways for chain
elongation within the Golgi apparatus of N-glycosidically-
linked oligosaccharides.[34]

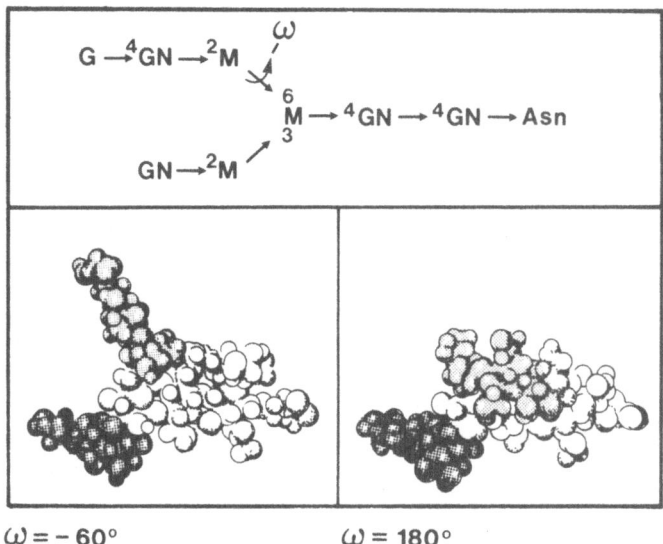

Fig. 9   Biantennary "glycopeptide" based on:

G   = glucose
GN = N-acetyl-2-deoxy glucoseamine
F   = fucose
M   = mannose
Asn = asparagine

The two different conformations[33] were generated
by rotating about the ω dihedral angle (cf. Fig. 4).

EXO-POLYSACCHARIDES

The large number of stereospecific molecules which can be formed from a relatively small number of carbohydrate monomers provides structural continuity, allows "recognition" and specific complexation in nature. The rules which govern this specificity can be stated[37]:

- The nature of the sugars present.
- The position and configuration of the glycoside linkage.
- The presence or absence of substituents.
- Conformational rigidity.

The numerous examples of regular complex copolysaccharides often involve familiar-looking material for cellulose chemists. Figure 10 shows two pneumococcal polysaccharides, Types III and VIII, also known as "specific soluble substance", which in the 1920's and early 1930's were shown to be antigenic although they were free of nitrogen and did not possess any of the properties of peptides. The knowledge achieved by the extensive studies on cellulose and carbohydrates in the first decades of this century was responsible for the early establishment of the chemical structures of Types III and VIII. The revolutionary work on bacterial transformation, in which Avery, MacLeod and McCarty in 1944 identified DNA as the genetic material[38], was done with pneumococci, the genetic marker being the production of a "specific capsular polysaccharide"[39]. These materials can claim to have been on the ground floor of the development of molecular biology and have since come into use as commercial polysaccharide vaccines.

Two aspects have to be considered when dealing with polysaccharides as antigens:

- Their capacity to induce antibodies in mammals i.e. immunogenicity.
- Their reactivity with antibodies i.e. antigenic specificity.

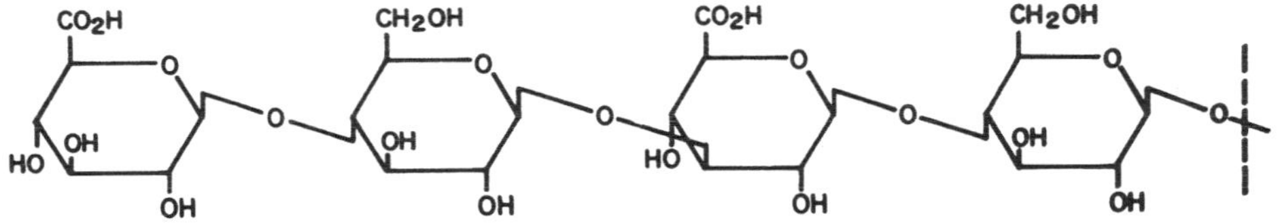

TYPE III

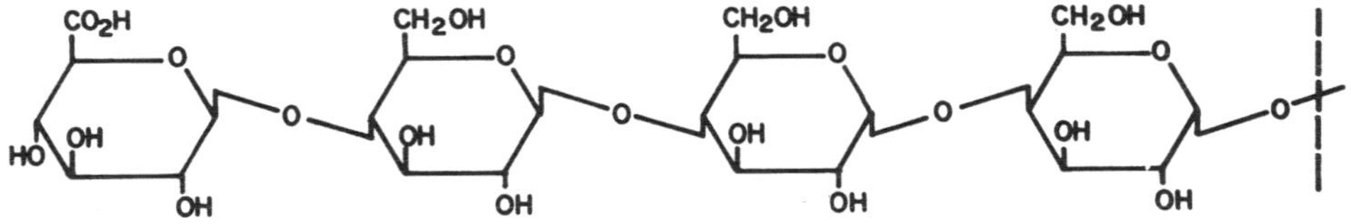

TYPE VIII

Fig. 10 Type III and Type VIII pneumococcal polysaccharides which are regular structures with a dimeric and tetrameric repeating unit, respectively. The structures are drawn from left to right corresponding to "non-reducing" end at the left. Both structures contain one glucuronic acid unit per repeat and two repeating units are shown for the Type III.

The oligosaccharide repeating unit is the chemical expression of immunological character and it may even be one or two specific sugars with their configuration/substitution providing the immunodominant characteristic. Because of their repetitive structure, bacterial polysaccharides have the same antigenic determinants repeated many times i.e. they are poly-functional with respect to reactions with antibodies.

Since it has been shown[40] that protective immunization can be performed with isolated polysaccharides and related artificial antigens, carbohydrate polymers have been in the forefront of immunological research. For the polymer chemist there is a major synthetic challenge and a new molecular property to exploit i.e. attachment of immunogenic and antigenic oligosaccharides to functional carrier macromolecules. The use of viral enzymes i.e. bacteriophage (bacterial virus) glycanases to depolymerize specifically[41] into the oligosaccharide repeating units is now well-established. The so-called penetrase enzymes are located in the viral tail and effectively reduce the capsular blanket, which is several hundred nm thick, into a homogeneous product of oligosaccharide repeating units of the given exopolysaccharide[42]. Thus antigenic oligosaccharides are available for attachment to synthetic polymer backbones or membranes as a first step toward manmade vaccines or their equivalents.

Microbiology has allowed polysaccharide gum producers to develop very unusual molecules which duplicate and surpass some of the commercially successful plant polysaccharides. For example, Xanthan gum shown in Fig. 11 has a cellulose backbone and regularly occurring carboxylated side chains[43]. Pyruvic acid reacted with the terminal carbohydrate of the side chains adds another carboxylate group. Unlike cellulose, individual chains are soluble in water, but do not show the typical polyelectrolyte viscosity drops with pH, ionic strength or temperature because an ordering factor intervenes encouraging stiffening of the chain as depicted in Fig. 12. This self-recognition/complexation behaviour is

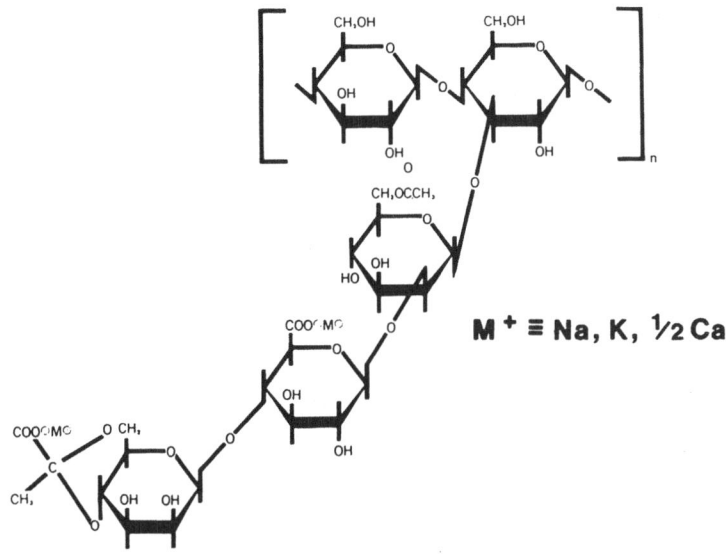

$M^+ \equiv Na, K, \frac{1}{2}Ca$

Fig. 11   Chemical repeating unit of sodium salt form of
          Xanthan Gum.

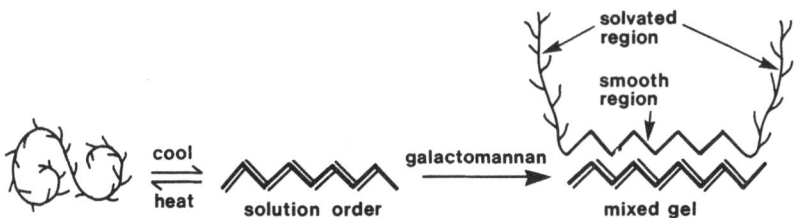

Fig. 12   Order-disorder transition of Xanthan Gum and
          complexation of the ordered region with a mannan
          segment in a galactomannan chain.

increasingly recognized in bacterial polysaccharides. It
provides high performance rheological characteristics by
enhancing the conformational rigidity in a novel manner.
The ordering phenomenon can be further amplified by
complexation with another polysaccharide such as Guar
gum[44], as shown in Fig. 12. This sort of synergism leads
to thermoreversible gelation in a system containing the two
polysaccharides[43].

Xanthan gum is a regular copolysaccharide not a
statistical one like partial derivatives of cellulose and
plant gums. While the regularly occuring side chain
renders Xanthan gum so different from cellulose, the
influence of a different glycosidic linkage in the backbone
is the most important fact in modifying polysaccharide
chain conformation.[45] Figure 13 shows the mimimum energy
conformations of Type III pneumococcal polysaccharide[46]
where the repeating unit is the disaccharide cellobiouronic
acid linked via $\beta(1\rightarrow3)$ linkages. The Type III chain
conformation is close to that of cellulose, being an
extended 3_1 helix. i.e., the interruption of $\beta(1\rightarrow4)$
linkages has not significantly perturbed the characteristic
stiffness of the cellulosic chain. It is a fact that the
$\beta(1\rightarrow4)$ linkage in any glycan has a dominant stiffening or
chain-extending effect[45a]. However, a sequence of $\beta(1\rightarrow3)$
links produces a totally new minimum energy conformation,
as may be seen in Fig. 14 by focusing on a single strand
which is a large amplitude helix with six glucoses per
turn. Furthermore, the conformational uniqueness of the
sequence of $\beta(1\rightarrow3)$ links has far-reaching effects on the
crystallization behaviour. Thus $\beta(1\rightarrow3)$ glucans form triple
helical structures and crystallize in the triplex form
[47,48,49] as shown in Figure 14. This triplex persists to
varying degrees in solution, as with collagen/gelatin and
leads to remarkable hydrodynamic[50,51] effects within
families of $\beta(1\rightarrow3)$ glucans. Commercially, these various
species are known as Curdlan[52], the pure $\beta(1\rightarrow3)$ glucan
and Scleroglucan[53] (or Schizophillan)[54] which is a
regular copolymer with a tetrasaccharide repeating unit as
shown in Formula I.

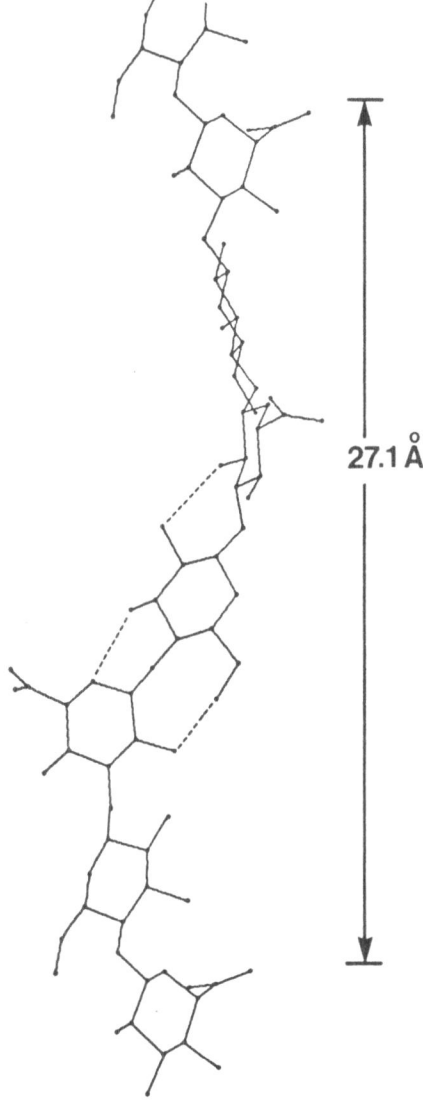

27.1 Å

Fig. 13 Minimum Energy conformation
         of Type III pneumococcal poly-
         saccharide which is a poly
         β(1→3) cellobiouronic acid
         i.e., the glycosidic linkage
         alternates from β(1→3) to
         β(1→4).

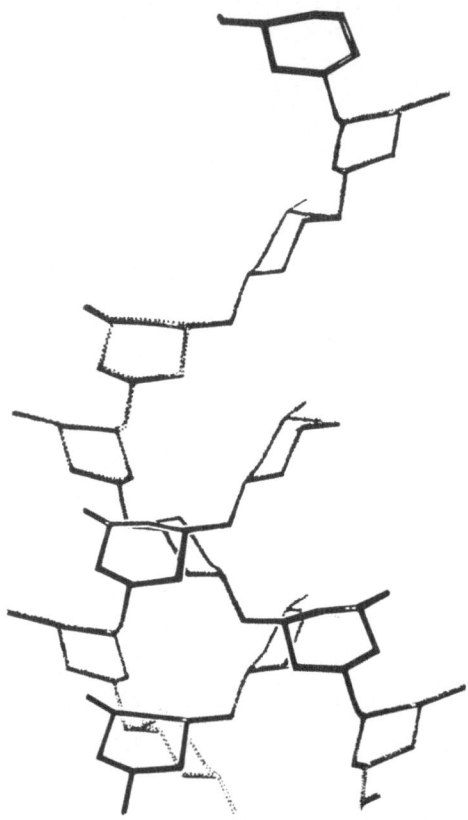

Fig. 14   Triple helical arrangement of $\beta\,(1 \rightarrow 3)$ glucan in the
          crystalline state.   A single strand corresponds to a
          large amplitude helix with six residues per turn.

$$\beta\text{-D-Glcp}$$
$$1$$
$$\uparrow$$
$$6$$

$$\left[\rightarrow 3)\text{-}\beta\text{-D-Glcp}(1 \rightarrow 3)\text{-}\beta\text{-D-Glcp}(1 \rightarrow 3)\text{-}\beta\text{-D-Glcp}(1\text{-} \right]_n \rightarrow \qquad (I)$$

While the former is only soluble in aqueous alkali, the latter dissolves readily in water and maintains an ordered conformation in this solvent[50]. As a result, its pseudoplastic solutions have good viscosity tolerance to a broad range of temperature, pH and salt concentrations. Overall, $\beta(1 \rightarrow 3)$ glucans provide high performance through unique self-complexing of a kind which has never truly been achieved by molecular designing of a synthetic polymer.

GARBAGE POLYESTERS

It is estimated[55] that some 10^{11} tons of carbohydrate polymers, mostly cellulose, are produced annually i.e., 25 tons per person on earth. Fortunately, carbohydrate-rich materials including urban refuse, agricultural residues and lignocellulosics are good fermentation substrates, hence microorganisms assure that biosynthesis and decay are in balance. That a major product of nature's biomass transformation should be a high molecular weight linear polyester is perhaps unexpected.

On second thought, since glyceric acid is the reference molecule for carbohydrates, why should nature refrain from stereoregular conversion of simple hydroxyacids into polyesters? Formula II shows the structural formula of poly-β-hydroxybutyrate (PHB) which is the main reserve material in bacteria when these are fed a nitrogen deficient carbohydrate rich diet.

$$\left[\begin{array}{ccccc} & H & & H & \\ & | & & | & \\ C & - & C & - & C & - & O \\ & | & & | & & \| & \\ & CH_3 & & H & & O & \end{array}\right]_n$$

(II)

This chiral polyalkanoate, based on D(-) β
hydroxybutyric acid monomer, was discovered in the 1920's
and remained a microbiological fact while polymer chemists
struggled with the likes of cellulose, natural rubber, and
silk. However, like starch in the plant world, PHB is
ubiquitous in the bacterial kingdom and serves exactly the
same purpose. Figure 15 shows how the concentration of PHB
in the cells builds up during the "log"phase of growth then
drops rapidly, presumably being metabolised in relation to
spore development[58]. During the growth cycle polymer
formation is readily followed by microscopic observations
on aseptically withdrawn samples which are stained with
Sudan black. PHB is in the form of 0.5 μm spherical
granules which can amount to 50-80% of the dry weight of
the cells. The polymer molecular weight can be 10^6, or
more, and a wide range of bacteria have been shown to
produce exactly the same material[59].

Numerous procedures have been described for isolating
either the granules themselves (95% PHB) or purified
polymer. Fig. 16 is a schematic outline of a patented[60]
isolation procedure for purified PHB. The thermoplastic
properties[61] of this purified material resemble
conventional polyolefins.

PHB can be thought of as a biomass transducer, in the
jargon of biotechnology. The polymer and its degradation
products are intermediate between carbohydrate materials
and alkanes. Once isolated, the PHB granules are readily
converted chemically to useful small molecules[62]. The
well-known thermal instability of β-hydroxyesters makes PHB
an ideal pyrolysis substrate leading to high yields of
crotonic acid. Even more striking is the fact that PHB is

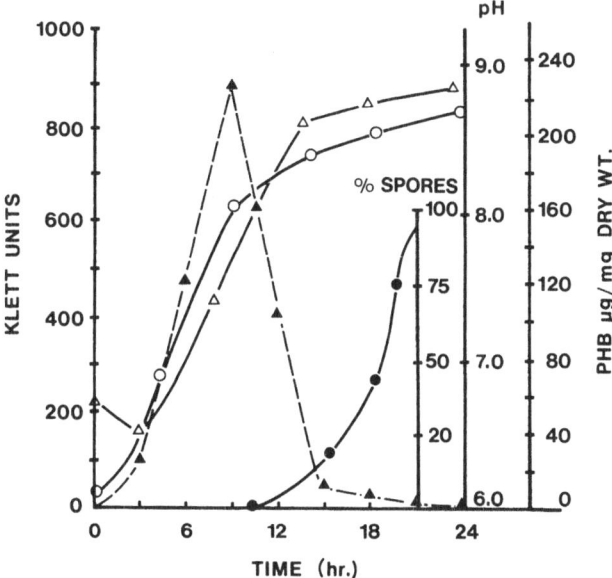

Fig. 15 Culture growth cycle for <u>B. cereus</u> grown at 37°C
under forced aeration where (o) bacterial growth
(Klett units are a measure of turbidity), (△)pH,
(▲)PHB, (•) spores.

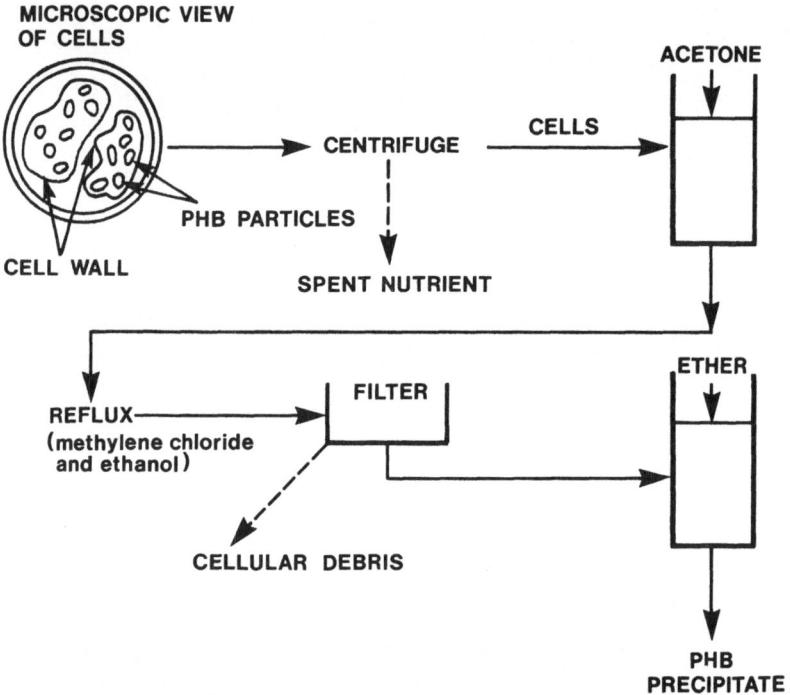

Fig. 16   J. N. Baptist's multistep procedure[60] for PHB
          extraction starting from bacterial cells.

just one of a family of microbial poly-β-hydroxyalkanoates
(PHA) all of which have potential as a source of small
molecule organic chemicals. Fig. 17 is meant to illustrate
some of the chemicals that have been isolated[63] from PHA.
Suitable microbial manipulation will ensure that one
polyester or the other is produced exclusively.

In spite of this susceptibility to pyrolysis, PHB can
be melt spun into fibers[64]. The properties of the fibers
including melting point are almost identical to those of
isotactic polypropylene[65]. The application as an
absorbable surgical material, first proposed by
Baptist[61], is under active development[64].

CONCLUSIONS

The ubiquity of carbohydrates in nature is the message
to be read in the expression "carbohydrate polymers".
Whether the term should be made to include polynucleotides
and natural polyalkanoates, such as PHB, is left to the
reader's choice.

The resurgence of interest in the carbohydrates is
illustrated by the recent review article which shows the
properties of cyclodextrins[66] as catalysts which provide
geometric control, as do enzymes, of some standard organic
reactions. Similarly, cell/cell interactions such as
recognition of specific plant cells by pathogens are
exciting new areas of research and commercial
application[67]. At times, the scope of polysaccharides
extends into unexpected areas such as the anti-tumour
properties of Schizophyllan [68].   Current wisdom suggests
that stimulation of the macrophages by certain
polysaccharides is the explanation for well-documented
tumour remission following polysaccharide injections.

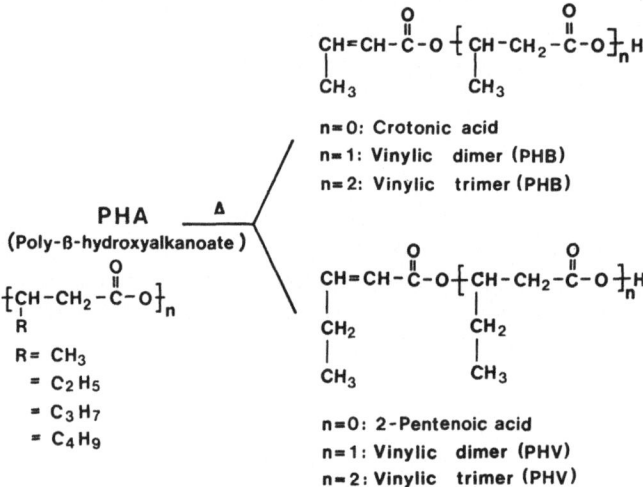

Fig. 17   Monomers and oligomers obtained by pyrolysis of PHA
The reaction scheme is shown for two different R
substituents[75]. The oligomers with a vinylic end
group are referred to as "dehydrates."

In a similar vein, targeted drugs[69], thanks to
carbohydrate labels on vesicle carriers, promises to open
a new era of medicinal treatment. In this way, medicaments
which might be toxic to the organism as a whole can be
directed exclusively to the infected organ.

Finally, the non-renewability of petroleum will
eventually provide an opportunity for chemicals from
biomass. Much of the transformation for this purpose will
be microbiological and the raw materials will be
carbohydrate polymers. Here the impact of the new biology
in the form of genetic engineering for optimized yield and
process promises exciting new concepts in chemical
production. Bioconversion processes already represent
economically feasible methods for producing basic organic
chemicals based on renewable raw materials.

REFERENCES

1.   Comprehensive Organic Chemistry, ed. by Derek Barton
      and W. David Ollis, Part 26, Carbohydrate Chemistry,
      Pergamon Press, N.Y., 5:685-831 (1979)
1a.   N. Sharon, Complex Carbohydrates, Addison-Wesley Pub.
      Co., Reading, Mass. (1975)
2.   V. J. Chapman, in Encyclopedia Britannica, 15th edn.,
      1:487-499
3.   A. S. Perlin and S. Suzuki, Canad. J. Chem. 40:50
      (1962)
4.   D. Gagnaire, R. H. Marchessault, Marc Vincendon,
      Tetrahedron Letters, 45:3953-3956 (1975)
5.   M. J. How, J. S. Brimacombe and M. Stacey, Advs. in
      Carbo. Chem. 19:303-415 (1964)
6.   I. Orskov and M. A. Fife-Asbury, Int. J. Systematic
      Bacteriol. 27:386 (1977)
7.   Future Sources of Organic Raw Materials : CHEMRAWN I,
      ed. by L. St. Pierre and G. R. Brown, Pergamon Press,
      N.Y. (1980)
8.   A. A. Demain, in: Annual Reports of Fermentation
      Processes, ed. by G. T. Tsao, Academic Press, New
      York, 4:193-208 (1980)
9.   P. R. Sundararajan and V.S.R. Rao, Biopolymers 8:305-
      312 (1969)

10. G. N. Ramachandran, C. Ramakrishnan and
    V. Sasisekharan in: Aspects of Protein Structure, ed.
    by G. N. Ramachandran, Academic Press, New York,
    pp.121-135 (1963)
11. A. Bourret, H. Chanzy and R. Lazaro, Biopolymers
    11:893-898 (1972)
12. B. Wunderlich in Macromolecular Physics, vol. 2,
    Academic Press, New York (1976)
13. A. Sarko and R. H. Marchessault, J. Polym. Sci.
    28C:317-331 (1969)
14. R. D. Preston, The Physical Biology of Plant Cell
    Walls, Chapman and Hall, London (1974)
15. J. Blackwell, Biopolymers 7:281 (1969)
16. R. H. Marchessault and Y. Deslandes, Carbohyd. Res.
    75:231-242 (1979)
17. Antony W. Burgess, J. Theor. Biol. 96:21-38 (1982)
18. I. Nieduszinsky and R. H. Marchessault, Biopolymers
    II:1335-1344 (1972)
19. C. Lelliott, E. D. T. Atkins, J. W. F. Juritz and
    A. M. Stephen, Polymer 19:363-367 (1978)
20. H. Chanzy, A. Grosrenaud, J. P. Joseleau, M. Dube and
    R. H. Marchessault, Biopolymers 21:301-319 (1982)
21. R. H. Marchessault, A. Buleon, Y. Deslandes and
    T. Goto, J. Coll. and Interface Sci. 71:375 (1979)
22. T. Painter, Pure and Appl. Chem. (in press)
23. Daniel C. Carter, John R. Ruble and G. A. Jeffrey,
    Carboh. Res. 102:59-67 (1982)
24. R. S. Werbowyz and D. Grey, Mol. Cryst. Liq. Cryst.
    34:97 (1976); ibid. Macromolecules 13:69 (1980).
25. Y. Onogi, J. L. White and J. F. Fellers, J. Pol. Sci,
    Polym. Phys. Ed. 18:663 (1980)
26. Bacterial Adherence, ed. by E. H. Beachey, Chapman
    and Hall, London (1980)
27. N. Sharon, Pure and Appl. Chem. (in press)
28. I. Goldstein, private communication
29. Jean Montreuil, Pure and Appl. Chem. 42:431-477 (1975)
30. Klaus Bock, Pure and Appl. Chem. (in press)
31. J. P. Carver and A. A. Grey, Biochemistry 20:6607-6616
    (1981)
32. I. A. Wilson, J. J. Skekel and D. C. Wiley, Nature
    289:366-372 (1981)
32a. Johann Deisenhofer, Biochemistry 20:2361 (1981)

33. Jean Robert Brisson, Ph.D. thesis: The Three-Dimensional Structure of Asparagine-Linked Glycopeptides, University of Toronto (1982)
34. Harry Schacter, Saroja Narasimhan, Noam Harpaz and Gregory D. Longmore in: Membranes and Transport, ed. by Anthony N. Mastonosi, Plenum Pub. Corp. vol. 1, pp. 255-262 (1982)
35. R. U. Lemieux, K. Bock, L. Delbaere, S. Koto and V. S. R. Rao, Can. J. Chem. 58:631 (1980)
36. Margaret Biswas and V. S. R. Rao, Carboh. Polymers 2:205 (1982)
37. R. H. Marchessault in: Milton Harris: Chemist, Innovator and Entrepreneur, ed. by Miklos M. Breuer, Am. Chem. Soc., Washington, D.C. (1982)
38. O. T. Avery, C. M. MacLeod and M. McCarty, J. Exp. Mod 79:137 (1944)
39. R. Dubos The Professor, The Institute and DNA, The Rockefeller University Press, New York (1976)
40. M. Heidelberger, C. M. MacLeod and M. M. DiLapi, J. Immunol. 66:145-149 (1951).
41. G. G. S. Dutton, Keith L. Mackie, Angela V. Savage, Dietlinde Rieger-Hug and Stephan Stirm, Carboh. Res. 84:161-170 (1980)
42. G. G. S. Dutton, A. V. Savage and M. Vignon, Can. J. Chem. 58, (1980)
43. I. W. Cottrell in: Fungal Polysaccharides, ed. by Paul A. Sandford and Kazuo Matsuda, A.C.S. Symposium Series 126, pp. 251-270 (1980)
44. H. R. Schuppner, U. S. Pat. 3,577,016 (1971)
45. D. A. Rees and W. E. Scott, J. Chem. Soc. B:469 (1971)
45a. R. H. Marchessault and Y. Deslandes, Carboh. Polymers, 1:31-38 (1981)
46. R. H. Marchessault, I. Imada, T. L. Bluhm and P. R. Sundararajan, Carboh. Res. 83:287-302 (1980)
47. Y. Deslandes, R. H. Marchessault and A. Sarko, Macromolecules 13:1466-1471 (1980)
48. E. D. T. Atkins, K. D. Parker, J. Polymer Sci., C28:69 (1968)
49. R. H. Marchessault and Y. Deslandes, Carboh. Res. 75:231-242 (1979)
50. T. Norisuye, T. Yanaki and H. Fujita, J. Pol. Sci., Phys. Ed. 18:547-558 (1980)

51. K. Ogawa, T. Watanabe, J. Tsurugi, S. Ono, Carboh. Res. 23:399 (1972)
52. T. Harada, Process Biochem. 9:21-25 (1974)
53. T. L. Bluhm, Y. Deslandes, R. H. Marchessault, S. Perez and M. Rinaudo, Carboh. Res. 100:117-130 (1982)
54. S. Kikumoto, T. Miyajima, K. Kimura, S. Okubo and N. Komatsu, J. Agr. Chem. Soc. Japan 45:162 (1971)
55. K. Hess in: Die Chemie der zellulose and Ohrer Beglieter, Leipzig: Akademischen Verlaggesellschaft, M.B.H. (1928)
56. M. Lemoine, Ann. Inst. Pasteur, 39:144 (1925)
57. C. Peaud-Lenoel and A. Kepes, Bull. Soc. Chim. Biol. 34:563-575 (1952)
58. R. Alper, D. G. Lundgren, R. H. Marchessault and W. A. Cote, Biopolymers 1:545-556 (1963)
59. D. G. Lundgren, R. Alper, C. Schnaitman and R. H. Marchessault, J. Bacteriol. 89:245-251 (1965)
60. J. N. Baptist, U. S. Pat. 3,036,959 and 3,044,942
61. J. N. Baptist and F. X. Werber, SPE Trans. 4:245 (1964)
62. S. Coulombe, P. Schauweker, R. H. Marchessault and B. Hauttecoeur, Macromolecules 11:279-281 (1978)
63. H. Morikawa and R. H. Marchessault, Can. J. Chem. 59:2306-2313 (1981)
64. Eric R. Howells, Chemistry and Industry pp 508-511 (1982)
65. R. H. Marchessault, S. Coulombe, H. Morikawa, K. Okamura and J. F. Revol, Can. J. Chem. 59:38-44 (1981)
66. Ronald Breslow, Science 218:532-537 (1982)
67. E. R. Morris, D. A. Rees, G. Young, M. D. Walkingshaw and A. Karke, J. Mol. Biol. 110:1 (1977)
68. K. Tabata, T. Ikumoto, T. Yanaki, W. Itoh and T. Kojima, Paper V-21, Abstracts, XIth International Carbohydrate Symposium, Vancouver, Canada Aug. (1982)
69. Marcia R. Mauk, Ronald C. Gamble and J. D. Baldeschwieler, Science 207:309-311 (1980)
70. H. Geyer, S. Stirm and K. Himmelspach, Med. Microbiol. Immunol. 165:271-288 (1979).
71. S. C. Charms and A. M. Stephen, Carbohydr. Res. 35:73 (1974)
72. I. W. Sutherland, J. Gen. Microbiol. 94:211-216 (1976)

73. A. J. Chakraborty, H. Friebolin, H. Niemann and S. Stirm, Carbohydr. Res. 59:523-530 (1977)
74. G. G. S. Dutton and T. E. Folkman, Carbohydr. Res. 80:147-161 (1980)
75. L. L. Wallen and W.K. Rohwedder, Environ. Sci. Technol. 8:576 (1974)

GENE SYNTHESIS — TOWARD A FUTURE WORLD

Marvin H. Caruthers

University of Colorado
Department of Chemistry
Boulder, Colorado

SYNOPSIS

Several recent advances in molecular biology have led to dramatic progress in genetic engineering. Among these are the development of useful plasmid vectors, the characterization and usage of restriction endonucleases, and the elucidation of methods for sequencing and synthesizing DNA. A discussion of these developments and the use of these methods for synthesizing, cloning and expressing an immune interferon gene will be presented.

INTRODUCTION

H. G. Khorana in 1967 had the vision and courage to propose that genes could be chemically synthesized.[1] As is usually the case in Khorana's laboratory, this bold proposal soon became an historical milestone in biochemistry when the first chemical synthesis of a gene was announced in 1970.[2] Unfortunately at that time, the necessary technologies for utilizing a synthetic gene (or even a natural gene) biologically in a controlled manner were simply unavailable. These technologies have since been developed and are referred to collectively as genetic engineering. Additionally, the tremendous effort and forebearance required to complete this task was soundly criticized as a "tour de force" in the popular scientific press.[3] Nevertheless as usually happens when research is being conducted at the very edge of our scientific capabilities, the necessary technology for manipulating and expressing synthetic as well as natural DNA soon became available and was

55

quickly integrated with Khorana's procedures in such a way that
the successful expression of several synthetic genes has now
been achieved.[4-10] Thus a concept originally put forward as an
(almost) unattainable goal, or attainable only after significant indulgence, is now used routinely for solving many biochemical problems and for generating a large variety of proteins. In this lecture I will review the various technologies
that have led to the rapid synthesis of genes and to the isolation of natural genes. I will then outline how we have used
these procedures for the synthesis and expression of an immune
interferon gene. Finally, I will speculate on the future role
of genetic engineering in the biotechnology area.

BASIC TOOLS FOR GENETIC ENGINEERING

Plasmids

     Plasmids are the molecular structures that serve as vehi-
cles for the expression of genes. They multiply independently
within host cells, are composed of circular, double-stranded
DNA, and are inherited in a regular manner. These subcellular
organelles are found in virtually all bacterial species and
generally represent between a fraction of one percent and two
or three percent of the cell's total DNA. Plasmids can be used
as vehicles for introducing nonbacterial genes into bacteria[11]
because they contain or can be constructed to contain genes
that render the host cell resistant to a wide variety of toxic
agents including antibiotics. A plasmid that is particularly
useful for the molecular cloning of foreign DNA into *E. coli* is
pBR322 (Fig. 1). This plasmid contains genes for resistance to
tetracycline and ampicillin. Thus when pBR322 is taken up by
E. coli, the cells are resistant to these antibiotics. This
method of conferring drug resistance provides a very powerful
tool for cloning nonbacterial DNA into *E. coli*. For example,
insertion of a foreign piece of DNA into the tetracycline gene
renders the tetracycline gene inactive. Therefore cells containing pBR322 with this foreign DNA insert are resistant to
ampicillin but sensitive to tetracycline. In contrast, cells
that fail to take up the vector are sensitive to both antibiotics and cells containing pBR322 without the DNA are resistant
to both. Therefore pBR322 (or any other plasmid) containing
two drug resistance genes provides a mechanism for cloning nonbacterial DNA into bacteria and has been used extensively for
this purpose.

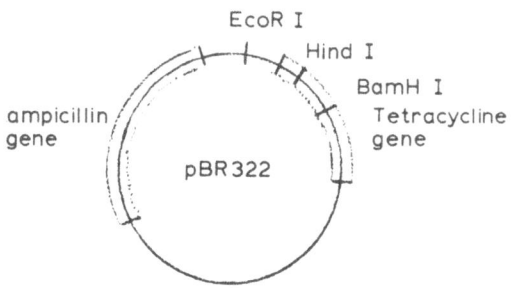

Fig. 1. A schematic representation of plasmid pBR322. The
tetracycline gene, ampicillin gene and the relative
location of certain selected restriction endonuclease
cleavage sites are shown.

Restriction Endonuclease Enzymes

Restriction endonucleases are found in a wide variety of
prokaryotes. These endonucleases recognize unique DNA sequences (usually four to six contiguous nucleotide base pairs)
and hydrolyze phosphodiester bonds on both strands at sites
usually within this sequence. Their biological role is to
degrade foreign DNA. However the cell's own DNA is not degraded because the nucleotide bases recognized by the restriction enzyme are methylated. A striking characteristic of these
cleavage sites is that they possess twofold rotational symmetry
which means that the sequence of base pairs is palindromic.
The cleavage sites are symmetrically positioned with respect to
the twofold axis. An example is shown in Fig. 2. The sequence
recognized by a restriction enzyme from *Bacillus amyloliquefaciens H* is G-G-A-T-C-C- (the *Bam*HI site). The -G-G- phosphodiester bond in each strand distal to the symmetry axis is
cleaved by this endonuclease. Currently, well over a hundred
restriction enzymes having different sequence specificity have
been purified and characterized.[12] Their names consist of a
three letter abbreviation for the host organism (e.g. *Bam* for
Bacillus amyloliquefaciens H) followed by a strain designation
and a Roman numeral identifying the restriction enzyme from
that particular bacteria (some bacteria have more than one
restriction enzyme).

```
5'--G-G-A-T-C-C--  -3'
3'--C-C-T-A-G-G--  -5'
         ↓BamH I

5'--G              +  pG-A-T-C-C-- -3'
3'--C-C-T-A-Gp                G-- -5'
```

Fig. 2. An illustration of the sequence specificty for the
         *Bam*HI restriction endonuclease. The symbol p repre-
         sents a 5'-phosphate.

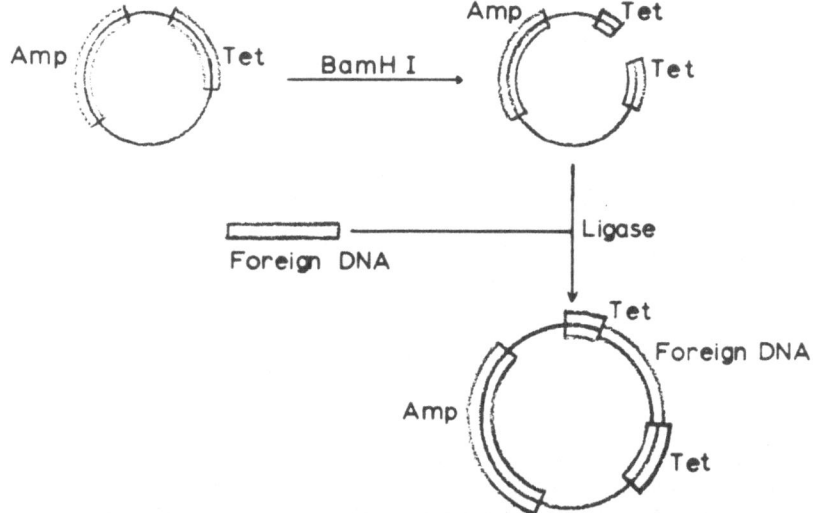

Fig. 3. An example illustrating the insertion of a foreign DNA
         segment into the tetracycline gene of pBR322 at the
         *Bam*HI site. Tet and Amp refer to tetracycline and
         ampicillin genes, respectively.

     Restriction enzymes in combination with DNA ligases and
kinases can be used to specifically insert a foreign DNA into a
plasmid. An example is illustrated in Fig. 3. The pBR322 plasmid can be cleaved at a unique *Bam*HI site within the tetracycline gene. The staggered cuts made by this enzyme generate
complementary single-stranded ends (cohesive ends). If a DNA

fragment to be inserted into this plasmid also has *Bam*HI sites,
the foreign DNA can first be cut with this enzyme and then
inserted into pBR322 using T4 DNA ligase to join the two DNAs at
their common, overlapping single strand regions. Once this
recombinant plasmid has been synthesized, transformation and
selection of *E. coli* cells resistant to ampicillin but sensitive to tetracycline is possible and leads to isolation of a
bacterial strain containing the required foreign DNA inserted
into a plasmid. A recent innovation in this area is the development of synthetic adaptors composed of oligodeoxynucleotides
(10-20 monomers each) containing one or more unique restriction
enzyme cleavage sites.[13] An appropriate choice of adaptors can
be used to interface nonhomologous restriction fragments with
various plasmids.

DNA Sequencing

There are two basic approaches for sequencing DNA. These
methodologies have been published in considerable detail.[14,15]
The first step in either approach is to generate a series of
deoxyoligonucleotide homologs from the segment of unknown
sequence. This series constitutes those deoxyoligonucleotides
having the same 5'-end but differing as to length by one nucleotide unit. An example of such a series is shown in Fig. 4 for
the deoxyoligonucleotide d(pApGpCpGpTpG). The two sequencing
approaches differ in how this series is generated. One method
involves four chemical procedures with each having unique specificity for cleaving DNA adjacent to only one or two of the four
nucleotide bases. The other method generates the same specificity using enzymes. One important concept shared by both
approaches is to generate all possible homologs by limited reaction at each base. Others involve ordering the homologs as to
size by polyacrylamide gel electrophoresis and introducing a
$[^{32}P]$-phosphate label into the molecule. The simultaneous generation of all possible homologues from separate reactions
specific for each base can be displayed and compared side-by-
side on the same gel. An example is shown in Fig. 4. By
reading the first radioactive band representing the most mobile
or smallest component (i.e. the band nearest the anode) and
proceeding to read toward the cathode in a zigzag fashion as
different bands are encountered, the sequence of the deoxyoligonucleotide can be determined. This method is extremely fast and
can be used to sequence deoxyoligonucleotides containing approximately 100 to 200 mononucleotides per experiment. By using
overlapping DNA segments generated by cleavage with different
restriction enzymes, thousands of contiguous base pairs can be

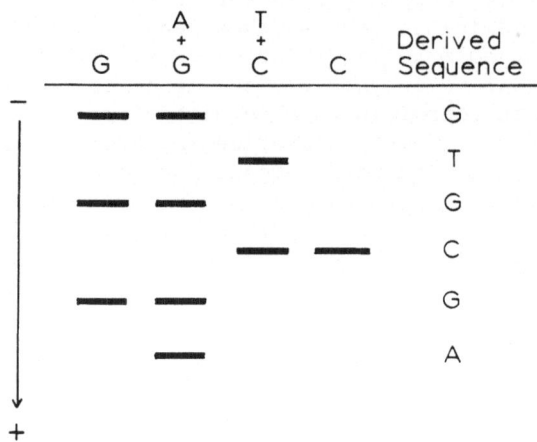

Figure 4. Gel electrophoresis of partially degraded DNA sam-
           ples. Samples marked G, A + G, T + C, and C designate
           different chemical reactions specific for G, A, T, and
           C, respectively.

sequenced. The development of this methodology therefore provides a mechanism for completely characterizing any cloned DNA
fragment.

DNA Synthesis

        The basic strategy we have developed for the synthesis of
a deoxyoligonucleotide [16,17] is outlined in Fig. 5. Thus mononucleotides are added sequentially to a nucleoside covalently
attached to an insoluble polymer support. Reagents, starting
materials and side products are then removed simply by filtration. After the addition of all mononucleotides, the deoxyoligonucleotide is chemically freed of blocking groups, hydrolyzed
from the support and purified to homogeneity by polyacrylamide
gel electrophoresis.

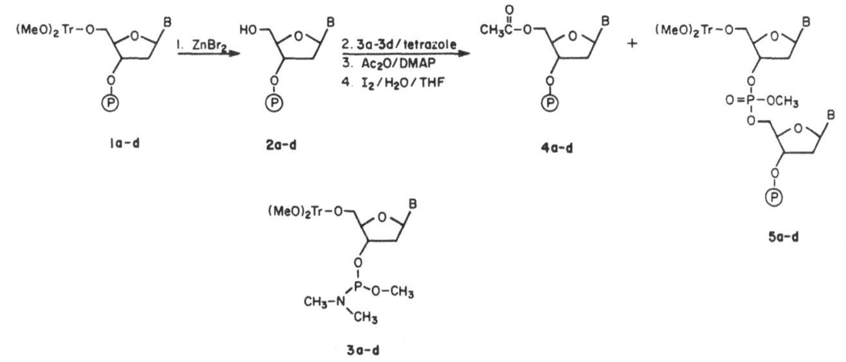

Fig. 5.   Steps in the synthesis of a dinucleotide. B refers to
          thymine, N-benzoylcytosine, N-benzoyladenine, or N-
          isobutyrylguanine for 1a-d, respectively. A similar
          nomenclature applies to 2a-d, 3a-d, 4a-d, and 5a-d.
          (MeO)$_2$Tr refers to the di-p-anisylphenylmethyl group.
          ⓟ refers to silica gel.

The insoluble polymer support is synthesized from silica
gel and an appropriately derivatized deoxynucleoside. We routinely start with high performance liquid chromatography (HPLC)
grade silica gel as a support matrix.[18] The initial synthesis
step involves refluxing 3-aminopropyl triethoxysilane with
silica gel in dry toluene for 3 h to form the silylamine. This
derivatized silica gel is then allowed to react with an appropriately protected nucleoside containing a 3'-p-nitrophenyl-
succinate ester. The product is the nucleoside covalently
linked to the support (compounds 1a-d).

The synthesis of a deoxyoligonucleotide proceeds by addition of mononucleotides to the polymeric support. The first
step is removal of the dimethoxytrityl group using ZnBr$_2$ in
nitromethane. Depurination, a troublesome problem with protic
acids, is not observed with this Lewis acid. The next step is
addition of a mononucleotide to 2a-d. We routinely use suitably protected deoxynucleoside 3'-phosphoramidites (compounds
(3a-d) which are ideal as intermediates in deoxyoligonucleotide
synthesis. These reagents are easy to prepare using standard
organic chemical procedures[19] and can be stored indefinitely as
stable white solids. When mixed with various weak acids such
as tetrazole in acetonitrile, these phosphoramidites are converted to the corresponding tetrazolides which react very

rapidly with compound 2a-d. The reaction is complete in less
than a minute but we usually continue the condensation for 5
minutes. Using these suitably protected deoxynucleoside phosphoramidites, we synthesize deoxyoligonucleotides containing
from 10 to 30 mononucleotides. The final steps involve first
acylation of non-reacted 5'-hydroxyl groups with acetic anhydride and then oxidation of the phosphite internucleotide
linkage to the corresponding phosphate using I_2 in water, 2,6-
lutidine and tetrahydrofuran.

Upon completion of a synthesis, the deoxyoligonucleotide is
freed of protecting groups and isolated by polyacrylamide gel
electrophoresis. Silica gel containing the synthetic deoxyoligonucleotide is first treated with triethylammonium thiophen-
oxide in order to remove the methyl groups from internucleotide
phosphotriesters. The next step is hydrolysis of the ester
joining the deoxyoligonucleotide to the support (concentrated
ammonium hydroxide for 3 h at 20°C). After centrifugation and
recovery of the supernatant containing the mixture of failure
sequences and the correct deoxyoligonucleotide, the N-benzoyl
groups from deoxycytidine and deoxyadenosine, and the N-iso-
butyryl group from deoxyguanosine are removed by warming the
concentrated ammonium hydroxide solution at 50°C for 12 h. The
final product is isolated free of failure sequences by a one
step procedure involving electrophoresis on 20% polyacrylamide
gel.

Several devices have been used to aid in the synthesis of
deoxyoligonucleotides. Initially we developed a semi-automatic
apparatus containing a pump, a column, and a series of reagent
bottles.[18] Other less complex devices can also be used. For
example we have successfully synthesized compounds in test
tubes and on simple sintered glass funnels. In the former case
each synthesis step is followed by a low speed centrifugation
and decantation in order to remove solvents and reagents. For
the latter case reagents and solvents are removed by filtration.
More recently we have made considerable progress toward developing a completely automatic microprocessor controlled machine.
This synthetic methodology can therefore be adapted to a wide
variety of manual, semi-automatic, or automatic machines and to
the requirements of a large number of molecular biologists and
chemists. So far we have used this approach to synthesize several hundred deoxyoligonucleotides for a large number of pro-
jects. The synthesis of the gene for immune interferon will be
described in order to illustrate how this chemistry can be used
as a successful part of a program in genetic engineering.

SYNTHESIS OF THE IMMUNE INTERFERON GENE

The immune interferon gene has been isolated, sequenced, and expressed.[20] One approach for studying the biological activity of immune interferon is to change the amino acid

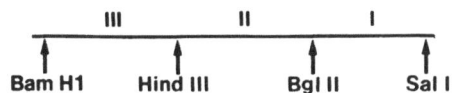

"Unique" Restriction Site Map of Gamma Interferon Gene

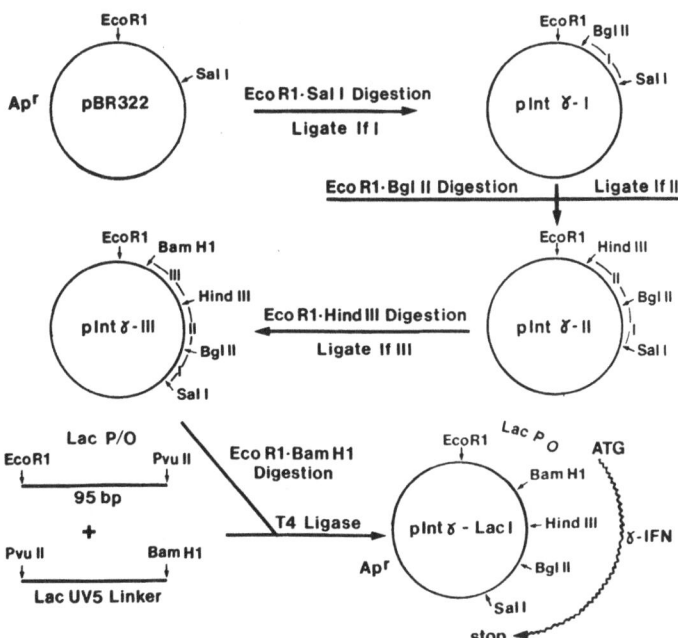

Fig. 6.   The synthesis plan for the immune interferon gene. The top part of the figure schematically illustrates the immune interferon gene containing unique restriction endonuclease cleavage site. The remainder of the figure outlines the synthetic strategy as discussed in the text.

sequence at specific sites and then determine how these changes alter the functions of the protein. This objective can best be accomplished by synthesizing an immune interferon whose DNA sequence (and consequently the protein sequence) is readily modified. We therefore devised a synthesis plan (top part of Fig. 6) such that several unique restriction endonuclease cleavage sites were included as part of the gene.[21] In this way, various portions of the gene such as between the *Hind*III and *Bgl*II sites could be enzymatically excised and replaced with new synthetic DNA duplexes coding for altered immune interferons. This overall plan therefore has considerable flexibility. We can study and modify one section of immune interferon without totally resynthesizing the gene. The remaining portion of Fig. 6 outlines our cloning and synthesis strategy. As can be seen by a careful examination of this figure, the gene was synthesized and cloned in three sections. Each section was composed of a gene section bounded by unique restriction endonuclease sites and a small synthetic DNA linker containing an *Eco*RI site. This strategy allowed us to always clone synthetic gene segments between the pBR322 *Eco*RI site and the newly generated, unique restriction endonuclease site. Finally the synthetic gene was joined to the UV5 *lac* promoter and interferon expressed at concentrations comparable to published amounts.[20] Current experiments are underway to modify the protein sequence of this gene and to synthesize the genes for other biologically interesting proteins using similar synthesis strategies.

EXPERIMENTAL SECTION

General Methods

     Thiophenol, 4-dimethylaminopyridine, anhydrous $ZnBr_2$, and iodine were purchased from Aldrich and used without further purification. Reagent grade acetic anhydride, triethylamine, and t-butylamine were used as received. Common solvents such as tetrahydrofuran, dioxan, acetonitrile, nitromethane, and methanol were stored over activated 4A molecular sieves and used without further purification. Dry acetonitrile was obtained by refluxing reagent grade solvent over CaH_2 for several hours and distilling just prior to usage. 2,6-Lutidine was obtained by refluxing reagent grade solvent over CaH_2 for one hour followed by distillation from 4A molecular sieves. 2,6-Lutidine was stored in the dark. Tetrahydrofuran was dried over sodium-benzophenone and used freshly distilled.

1-H-Tetrazole (Aldrich) was sublimed at 110-115°C at 0.05 mm Hg
prior to use.

All solution transfers involving dry reagents were completed with clean syringes dried in ovens at 50°C.

Synthesis of the Support

The initial step was synthesis of the succinilated deoxynucleosides. All four were prepared using the same general
procedure. To a solution of 5'-demethoxytritylnucleoside
(5 mmol) in anhydrous pyridine was added 4-dimethylaminopyri-
dine (0.61 g; 5 mmol) and succinic anhydride (6.6 g; 6 mmol).
The reaction was monitored by tlc(acetonitrile:water; 9:1) and
was usually complete after 12 h at 20°C. Occasionally a second
portion of succinic anhydride (0.1 g; 1 mmol) was added. The
reaction was next quenched with water (0.1 ml) for 10 min at
20°C. The reaction mixture was evaporated *in vacuo*, and then
co-evaporated twice with dry toluene (2 X 20 ml). The residue
was redissolved in dichloromethane (40 ml) and the solution was
washed successively, once with 10% citric acid (10 ml) and
twice with water (2 X 10 ml). The organic solution was dried
over anhydrous sodium sulfate, and evaporated *in vacuo*. The
residue was redissolved in 10 ml dichloromethane (containing 5%
pyridine) and the product precipitated into pentane:ether (200
ml; v/v). The precipitate was dried *in vacuo* (yield: 70-85%).
In order to derivatize silica gel containing an amino function,
the succinylated nucleoside (1 mmole) was dissolved in dioxane
(4 ml) containing dry pyridine (0.2 ml) and then *p*-nitrophenol
(140 mg; 1 mmol) was added. A solution of dicyclohexylcarbodi-
imide (220 mg) in anhydrous dioxane (1 ml) was added and the
reaction was monitored by tlc (silica gel plate; benzene:diox-
ane,3:1). The reaction was virtually complete after 2 h at
room temperature. Dicyclohexylurea was removed by centrifugation, and the supernatant containing the desired product was
used directly in the condensation reaction.

Deoxynucleosides were attached to the support using the
following general procedure. HPLC grade silica gel (12 g,
Fractosil 200, Merck) was exposed to a 15% relative humidity
(saturated LiCl) for at least 24 h. The silica was then
treated with 3-triethoxysilylpropylamine (13.8 g, 0.01 M in dry
toluene) for 12 h at 20°C and 18 h at reflux. It was isolated
by centrifugation, washed successively (3 times each) with toluene, methanol and 50% aqueous methanol. The silica was then
shaken with 50% aqueous methanol (200 ml) at 20°C for 18 h.

After isolation by centrifugation, the silica was washed with
methanol and ether and dried *in vacuo*. The dried silica was
suspended in anhydrous pyridine and treated with trimethylsilyl
chloride (15 ml) for 12 h at 20°C. After isolation by centrifugation, the silica was washed four times with methanol, twice
with ether, and then dried *in vacuo*. The dry silica (3 g) was
suspended in DMF (5 ml) and a solution of the 5'-dimethoxytri-
tylnucleoside-3'-*p*-nitrophenylsuccinate in dioxane and 1 ml of
triethylamine was added. The suspension was shaken at 20°C for
4 h. Ninhydrin test at this stage indicated the existance of
free amino groups on the resin. To cap these groups, acetic
anhydride (0.6 ml) was added and the mixture was shaken for
another 30 min, after which time a negative ninhydrin test was
obtained. The silica was isolated by centrifugation, washed
successively (3 times each) with DMF, 95% ethanol, dioxane and
ethyl ether, and then dried *in vacuo*. Analysis for the extent
of dimethoxytritylnucleoside attached to the support was done
spectrophotometrically. An accurately weighed sample of silica
(10-15 mg) was treated with 0.1 M *p*-toluenesulfonic acid in
acetonitrile and the optical density of the supernatant obtained after centrifugation was measured at 498 nm. (The
extinction coefficient of dimethoxytritanol is 7.0 X 10^4.) The
extent of derivation to form 1a-d was found to be the following: (MeO)$_2$TrdT, 62 μmol/g; (MeO)$_2$TrdibG, 56 μmol/g;
(MeO)$_2$TrdbzA, 65 μmol/g; (MeO)$_2$TrdbzD, 68 μmol/g.

Synthesis of Deoxynucleosidephosphoramidites

The careful preparation of compounds 3a-d is of critical
importance. The synthesis begins with the preparation of
chloro-N,N-dimethylaminomethoxyphosphine [CH$_3$OP(Cl)N(CH$_3$)$_2$]
which is used as a monofunctional phosphitylating agent. A 250
ml addition funnel was charged with 100 ml of precooled anhydrous ether (-78°C) and precooled (-78°C) anhydrous dimethylamine (45.9 g, 1.02 mol). The addition funnel was wrapped with
aluminum foil containing dry ice in order to avoid evaporation
of dimethylamine. This solution was added dropwise at -15°C
(ice-acetone bath) over 2 h to a mechanically stirred solution
of methoxydichlorophosphine (47.7 ml, 67.32 g., 0.51 mol) in
300 ml of anhydrous ether. The addition funnel was removed
and the 1 liter, three-necked round bottom flask was stoppered
with serum caps tightened with copper wire. The suspension was
mechanically stirred for 2 h at room temperature. The suspension was filtered and the amine hydrochloride salt was washed
with 500 ml anhydrous ether. The filtrate and washings were
combined and ether was distilled at atmospheric pressure. The

residue was distilled under reduced pressure. The product was collected at 40-42°C at 13 mm Hg and was isolated in 71% yield (51.1 g, 0.36 mol). d^{25}= 1.115 g/ml. ^{31}P-N.M.R., δ = -179.5 ppm (CDCl$_3$) with respect to internal 5% v/v aqueous H$_3$PO$_4$ standard. ^{1}H-N.M.R. doublet at 3.8 and 3.6 ppm J$_{P-H}$ = 14 Hz (3H, OCH$_3$) and two singlets at 2.8 and 2.6 ppm (6H, N(CH$_3$)$_2$). The mass spectrum showed a parent peak at m/e = 141.

Compounds 3a-d were prepared by the following procedure. 5'-O-Di-p-anisylphenylmethyl nucloside (1 mmol) was dissolved in 3 ml of dry, acid free chloroform and diisopropylethylamine (4 mmol) in a 10 ml reaction vessel preflushed with dry nitrogen. [CH$_3$OP(Cl)N(CH$_3$)$_2$] (2 mmol) was added dropwise (30 sec) by syringe to the solution under nitrogen at room temperature. After 15 min the solution was transferred with 35 ml of ethyl acetate into a 125 ml separatory funnel. The solution was extracted four times with an aqueous, saturated solution of NaCl (80 ml). The organic phase was dried over anhydrous Na$_2$SO$_4$ and evaporated to a foam under reduced pressure. The foam was dissolved with toluene (10 ml) (3d was dissolved with 10 ml of ethyl acetate) and the solution was added dropwise to 50 ml of cold hexanes (-78°C) with vigorous stirring. The cold suspension was filtered and the white powder was washed with 75 ml of cold hexanes (-78°C). The white powder was dried under reduced pressure and stored under nitrogen. Isolated yields of compounds 3a-d were 90-94%. The purity of the products was checked by ^{31}P-N.M.R. Compounds 3a-d are characterized as two peaks between -146 and -145 ppm. Various impurities are sometimes observed with peaks between 0 and +10 relative to phosphoric acid. These impurities do not appear to inhibit the condensation reactions.

Outline of the Synthesis Cycle

The appropriately derivatized deoxynucleoside attached covalently to silica gel (compound 1a, 1b, 1c, or 1d) was treated with a saturated solution of anhydrous ZnBr$_2$ in nitromethane:methanol (95:5) for 4 min. The support was then washed with nitromethane followed by methanol. Before the condensation step the silica gel was carefully washed several times with dry acetonitrile under an inert atmosphere (N$_2$). Stock solutions of sublimed tetrazole and appropriately protected 2'-deoxynucleoside-3'-N,N-dimethylaminomethoxyphosphines were prepared. Usually these stock solutions were sufficient for at least three condensations. For each µmole of deoxynucleoside attached covalently to silica gel, tetrazole

(60 mole) and the deoxynucleoside-3'-N,N-dimethylaminomethoxy-phosphine in the condensation mixture was about 0.1 M. The condensation reaction was stopped after 5 min. Immediately following the condensation reaction, the silica gel was washed with a tetrahydrofuran:water:2,6-lutidine solution (2:2:1) for 3 min. Oxidation of trivalent phosphorus to pentavalent phosphate (5a-d) was completed using a 0.2 M solution of iodine in tetrahydrofuran:water:2,6-lutidine (2:2:1) for five min. The silica gel was washed with methanol followed by tetrahydrofuran. Acylation of unreactive hydroxyl groups was completed by adding first a solution of 4-dimethylaminopyridine in dry tetrahydrofuran (2 ml; 6.5% w/v) and then a solution of acetic anhydride in 2,6-lutidine (0.4 ml; 1:1) to the support. After 5 min, this acylation solution was removed and the silica gel was washed with methanol and nitromethane. This step completes one synthesis cycle.

Isolation of Synthetic Deoxyoligonucleotides

After completion of the appropriate synthesis cycles, deoxyoligonucleotides free of protecting groups and side products were isolated using the following procedure. Silica gel containing the deoxyoligonucleotide was first treated with a solution containing thiophenol:dioxane:triethylamine (1:2:2) for 90 min at room temperature. This deprotection step removes the methyl phosphotriester protecting group. The support was next treated with concentrated ammonium hydroxide for 24 h at 60°C. The liquid phase was evaporated to dryness *in vacuo* and the residue was treated with a solution of t-butylamine in methanol (1:1) for 24 h at 60°C in a screw cap vial. The reaction mixture was evaporated to dryness *in vacuo* and then loaded on a Sephadex G-50 column. Fractions containing deoxyoligonucleotides were pooled and the DNA isolated by ethanol precipitation. The crude DNA pellet was dissolved in formamide and loaded on a 20% denaturing polyacrylamide gel. After electrophoresis, the band containing deoxyoligonucleotide product was visualized using an ultraviolet lamp. The gel slice containing the product was eluted and desalted using standard procedures.[22] Usually the product is the major u.v. light absorbing band on the gel.

LOOKING TOWARD THE FUTURE

Genes coding for unique proteins can now be isolated from natural chromosomes or chemically synthesized. Moreover by a

series of biochemical and microbiological manipulations in
plasmids, these genes can be cloned and expressed in appropriate micro-organisms. Clearly this technology will be used to
express a large number of proteins having potential benefit as
pharmaceutical reagents. Already human insulin [6,10] human
growth hormone,[7] and various human interferons[8,9] have been
cloned and expressed from natural or synthetic genes. This
technology should allow us to pursue several other directions
as well. For example, the amino acid sequence of known enzymes
can now be modified by chemically synthesizing the enzyme gene
in modified form. Perhaps in this manner a new field of industrial catalysts can be generated having specificity and potential catalytic activity far in excess of the standard inorganic, organic, and organometallic catalysts currently being
used. Perhaps as well, this technology will prove useful in
designing new drugs that can be used to replace natural but
unstable lymphokines and hormones. This should be possible
because these natural proteins can now be produced in large
quantities using genetic engineering procedures. Consequently
the tertiary structure can be deduced and used to determine the
structure parameters of potentially interesting new drugs that
may substitute for the natural protein. As is usually the case
with new technologies, the real merits and advantages of genetic engineering have undoubtedly eluded our projections. Only
the future will show us the way.

ACKNOWLEDGEMENTS

    The development of the chemistry outlined in this manu-
script was supported by NIH (GM21120 and GM25680) in the
author's laboratory. This research was completed by a group of
extremely capable and imaginative colleagues whose names appear
on the cited references. The synthesis of the immune interferon gene was completed in the laboratories of AMGen Development Inc., Boulder, Colorado, and will be published in detail
elsewhere.

REFERENCES

1. H. G. Khorana, "Progress Towards the Synthesis of the
    Yeast Ala tRNA Gene," in Proceedings of the Seventh Inter-
    national Congress of Biochemistry, Tokyo, August 19-25,
    1967, p. 17.

2. K. L. Agarwal, H. Büchi, M. H. Caruthers, N. Gupta, H. G.
    Khorana, K. Kleppe, A. Kumar, E. Ohtsuka, U. L. RajBhan-
    dary, J. H. van de Sande, V. Sgaramella, H. Weber, and T.
    Yamada, Nature 227, 27 (1970).
3. Cell Biology Correspondent, "Is Enough Too Much," Nature
    New Biology 241, 33 (1973).
4. H. G. Khorana, Science 203, 614 (1979).
5. K. Itakura, T. Hirose, R. Crea, A. D. Riggs, H. L.
    Heyneker, F. Boliver, and H. W. Boyer, Science 198, 1056
    (1977).
6. D. V. Goeddel, D. G. Kleid, F. Bolivar, H. L. Heyneker,
    D. G. Yansura, R. Crea, T. Hirose, A. Kraszewski, K.
    Itakura, and A. Riggs, Proc. Natl. Acad. Sci. USA 76,
    106 (1979).
7. D. V. Goeddel, H. L. Heyneker, T. Hozumi, R. Arentzen, K.
    Itakura, D. G. Yansura, M. J. Ross, G. Miozzari, R. Crea,
    and P. Seeburg, Nature 381, 544 (1979).
8. M. D. Edge, A. R. Greene, G. R. Heathcliffe, P. A. Mea-
    cock, W. Schuch, D. B. Saclon, T. C. Atkinson, C. R. New-
    ton, and A. F. Markham, Nature 292, 756 (1981).
9. S. Nagata, H. Taira, A. Hall, L. Johnsrud, M. Streali, J.
    Ecsödi, W. Boll, K. Cantrell, and C. Weissman, Nature 284,
    316 (1980); D. V. Goeddel, E. Yelverton, A. Ullrich, H. L.
    Heyneker, G. Miozzari, W. Holmes, P. Seeburg, T. Dull, L.
    May, N. Stebbing, R. Crea, S. Maeda, R. McCandliss, A.
    Sloma, J. Tabor, M. Gross, P. Famillette, and S. Pestka,
    Nature 287, 411 (1980).
10. H. M. Hsiung, W. L. Sung, R. Brousseau, R. Wu, and S. A.
    Narang, Nucleic Acids Research 8, 5753 (1980).
11. J. F. Morrow, S. N. Cohen, A. C. Y. Chang, H. W. Boyer, H.
    M. Goodman, R. B. Helling, Proc. Natl. Acad. Sci. USA 71,
    1743 (1974).
12. R. J. Roberts, Directory of Restriction Endonucleases, in
    Methods in Enzymology, Vol. 68, R. Wu, ed., Academic Press
    New York, NY 1979, p. 27.
13. R. J. Rothstein, L. F. Lau, C. P. Bahl, S. A. Narang, and
    R. Wu, Synthetic Adaptors for Cloning DNA, in Methods in
    Enzymology, Vol. 68, R. Wu, ed., Academic Press, New York,
    NY 1979, p. 98.
14. A. J. H. Smith, DNA Sequence Analysis by Primed Synthesis,
    in Methods in Enzymology, Vol. 65, L. Grossman and K. Mol-
    dave, eds., Academic Press, New York, NY 1979, p. 560.
15. A. M. Maxam and W. Gilbert, Proc. Natl. Acad. Sci. USA 74,
    560 (1977).

16. M. H. Caruthers, S. L. Beaucage, C. Becker, W. Efcavitch, E. F. Fisher, G. Galluppi, R. Goldman, P. deHaseth, F. Martin, M. Matteucci, and Y. Stabinsky, New Methods for Synthesizing Deoxyoligonucleotides, in Genetic Engineering Vol. 4, J. K. Setlow and A. Hollaender, Eds. Plenum Press, New York, 1982, p. 1.
17. M. H. Caruthers, Chemical Synthesis of Oligodeoxynucleotides Using the Phosphite Triester Intermediates, in Chemical and Enzymatic Synthesis of Gene Fragments, A Laboratory Manual, H. G. Gassen and A. Land, eds., Verlag Chemie, Weinheim, Federal Republic of Germany, 1982, p. 71.
18. M. D. Matteucci and M. H. Caruthers, J. Am. Chem. Soc. 103, 3185 (1981).
19. S. L. Beaucage and M. H. Caruthers, Tetrahedron Lett. 22, 1859 (1981).
20. P. W. Gray, D. W. Leung, D. Pennica, E. Yelverton, R. Najarian, C. Simonsen, R. Deryuck, P. Sherwood, D. Wallace, S. Berger, A. Levinson, and D. Goeddel, Nature 295, 503 (1982).
21. Stabinsky, Y., et al., "Total Synthesis of Immune Interferon," in preparation.
22. Maxam, A. M. and Gilbert, W., in Methods in Enzymology, Vol. 65, L. Grossman and K. Moldave, eds., Academic Press, New York, 1979, p 499.

POLYMERIC MONOLAYERS AND LIPOSOMES AS MODELS FOR

BIOMEMBRANES AND CELLS

K. Dorn and H. Ringsdorf

University of Mainz
Institute of Organic Chemistry
D-6500 Mainz, West-Germany

SYNOPSIS

Conventional model membrane systems based on natural lipids usually lack long term stability. Polymer chemistry can help to overcome this disadvantage introducing polymerizable lipids into these systems. Polymerizable lipids used are diacetylenes, butadienes, acrylates or lipids which can undergo polycondensation.

The polymerization behaviour of these membrane systems can be characterized in monolayers and liposomes using surface pressure-area diagrams or DSC investigations respectively. Decreased membrane permeability and improved stability towards the addition of detergents of polymerized vesicles can be verified by leakage measurements of entrapped 6-carboxyfluorescein.

To introduce biological specifities into these stabilized systems polymerizable glycolipids are synthesized to provide surface recognition properties. Both monomeric and polymeric vesicles made of these lipids are agglutinated by lectins.

Further biological modifications are possible by adding natural lipids or membrane proteins (such as ATPase) to polymerizable membranes. Membrane protein activity can be retained after polymerization. The natural, non-polymerizable lipid component can be selectively degraded (for instance by phospholipase A2) to open up the previously stable compartment. The polymerization behaviour of

diacetylene and butadiene lipids strongly depends on the
miscibility with the natural component. Means to characterize
these mixed systems are again DSC and electron microscopy,
the latter evaluating patch formation using the ripple
structure of phospholipids.

Besides this "synthetic" route, the combination of
natural and polymerizable membrane components is principally
possible via the fusion of cell membranes with polymerizable
vesicles. Fusion of cells is possible using dielectrophoresis
and dielectric breakdown. Since it was possible to prepare
"giant" vesicles (visible under the light microscope) these
techniques were also successfully applied to vesicle-vesicle
fusion. Investigations on cell-vesicle fusion are currently
under way.

INTRODUCTION

This contribution will deal with the problem of what
can polymer chemistry contribute to improve the usefulness
of model systems for biomembranes and cells. Conventional
membrane model systems such as vesicles or black lipid
membranes based on natural lipid components usually lack
long term stability. So, the first goal on the way to a
better biomembrane model is to improve membrane stability.
However, besides having increased stability the model
system should also still be able to perform biological
membrane functions such as selective permeability, surface
recognition, membrane protein activity, and membrane fusion.
How polymer chemistry can provide means to realize stable
membrane model systems exhibiting these abilities will be
discussed in the following.

POLYMERIC MEMBRANE MODELS

Looking at the Singer-Nicolson model (1) biomembranes
mainly consist of lipids and proteins, the latter being
either partially (peripheral proteins) or completely
(integral proteins) embedded into the lipid matrix. These
proteins contribute a major part to the stability of natural
membranes (2). The use of these proteins, however, to
stabilize an artificial membrane system via incorporation is
rather limited and not yet realizable. The stabiliztation of
the lipid matrix itself, therefore, appears to be a more
convenient method also providing a greater potential use.

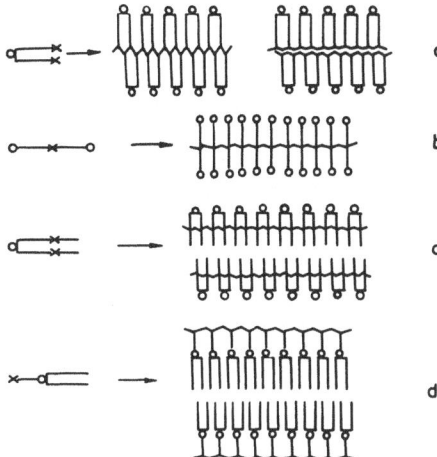

Fig. 1: Possible ways to synthesize polymeric model membranes.
(x = polymerizable group). a)-c) Polymerization with
preservation of headgroup properties; d) polymerization with
preservation of chain mobility. For examples of corresponding monomers, see table 1.

This can be achieved introducing lipids carrying polymerizable functionalities and polymerizing these compounds
in a membrane-like orientation, i.e. monolayers, planar
bimolecular lipid membranes (BLMs) or liposomes (3).

Principally there are two different possibilities to
place a polymerizable group into a lipid molecule (Fig.1).
The polymerizable functionality can either be located in
the hydrophobic moiety of the lipid (a-c) or be fixed to the
hydrophilic headgroup. Polymerizing lipids in the hydrophobic region will lead to polymeric membranes with
decreased alkyl chain mobility compared to the monomer,
however, the hydrophilic headgroup will be unchanged.
Having the polymerizable group fixed to the hydrophilic
portion of the lipid will not allow the use of biological
headgroups, however, the polymer chain will not disturb the

Table 1: Several examples of polymerizable, liposome-forming
lipid analogues (cf. Fig. 1).

Type	Compound	Ref.

a)   X———————◯   X————————◯
      X———————

(1) $CH_2=CH-(CH_2)_8COO-(CH_2)_2$
                     $\overset{+}{N}$ $\begin{smallmatrix} CH_3 \\ \\ CH_3 \end{smallmatrix}$ Br^- (3g)
  $CH_2=CH-(CH_2)_8COO-(CH_2)_2$

(2) $CH_2=C(CH_3)-CO-NH-(CH_2)_{10}COO-(CH_2)_2$
                               $\overset{+}{N}$ $\begin{smallmatrix} CH_3 \\ \\ CH_3 \end{smallmatrix}$ Br^- (17)
  $CH_2=C(CH_3)-CO-NH-(CH_2)_{10}COO-(CH_2)_2$

(3) $CH_2=C(CH_3)-COO-(CH_2)_{11}COO-CH_2$
   $CH_2=C(CH_3)-COO-(CH_2)_{11}COO-CH$ (19)
                      $CH_2-O-PO_3^--(CH_2)_2-\overset{+}{N}(CH_3)_3$

(4) $CH_2=C(CH_3)-COO-(CH_2)_{11}$
                       $\overset{+}{N}$ $\begin{smallmatrix} CH_3 \\ \\ CH_3 \end{smallmatrix}$ Br^- (36)
     $CH_3-(CH_2)_{15}$

(5) $CH_2=C(CH_3)-COO-(CH_2)_{11}-COO-(CH_2)_6$
                                    $\overset{+}{N}$ $\begin{smallmatrix} CH_3 \\ \\ CH_3 \end{smallmatrix}$ Br^- (27)
          $CH_3-(CH_2)_{17}$

(6) $CH_3-(CH_2)_{14}-COO-CH_2$
   $CH_2=C(CH_3)-COO-(CH_2)_{11}COO-CH$ (19)
                    $CH_2-O-PO_3^- -(CH_2)_2-\overset{+}{N}(CH_3)_3$

b)   ◯———————X————————◯

(7) $HOOC-(CH_2)_8-C\equiv C-C\equiv C-(CH_2)_8-COOH$ (17)

(8) $HO-(CH_2)_9-C\equiv C-C\equiv C-(CH_2)_9-OH$ (17)

(9) $H_2O_3P-O-(CH_2)_9-C\equiv C-C\equiv C-(CH_2)_9-O-PO_3H_2$ (26)

c)

(10) $CH_3-(CH_2)_{12}-C\equiv C-C\equiv C-(CH_2)_8-COO-CH_2$
$CH_3-(CH_2)_{12}-C\equiv C-C\equiv C-(CH_2)_8-COO-CH$ (18)
$CH_2-O-PO_3^--(CH_2)_2-\overset{+}{N}(CH_3)_3$

(11) $CH_3-(CH_2)_{12}-CH=CH-CH=CH-COO-CH_2$
$CH_3-(CH_2)_{12}-CH=CH-CH=CH-COO-CH$ (18)
$CH_2-O-PO_3^--(CH_2)_2-\overset{+}{N}(CH_3)_3$

(12) $CH_3-(CH_2)_{12}-C\equiv C-C\equiv C-(CH_2)_8-COO-(CH_2)_2$
$\overset{+}{N}$ CH_3 Br^- (3a)
$CH_3-(CH_2)_{12}-C\equiv C-C\equiv C-(CH_2)_8-COO-(CH_2)_2$ CH_3

(13) $CH_3-(CH_2)_{12}-C\equiv C-C\equiv C-(CH_2)_9-O$
$\overset{O}{\underset{\|}{P}}-OH$ (21)
$CH_3-(CH_2)_{12}-C\equiv C-C\equiv C-(CH_2)_9-O$

(14) $CH_3-(CH_2)_{12}-C\equiv C-C\equiv C-(CH_2)_8-COO-(CH_2)_2$ (22)
$N-(CH_2)_2-SO_3H$
$CH_3-(CH_2)_{12}-C\equiv C-C\equiv C-(CH_2)_8-COO-(CH_2)_2$

(15) $CH_3-(CH_2)_{12}-C\equiv C-C\equiv C-(CH_2)_8-COO-(CH_2)_2$
O (3a)
$CH_3-(CH_2)_{12}-C\equiv C-C\equiv C-(CH_2)_8-COO-(CH_2)_2$

(16) $CH_3-(CH_2)_{16}-COO-CH_2$
$CH_3-(CH_2)_{12}-C\equiv C-C\equiv C-(CH_2)_8-COO-CH$ (3c)
$CH_2-O-PO_3H_2$

(continued)

Table 1 (Continued)

---

d)

$$\text{——————————————} \bigcirc \text{——————X}$$

(17) $CH_3-(CH_2)_{17}-O-CH_2$
      $CH_3-(CH_2)_{17}-O-CH$ (17)
                        $CH_2-O-CO-(CH_2)_5-NH-CO-C(CH_3)=CH_2$

(18) $CH_3-(CH_2)_{17}-CO-CH_2$
      $CH_3-(CH_2)_{17}-CO-CH-NH-CO-C(CH_3)=CH_2$ (17)

(19) $CH_3-(CH_2)_{17}$ \
                           $\overset{+}{N}$ — CH_3
      $CH_3-(CH_2)_{17}$ / \ $(CH_2)_3-NH-CO-C(CH_3)=CH_2$ (27)

(20) $CH_3-(CH_2)_{10}-COO-(CH_2)_2$ \
                                        $\overset{+}{N}$ — CH_3 Br^-
      $CH_3-(CH_2)_{10}-COO-(CH_2)_2$ / \ $CH_2-CH=CH_2$ (3g)

---

alkyl chain mobility. Lipids of all four types have been
synthesized and table 1 shows a selection of a few examples
of each type (see also (4)).

     Besides polymerization another type of polyreaction,
the polycondensation, can be used to stabilize membrane
systems. Recently, Fukuda et al. (5) described polyamide
formation in monolayers. Long chain esters of glycine and
alanine were polycondensed to yield non-oriented polyamide
films of polyalanine and polyglycine. In analogy to this
reaction it is possible to prepare stabilized, oriented
membrane systems via polycondensation using long chain
α-amino acid esters (6).

$CH_3-(CH_2)_{15}-CH-CO-O-CH_3$ $CH_3-(CH_2)_{15}-CH-CO-O-(CH_2)_{21}-CH_3$
                 | |
                NH_2 NH_2
                (21)   (22)

$$CH_3-(CH_2)_{23}-\underset{\underset{NH_2}{|}}{CH}-CO-O-CH_3$$

(23)

$$CH_3-(CH_2)_{23}-\underset{\underset{NH_2}{|}}{CH}-CO-O-(CH_2)_{21}-CH_3$$

(24)

This polycondensation can also be carried out with a 1:1 mixture of an amphiphilic diamine (25) and an amphiphilic diester (26).

$$CH_3-(CH_2)_{17}-\underset{\underset{}{|}}{NH}-CH_2$$
$$CH_3-(CH_2)_{17}-NH-CH_2$$

(25)

$$CH_3-(CH_2)_{17}\diagdown_{\diagdown}\overset{CO-O-C_2H_5}{\underset{CO-O-C_2H_5}{\diagup}} C$$
$$CH_3-(CH_2)_{17}\diagup$$

(26)

These stabilized membrane systems on the basis of polyamides are biodegradable and therefore have a great potential use for in vivo applications.

The knowledge of bulk properties of polymerizable lipids is not sufficient to evaluate their ability to form membranes or to predict their polymerization behaviour in these oriented systems. Their interaction with an aqueous phase

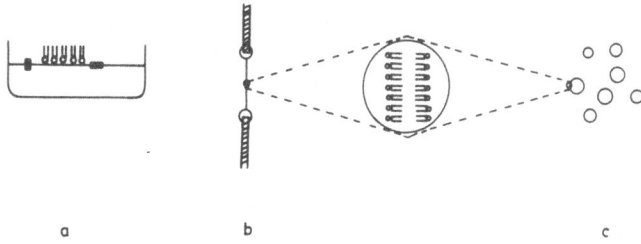

a   b   c

Fig. 2: Orientation of amphiphilic compounds in model membranes: a) monolayer; b) bimolecular lipid membrane (BLM); c) liposomes. Between b) and c) a cross section through the BLM or liposome is shown.

results in totally different properties which can be
described in terms of amphiphilic behavior, hydrophilic-
lipophilic balance, liposome formation or formation of
lyotropic phases (7,8). For thermodynamic reasons lipids
undergo self-organization when brought into contact whith
an aqueous environment. Several methods have been developed
to investigate these aggregation phenomena. Membrane models
most commonly used are monolayers (9), bimolecular lipid
membranes (BLMs) (10), and liposomes (11) as shown
schematically in fig. 2.

In this contribution the characterization of membrane
properties and polymerization behaviour of these polymerizable lipids in monolayers and vesicles will be
discussed.

POLYREACTIONS IN MONOLAYERS

Amphiphilic compounds such as lipids are able to form
monomolecular films when applied to a gas-water interface.
These two-dimensional systems can be characterized
recording phase diagrams in variation of surface-pressure,
area, or temperature. These surface pressure-area diagrams
are usually measured on a Langmuir film balance schematically
illustrated in fig. 3.

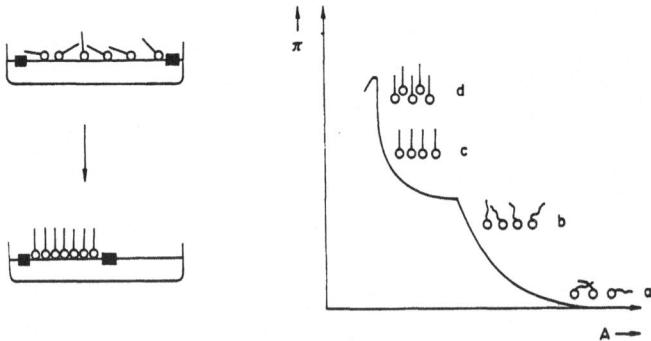

Fig. 3: Left: Langmuir film balance, schematic; right:
surface pressure-area diagram of a monolayer at the gas
water interface (a-d , see text). p=pressure, A=area

Different phases can be observed in the monolayer by
analogy to pV-diagrams of three-dimensional systems: at low
pressures (large areas) a gas-analogue state (a) is present
with low interaction of the lipid molecules spread on the
gas-water interface. Decreasing the area results in pressure
increase and the formation of a liquid-analogue phase (b)
with headgroup contact but high mobility in the hydrophobic
chains. Further compression yields a condensed or solid-
analogue phase with head-packing (c) or chain-packing (d).
Further decrease in surface area results in the collapse
of the film; beyond this collapse point there is no well-
defined system and the high orientation of the amphiphilic
molecules is lost.

Besides studying the influence of parameters such as
temperature, headgroup size and charge, alkyl chain length,
and pH, surface pressure-area diagrams also allow the
observation of polyreactions in such systems as shown in
fig. 4.

The orientation of the monomer units during polymerization of the monolayer remains unchanged; the reaction
results in a highly oriented stable film. UV-initiated
monolayer polymerization has been intensively studied in
recent years (12-15). Normally, the reactions involve
contraction of the film; the surface pressure-area diagrams
of the polymers exhibit a smaller occupied area, a steeper
slope, and a higher collapse pressure (13). Hence, the
polyreaction can be followed by measuring the film contraction
vs. time at constant surface pressure as shown in fig. 5
for the diacetylene lipid (12).

In case of the diacetylene lipids the polymerization
can also be followed spectroscopically by a special Langmuir

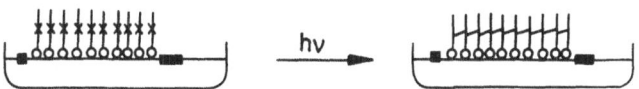

Fig. 4: Schematic presentation of the preparation of polymeric monolayers by UV irradiation, orientation of the
molecules being preserved (x=polymerizable group).

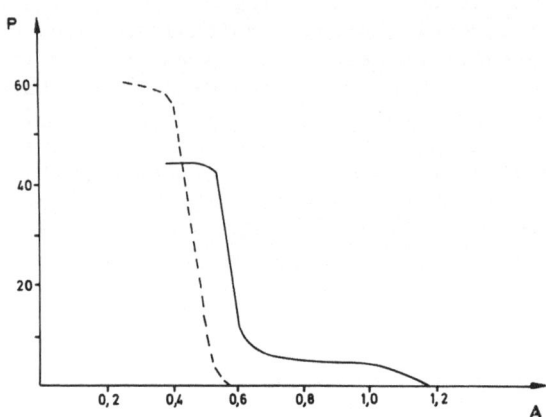

Fig.5: Surface pressure-area diagram of monomeric and
polymeric ammonium lipid (12) (table 1). (———)=monomer,
(- - -)=polymer; p=surface pressure (in mN/m); A=area
(in nm²/molecule).

balance built to fit into a commercial UV spectrometer (15).
This is possible, since polydiacetylenes exhibit a high
extinction coefficient in the visible region (blue/red).
The polyreaction of diacetylenes is topochemically controlled
(16), i.e. it is only possible in the solid-analogue state.

In contrast to this, methacrylic and diene derivatives
of lipids are not only polymerizable in the solid-like, but
also in the liquid-analogue phase (17,18). In addition, the
polymerized methacrylate and diene systems exhibit a higher
mobility due to non-conjugated and saturated polymer chains
and are therefore more suitable for the formation of
flexible membranes.

Similar results are obtained on the polycondensation of
long chain α-amino acid esters in monolayers. As shown in
fig. 6 for compound (21) the polymer (b) exhibits a lower
surface area per molecule compared to the monomer (a). The
polypeptide formation in the oriented monomer film is
shown in fig. 7.

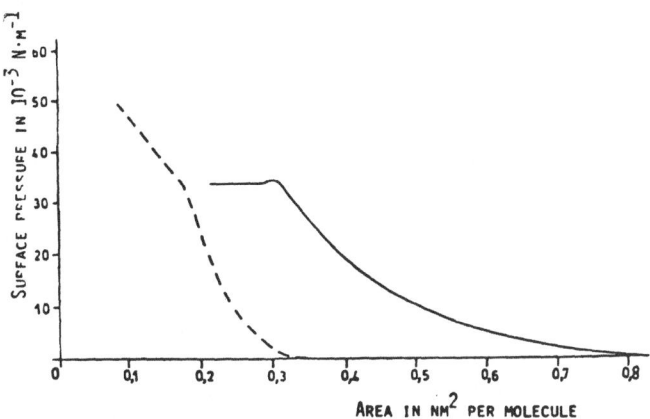

Fig.6: Surface pressure–area diagram of (21); (——)=monomer; 20°C, pH=8.5; (- - -)= after 24 h at 20°C; pH=8.5; condensation pressure 2 mN/m.

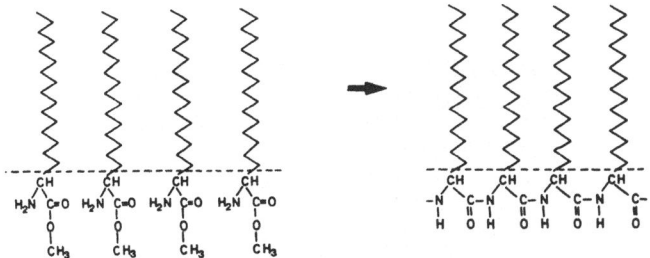

Fig. 7: Reaction scheme for the polycondensation of (21) in monolayers.

Polyamide formation can be demonstrated by FT–IR spectroscopy (disappearance of ester and formation of amide bands) and gel permeation chromatography of the polypeptides

(6). From the elution volumes, molecular weights between 2,000 and 10,000 can be estimated.

These investigations show that polyreactions in oriented planar systems are possible and lead to highly ordered and stable model membranes. It remains to be seen whether polymerization is possible in spherical bilayers such as liposomes and whether the vesicles thus formed exhibit a higher stability than the corresponding low molecular systems.

MEMBRANE STABILITY OF POLYMERIC LIPOSOMES

Liposomes are the closest approach to biomembranes; they are spherical structures having an aqueous interior and one or several lipid double-layers (fig. 8).

The most common methods to prepare liposomes are the ultrasonication of lipid suspension in water, the injection of etheral solutions of lipids into water, the dialysis of surfactant-lipid mixtures, or the hand-shaking of lipid films on glass surfaces in water.

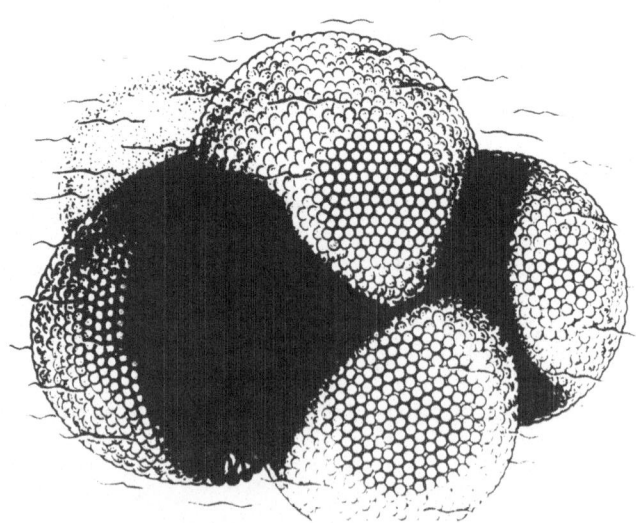

Fig.8: Schematic representation of unilamellar liposomes (i.e. consisting of one bilayer).

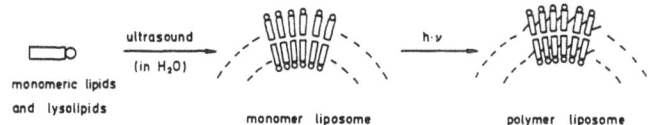

Fig. 9: Schematic representation of the polymerization of lipids in vesicular solution.

        The polymerizable lipid analogues in table 1 have mainly been transferred into liposomal solutions by ultrasonication of their crystal suspension in water (17,20). Small vesicles with a relatively narrow size distribution and mainly a single bilayer are formed after long sonication times. The monomeric liposomes are relatively unstable; like vesicles from natural lipids, their solutions turn turbid after some days.

        The monomeric vesicle solutions of methacrylic, diene, and diacetylene lipids can be polymerized by UV irradiation as schematically shown in fig. 9. In the case of the diacetylene lipids the polyreaction can be observed visually by the color change mentioned above. Electron microscopy can provide direct evidence that the polymerized systems still posses liposomal structure (3,20,23).

        Polymerization of liposomes affects their membrane stability. In contrast to monomeric liposomes the polymerized membrane systems remain stable for weeks without precipitation. Entrapped substances are released much slower from polymeric vesicles than from the corresponding monomers. This has been studied in the case of the diene lipid (11): entrapped 6-carboxyfluorescein (6-CF) in high concentration exhibits self-quenching; release into the surrounding aqueous medium results in strong fluorescence due to dilution (24). At room temperature vesicles made from DPPC (dipalmitoylphosphatidylcholine) are below the phase transition temperature showing 8% release after 40 hours (fig. 10). At room temperature vesicles made from the monomeric diene lecithin (11) are above their phase transition and therefore release the dye much more rapidly. Polymerized vesicles of the same compound, however, show no significant release after 40 hours.

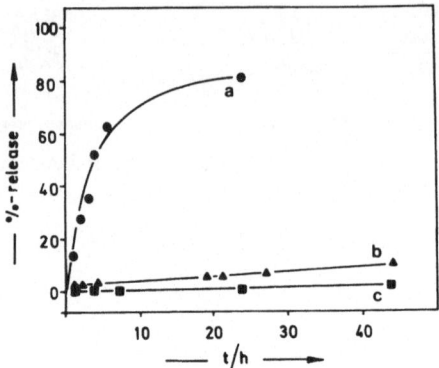

Fig. 10: Release of entrapped 6-carboxyfluorescein from
liposomes of monomeric (a) and polymeric (c) (11). For
comparison: DPPC (b) (dipalmitoylphosphatidylcholine).

An even more striking effect is produced by the
addition of SDS (sodium dodecylsulfate) as shown in fig. 11.
While the monomeric butadiene (11) and DPPC vesicles are
destroyed at low SDS concentrations, the polymerized vesicles
are stable up to 0.002 mol/l SDS.

Polymeric vesicles also show increased stability vs.
the addition of organic solvents (20).

This increased stability, however, is combined with the
presence of a polymer chain introducing increased viscosity
and thus reduced flexibility into the membrane system. How
does this reduced membrane affect one of the most vital
biomembrane properties, i.e. the phase transition temperature?

Whether polymerized artificial biomembrane systems are
too rigid to show a phase transition strongly depends on the
type of polymerizable lipid involved in membrane polymerization.

Especially in the case of the diacetylene lipids a loss
of phase transition can be expected due to the formation of
the rigid conjugated polymeric backbone. This was demonstrated
by DSC measurements with the diacetylene containing taurine

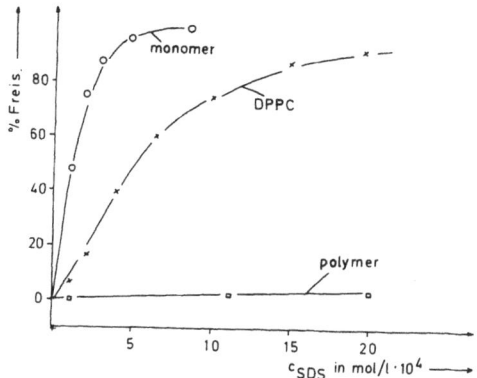

Fig.11: Release of enclosed 6-carboxyfluorescein vs. concentration of added SDS for the butadiene lecithin (11)

derivate (14). Fig. 12 illustrates the phase transition behaviour of (14) as a function of the polymerization time. The pure monomer liposomes show a transition temperature of 53°C. During the polymerization a decrease in phase

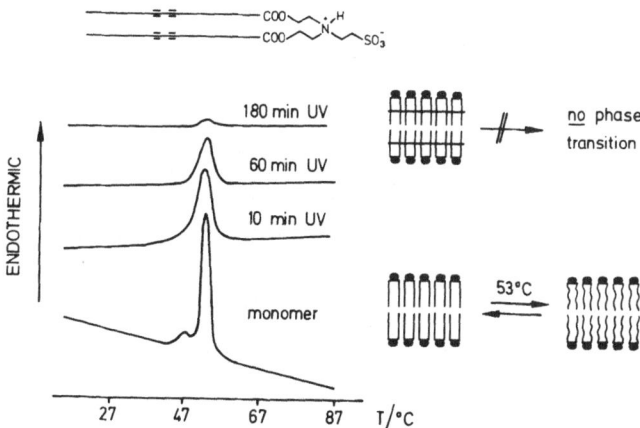

Fig.12: Measurement of phase transition of liposomes of (14) by DSC after different irradiation times.

transition enthalpy indicates a restricted mobility of the polymerized hydrocarbon core. Moreover, the phase transition finally disappears after complete polymerization of the monomer (34).

In contrast to this the phase transition of polymeric liposomes is retained if the polymer chain is more flexible or located on the surface of the vesicles instead in the hydrophobic core. The polymerized vesicles of the methacrylamide (19), for instance, show a phase transition temperature which is even slightly lower than the one of the corresponding monomer vesicles (27) as shown in fig. 13. This could be explained by a disordering influence of the polymer chain on the headgroup packing.

In analogy to the polycondensation of long chain α-amino acid esters in monolayers the formation of polypeptide vesicles from these compounds is also possible (6). In contrast to the polymerization reactions discussed above, however, these polycondensation reactions are not catalyzed at all and simply occur due to the high packing density of the reactive groups and their orientation in the bilayer. In addition it can be expected that the polypeptide vesicles are enzymatically cleavable, thus providing stable, but biodegradable vesicles.

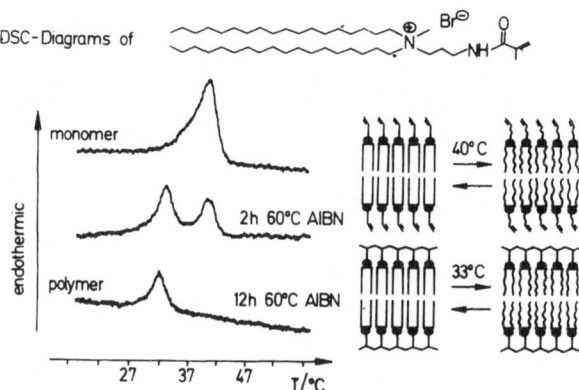

Fig.13: Phase transition temperature change during polymerization of (19). a=monomer, b=partially polymerized, c=polymerized.

MIXED SYSTEMS OF POLYMERIZABLE AND NATURAL LIPIDS

As mentioned in the beginning, increased membrane
stability is not sufficient to build better biomembrane
models. The stabilized systems have also to be able to
perform biological membrane properties such as selective
permeability. Since the polymerized systems discussed so far
combine increased stability with significantly decreased
permeability, methods have to be found to selectively open
up these stabilized membrane systems. A basis for this could
be the incorporation of labile components into the polymerizable system which could be destabilized, for instance,
by variation of pH (28), photochemical isomerization (29)
or enzymatic hydrolysis.

Such an enzymatic process, for example, is the hydrolysis
of a natural phospholipid membrane component by phospholipase
A2, as schematically shown in fig. 14. This enzyme cleaves
the ester bond in position two of a phospholipid producing
a lysophospholipid and a fatty acid which are both water
soluble, thus destroying the membrane.

First experiments investigating the action of phospholipase A2 on mixed membrane systems have been carried out
in monolayers measuring the area contraction under constant
surface pressure vs. time. As can be seen in fig. 15. the
main parameter influencing enzyme activity is the miscibility

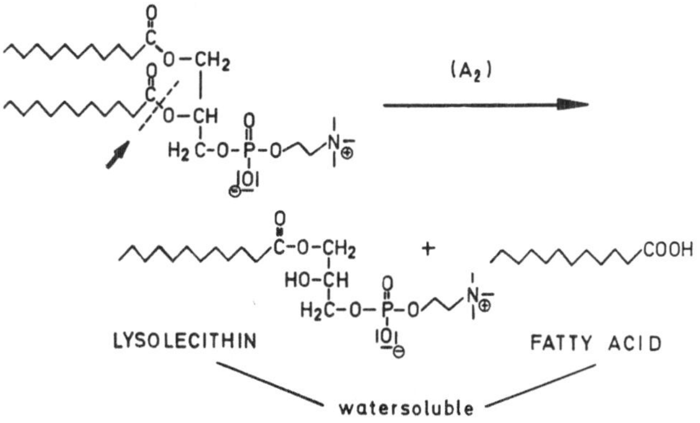

Fig.14: Enzymatic hydrolysis of a phospholipid by
phospholipase A2.

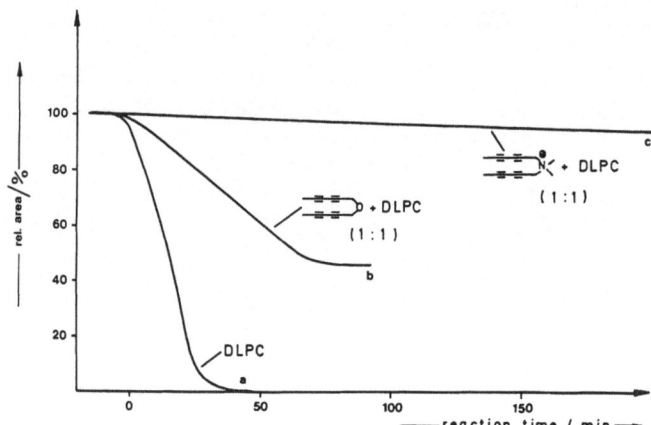

Fig.15: Relative area contraction vs. time during the action
of phospholipase A2 on mixed monolayers. a=pure DLPC,
b=immiscible system, c=miscible system.

of the lipid components. For a 1:1 mixture of DLPC (dilauroyl-
phosphatidylcholine) with an immiscible polymerizable
diacetylene lipid (b) the hydrolysis rate is slowed down
compared to a pure DLPC monolayer (a). In a miscible system
(c), however, the hydrolysis is slowed down to a much greater
extend. This could be explained by the fact that each DLPC
molecule is surrounded by non-hydrolysable lipid and therefore the action of phospholipase A2 might be sterically
hindered. This influence of miscibility on the hydrolysis
rate is magnitudes higher than the effect of the monolayer
being polymerized or not. Therefore, in trying to transfer
these hydrolysis investigations onto vesicle systems the
miscibility of lipids in vesicles has to be characterized.

     Besides using differential scanning calorimetry (30)
miscibility properties of vesicular systems can also be
characterized using freeze-fracture electron microscopy.
Here the "ripple" structure of phospholipids with saturated
alkyl chains, schematically indicated in fig. 16, is used
to indicate patch formation (immiscibility) in a mixed

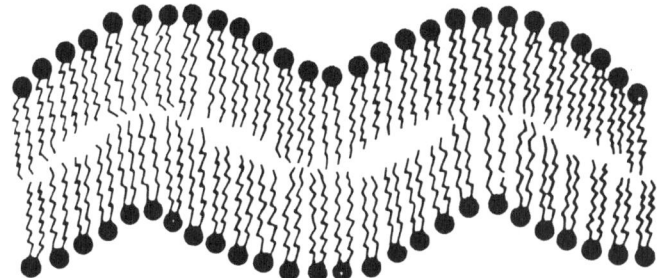

Fig.16: Schematic representation of "ripple" structure of
saturated phospholipids (17).

phospholipid membrane system. First experiments with the
system diacetylene lecithin (10)/DMPC (dimyristoylphospha-
tidylcholine) (31) indicate the existence of patches or
ripple-structured phospholipid next to unstructured areas
of diacetylene lipid as shown in fig. 17.

These miscibility experiments provide the basis for
further investigations on the selective opening of stable
vesicle compartments via enzymatic lipid hydrolysis.

SURFACE RECOGNITION OF POLYMERIC LIPOSOMES

In order to mimic basic biomembrane interactions an
artificial membrane system has to be susceptible to surface
recognition, for instance, by proteins. In natural membranes
these recognizing functionalities are usually located in
specific headgroups of membrane components.

For first simple recognition experiments in polymeric
membrane systems sugars were chosen as recognizable headgroups. Sugars can be recognized by lectins. Lectins are
proteins with specific binding sites for sugars and are
mainly found in plants. Concanavalin A (Con A), for instance,
is specific for mannose or glucose.

If a sugar moiety is part of an aggregate such as a
vesicle or a cell, on the addition of a lectin these
aggregates are agglutinated and precipitation occurs (32,33).

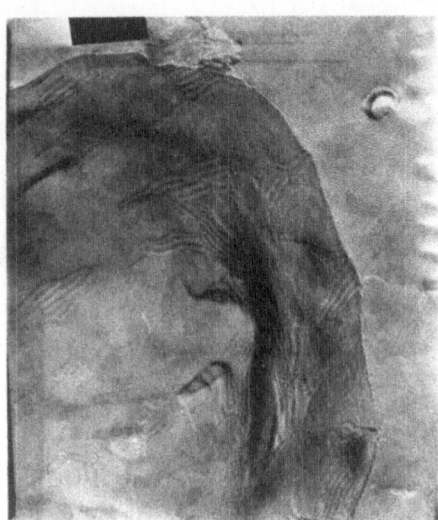

Fig.17: "Ripple" structure patch formation in (10)/DMPC
vesicles; 1:1 mixture.

In return, however, on addition of a low molecular sugar
these agglutinated aggregates can be redispersed by
competitive replacement. With these interactions in mind,
glycolipids with different sugar headgroups were synthesized
(fig. 18) carrying polymerizable diacetylene functionalities
in the hydrophobic chain (26).

On sonication these glycolipids form clear, colorless
liposomal solutions which can be polymerized to clear red
solutions, as already known for various other types of
diacetylene lipids. On addition of Con A agglutination
occurs within a few seconds forming a red precipitate next
to a clear supernatant. On addition of α-methylmannopyrano-
side the precipitate is redissolved resulting in the
original clear polymeric vesicle solution. These experiments
show that polymeric vesicle surfaces can be recognized by
natural proteins.

Presently attempts are made to incorporate lectins into
polymerized liposomal membranes to mimic vesicle-vesicle
recognition between lectin- and sugar carrying systems. The
possibility of incorporation of a membrane protein into
polymeric vesicles has already been demonstrated and will be
discussed in the following.

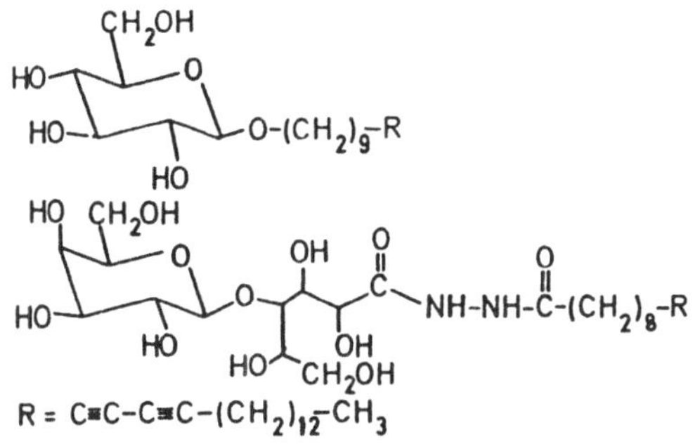

Fig.18: Polymerizable diacetylene glycolipids.

INCORPORATION OF MEMBRANE PROTEINS INTO POLYMERIC LIPOSOMES

Biological membranes are pictured as very selective barriers separating different biochemical reaction compartments. This high performance transport specifity depends solely on the presence of membrane proteins embedded in the lipid matrix On the other hand, however, most membrane proteins cease to function in the absence of lipids. Therefore, in order to introduce biological transport abilities into artificial membrane systems, protein-lipid interactions are of vital interest, i.e., how does a polymeric environment influence the activity of membrane proteins?

As an example of an asymmetric integral membrane protein the synthetase complex (ATPase) was incorporated into liposomes of the diacetylene taurine (14) (34). The protein consists of a hydrophobic membrane integrated part (F_0) and a water soluble moiety (F_1) carrying the catalytic site of the enzyme.

The incorporation of the ATPase into monomeric and polymeric diacetylene liposomes is achieved by simple incubation. However, there is a remarkable difference in the ATPase activity depending on the proteoliposome preparation technique (fig. 19). If ATPase is incorporated into already

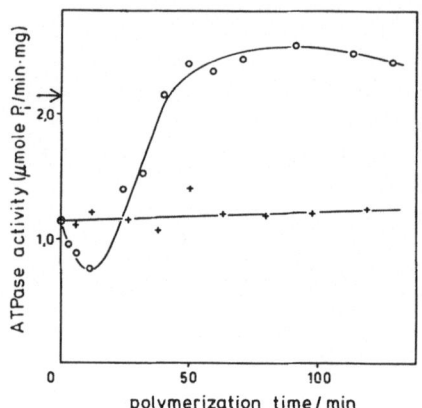

Fig.19: ATPase activity of ATP synthetase incorporated into
liposomes of (14): (o) incubation of the enzyme with
polymerized liposomes; (+) incubation of the enzyme with
monomeric liposomes followed by polymerization; (→) activity
of ATP synthetase in soybean lecithin liposomes.

partially polymerized liposomes the hydrolytic activity
exhibits a significant dependence on further polymerization
time. A minimum after 10 min polymerization is followed by
a strong increase (even above the activity in soybean lecithin
liposomes) upon prolonged irradiation. The activity minimum
corresponds with a red shift in the VIS-spectra of the
diacetylene polymer which is associated with a conformation
change in the polymer. Incubation of ATP synthetase complex
with monomeric liposomes followed by irradiation, however,
results only in a slight increase in hydrolytic activity.

      The activity increase in the polymeric liposomes compared
to the monomer can be ascribed to a structural change in the
bilayer organization during polymerization. DSC data indicate
residual "monomeric domains" in the polymerized liposomes,
so that the ATPase is most probably embedded in these domains,
stabilized by the polymer matrix as schematically shown in
fig. 20.

      These polymeric proteoliposomes have a considerably
higher long term stability and activity than, for instance,

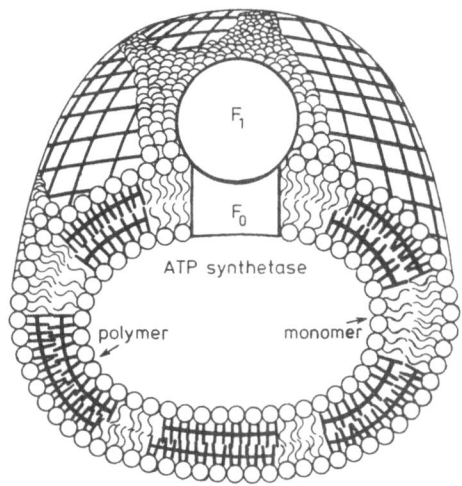

Fig. 20: Schematic representation of ATP synthetase
incorporated into a partially polymerized liposome.

ATPase containing soybean lecithin liposomes. Their lifetime
is practically only determined by the bacterial growth rate
in the particular sample.

FUSION OF POLYMERIZABLE VESICLES.

    In contrast to this "synthetic" route to polymeric cell
models discussed above, which involves putting together all
kinds of natural and synthetic membrane forming substances,
biomembranes with polymeric components can also be created
combining naturally formed cell membranes with artificial
membrane systems via fusion.

    Before trying to fuse polymerizable vesicles with
natural cells, investigations on vesicle-vesicle fusion are
necessary. A very selective fusion method recently described
by Zimmermann (35) was applied to these fusion experiments.

    Essentially, this method is based on subjecting cell
membranes, in which high intensity electrical fields occur
naturally, to a short external electric field pulse of
comparable intensity. Under these conditions the membrane
breaks down locally and becomes permeable. This process is

reversible, i.e., the membrane reconstitutes its original
properties in time intervals which can be experimentally
controlled.

If two cells adhere to each other during the field pulse,
electrical breakdown occurs in the contact zone between the
two·cells. If the membrane contact is close enough (1-2 nm),
lipid molecules are able to diffuse from one membrane into
the other forming bridges between both membranes. After
turning off the field pulse, the fusion of the two cells
into one sphere is energetically favoured. Thus, close
contact of membranes has to be achieved before fusion can be
induced by an electrical field pulse.

If cells are placed in an inhomogeneous electrical field
they start to migrate in the direction of greater field
intensity due to an induced dipole. They align themselves
like "strings of pearls" along the field line. This effect is
know as dielectrophoresis (36). Now, fusion of the cells
aligned is possible via an electrical field pulse.

The dipole induced in the cells is proportional to their
volume, i.e. only relatively large particles can be aligned
along field lines. So, in order to use this method for
vesicle-vesicle fusion, "giant" vesicles (50-100 µm) have
to be prepared (37). Using these large vesicles the fusion
process can be followed in a phase-contrast light microscope
as shown in fig. 21 for vesicles made from a 1:1 mixture of
cholesterol and a polymerizable butadiene lipid with a
cationic headgroup (38).

The vesicles are oriented between two electrodes of
100 µm distance applying a field strength of 2-4 kV/cm. The
diameter of the two large vesicles is about 40 µm (fig.21 a).
Fusion is initiated by a field pulse of 30-90 kV/cm and
20-50 µs duration. The intermingling of membrane lipids
occurs within a fraction of a second. The membrane boundary
between the two vesicles disappears forming an oval as shown
in fig. 21 b. A special liposome is formed after turning off
the external field. If the vesicles are unilamellar the whole
fusion process is completed within one second. In the case of
multilamellar vesicles the fusion time is extended to about
3-6 seconds. Fusion times for cells are usually in the
range of minutes (35).

Preliminary experiments to fuse the described "giant"

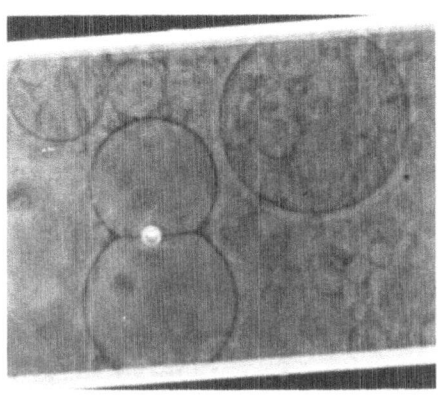

Fig. 21: Phase-contrast photographs of a fusion of LUVs (large unilamellar vesicles) prepared form a 1:1 mixture of a butadiene lipid with a cationic dimethylammonium bromide headgroup and cholesterol. Vesicle diameter: 37 and 45 μm respectively;

a) orientation in an ac field of about 2 kV/cm;

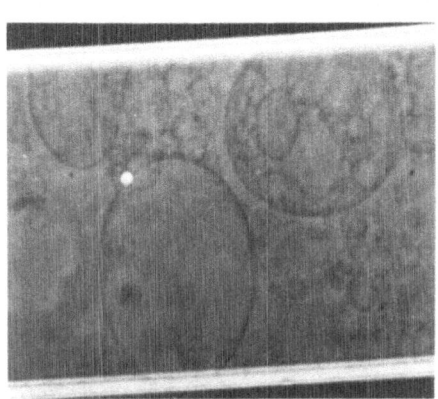

b) elongated fused liposome, one second after application of a 30 μs, 140 V/cm field pulse;

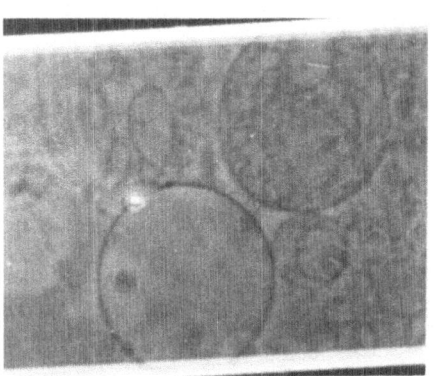

c) spherical new vesicle after turning off ac field; new diameter: 51 μm.

vesicles with Friend cells show problems in the simultaneous
orientation of vesicles and cells in the 0.3 M mannite
sólution necessary as an isotonic medium for the cells. The
induced dipoles of cells are much larger compared to vesicles,
leading primarily to cell-cell contacts and thus to cell-
cell fusion (39).

Vesicle-vesicle fusion experiments with polymerized
vesicles are presently under way. First investigations on
calcium-induced fusion with polymerized vesicles of the
diacetylene taurine (14) (40) indicate that the fusion
process is expected to be slowed down considerably compared
to the corresponding unpolymerized vesicles.

REFERENCES:

1) S.J.Singer, G.L. Nicolson, Science, 175, 720 (1972)
2) S.J.Singer in: Cell Membranes, G. Weissmann, R.Clairborne
    ed., HP Publishing Co., New York 1975, p. 44
3) a) D.Day, H.-H.Hub, H. Ringsdorf, Isr.J.Chem., 18, 325
    (1979). b) S.L.Regen, B. Czech, S.J.Singh, J.Am.Chem.Soc.,
    102, 6638 (1980). c) H.-H. Hub, B.Hupfer, H.Koch,
    H.Ringsdorf, Angew.Chem.Int.Ed.Engl., 19, 938 (1980).
    d) D.S. Johnston, S. Sanghera, M. Pons, D.Chapman,
    Biochim.Biophys.Acta, 602, 57 (1980). e) W.Curatolo,
    R.Radhakrishnan, C.M.Gupta, H.G.Khorana, Biochemistry,20,
    1374 (1981). f) D.F.O'Brien, T.H.Whitesides, R.T.Kling-
    biel, J.Polym.Sci.Polym.Lett., 19, 95 (1981). g) P.Tundo,
    D.J. Kippenberger, P.L. Klahn, N.E.Pietro, T.C. Tao,
    J.H. Fendler, J.Am.Chem.Soc., 104, 456 (1982). h) R.Benz,
    W. Prass, H. Ringsdorf, Angew.Chem.Suppl., 1982, 869.
4) H.Fendler in: Surfactants in Solution, K.L.Mittal ed.,
    Plenum Press (Proceedings of the International Symposium,
    1982, Lund, Sweden) in press.
5) K. Fukuda, Y. Shibasaki, H. Nakahara, J.Macromol.Sci.,
    Chem., 15, 999 (1981).
6) T. Folda, L. Gros, H. Ringsdorf, Makromol.Chem.Rapid
    Commun., 3, 167 (1982).
7) E. Sackmann, Ber.Bunsenges.Phys.Chem., 78, 929 (1974).
8) D. Small, Pure Appl.Chem., 53, 2095 (1982)
9) G.L.Gaines: Insoluble Monolayers at Liquid-Gas Interfaces.,
    Interscience, New York 1966.

10) H.C. Tien: Bimolecular Lipid Membranes, Theory and
    Practice. Marcel Dekker, New York 1974.
11) H.C. Huang, Biochemistry, 8, 344 (1969).
12) D. Naegele, H. Ringsdorf in H.G. Elias: Polymerization in
    Oriented Systems. Midland Macromolecular Monographs Vol.3,
    Gordon and Beach, New York 1977, p. 79
13) R. Ackermann, O. Inacker, H. Ringsdorf, Kolloid Z.Z.Polym.,
    249, 1118 (1979).
14) a) S.A. Letts, T. Fort, Jr., J.B. Lando, J.Colloid
    Interface Sci., 56, 64 (1976). b) M. Hatada, M. Nishii,
    J.Polym.Sci., 15, 927 (1977). c) D. Naegele, H.Ringsdorf,
    B. Tieke, G. Wegner, D. Day, J.B. Lando, Chem.-Ztg., 100,
    426 (1976). d) D. Day, H. Ringsdorf, J.Polym.Sci.Polym.
    Lett. Ed., 16, 205 (1978). e) H. Ringsdorf, Am.Chem.
    Soc.Div. Org. Coatings and Plast.Chem. Preprints,
    42, 379 (1980).
15) a) D. Day, H. Ringsdorf, Makromol.Chem., 180, 1059 (1979).
    b) H.-H. Hub, Ph. D. Thesis, University of Mainz 1981.
16) a) G. Wegner, Makromol.Chem., 154, 35 (1972). b) R.H.
    Baughman, J.Appl.Phys., 43, 4362 (1972).
17) A. Akimoto, K. Dorn, L. Gros, H. Ringsdorf, H. Schupp,
    Angew.Chem.Int.Ed.Engl., 20, 90 (1981).
18) B. Hupfer, H. Ringsdorf, H. Schupp, Makromol.Chem., 182,
    247 (1981).
19) S.L. Regen, A. Singh, G. Oehme, H. Singh, Biochim.
    Biophys.Res.Commun., 101, 131 (1981).
20) H.-H.Hub, B.Hupfer, H. Koch, H. Ringsdorf, Angew.Chem.,
    92, 962 (1980).
21) K. Aliev, unpublished
22) H. Koch, H. Ringsdorf, Makromol.Chem., 182, 255 (1981)
23) E. Lopez, D.F.O'Brien, T.H. Whitesides, J.Am.Chem.Soc.,
    104, 305 (1982).
24) J.N. Weinstein, S. Yoshikami, L. Heukart, R. Blumenthal,
    W.A. Hagins, Science, 195, 489 (1977).
25) H. Schupp, Ph.D. Thesis University of Mainz 1981
26) H. Bader, H. Ringsdorf, J. Skura, Angew.Chem.Int.Ed.Engl.,
    20, 91 (1981).
27) K. Dorn,Ph.D. Thesis University of Mainz 1983.
28) M.B. Yatvin, W. Kreutz, B. A. Horwitz, M. Shinitzky,
    Science, 210, 1253 (1980).
29) R. Kano, Y. Tanaka, T. Ogawa, M. Shimomura, J. Okahata,
    T. Kunitake, Chem.Lett., 1980, 421.
30) a) R. Bueschl, B. Hupfer, H. Ringsdorf, Makromol.Chem.
    Rapid Commun., 3, 589 (1982). b) D.F. O'Brien, E.Lopez,
    T.H. Whitesides, Biochim.Biophys. Acta, 1982, in press
31) R. Bueschl, H. Ringsdorf, E. Sackmann, in prep.

32) I.J. Goldstein, ed.: <u>Carbohydrate-Protein Interactions</u>,
    ACS Symposium Series, Vol. 88 Washington D.C. 1979.
33) J.H. Fendler, <u>Acc.Chem.Res.</u>, <u>13</u>, 7 (1980).
34) N. Wagner, K. Dose, H. Koch, H. Ringsdorf, <u>FEBS Lett.</u>,
    <u>132</u>, 313 (1981).
35) U. Zimmermann, P. Scheurich, G. Pilwat, R. Benz,
    <u>Angew.Chem.Int.Ed.Engl.</u>, <u>20</u>, 325 (1981).
36) P. Scheurich, U. Zimmermann, H. Schnabl, <u>Plant Physiol.</u>,
    <u>67</u>, 849 (1981).
37) H.-H. Hub, U. Zimmermann, H. Ringsdorf, <u>FEBS Lett.</u>, <u>140</u>,
    254 (1982).
38) R. Bueschl, H. Ringsdorf, U. Zimmermann, <u>FEBS Lett.</u>, <u>150</u>,
    38 (1982)
39) R. Bueschl, personal communcation
40) H. Koch, Ph.D. Thesis, University of Mainz 1983

POLYMER CARBON

THE START INTO A NEW AGE OF POLYMER APPLICATION

E. Fitzer

University of Karlsruhe
Kaiserstraße 12, D-7500 Karlsruhe, FRG

SYNPOSIS

Polymer carbon comprises a group of materials consisting only of elemental carbon, but differing in mechanical properties from those of graphite by high strength, resistivity against shear stress and extraordinary stiffness. These properties can be explained by the high bonding strength between sp^2-hybridized carbon atoms and the special microstructure with parallel arrangement of polyaromatic sheets, however, without any crystallographic order in direction perpendicular to the sheets. The relation to the MARCK-BERNAL structure of graphite (1924) and that of amorphous carbon, as described by DEBYE - SCHERRER in 1917 and as defined today is discussed.

Polymer carbon is made from non melting polymers by thermal degradation and conserves exactly the same morphology as that of the precursor polymer in spite of volume shrinkage during pyrolysis. Polymers without preferred orientation of the macromolecules yield in isotropic polymer carbon - so called glass like carbon - the structure of which reminds one that of polymers with planar macromolecules. Carbon fibres with preferred orientation of the polyaromatic, graphite like carbon sheets can be obtained from polymer fibres in which the macromolecules have been oriented during melt spinning,by subsequent stretching of the polymer fibres or even by hot working of isotropic carbon fibres at graphitization temperatures. These various methods and their influence on the resulting properties of the carbon fibres are explained.

The paper treats finally the capacity of polymer carbon in form of carbon fibres for the reinforcement of organic polymers. With such advanced composites first time polymer materials are created, the properties of which exceed those of metals in strength, stiffness, resistivity against corrosion and fatigue, the weight of which is but half an order of magnitude less than that of steel. All-carbon composites - these are carbon fibres in a matrix of isotropic polymer carbon - became the high temperature material par excellence for application under most extreme conditions.

Although these composites with polymer carbon are already widely used in critical structural parts in aerospace technology, a broad industrial application in place of metals still needs improved polymer matrices, development of more economic fabrication methods for the composites, sophisticated design for special tailoring of the material and further cost reduction for the carbon fibres. Nevertheless, material scientists, trusting on further development of polymers, dream of the vision of a future "carbon age".

THE NATURE OF POLYMER CARBON

Before X-ray structural analysis became available, chemists classified 3 types of elemental carbon: diamond, graphite, and the so-called amorphous carbon.

The structure of diamond was already determined as a tetrahedral structure in 1913 by W.H. BRAGG and W.L. BRAGG (1). We know today that this arrangement of the carbon atoms indicates sp_3-hybridization of the binding electrons of carbon as in non aromatic carbon hydrogen compounds. The strong covalent C/C-bonds in four equivalent directions cause the well known isotropic strength of diamond.

The exact description of the graphite structure, however, turned out to be more difficult (Fig.1).

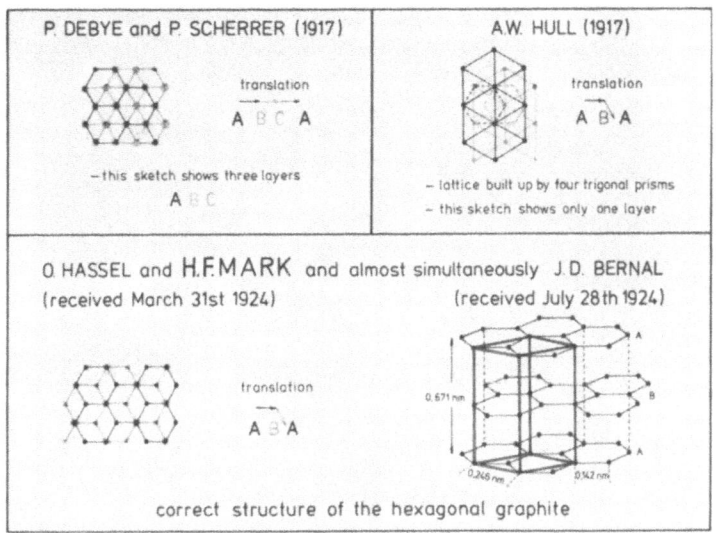

Fig. 1 : History of research on the graphite
structure (2,3,4,5)

In 1917, not only P. HULL (2) but also A.W. DEBYE
and P. SCHERRER (3) failed, although the last cited authors have already understood the layer
structure of graphite, but have assumed a wrong
sequence for the layer stacking. The crystalline
structure of the hexagonal form of graphite was
described for the first time correctly in 1924 (4)
by our today's recipient of the Division of Polymer Chemistry award, Prof.Dr.Hermann F. MARK and
O. HASSEL, and independently some months before
by J.D. BERNAL (5). The graphite structure is re-
fered today mostly as BERNAL structure. One should
and we will use in the future the classification
MARK-BERNAL - structure. Today's chemists will explain this layer structure by perfect sp_2-hybridi-
zation where the $\overline{\pi}$ -electrons are completely de-
localized.

     The nature of amorphous carbon was described
by P. DEBYE and P. SCHERRER in 1917 (3) already

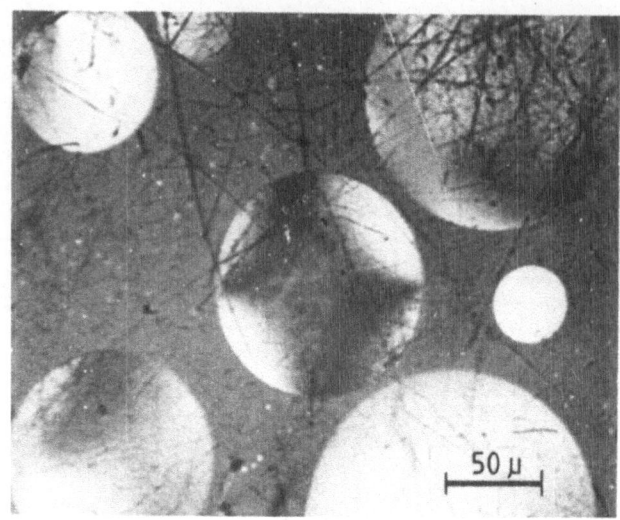

Fig. 2 : Mesophase spherolites formed in pitch
          by heat treatment

as planar carbon molecules similar as in polyaromatics. They have found such C_x molecules in all
different types of amorphous carbon, a finding
which in principle is still valid today. Only the
size of these structural units, assumed as 8 to
10 carbon atoms by DEBYE-SCHERRER, does not fit to
today's knowledge of the structure of the pyrolysis
residues of carbonaceous compounds.

    As known since hundred years, graphite can be
made synthetically from structurally disordered
carbons by heat·treatment at temperatures around
3000 K, where graphitization occurs. Preconditions
for such solid state transformation from a disordered layer structure to the 3-dimensional crystalline order, however, is a structural preorientation of the carbon layers, a "pregraphitic"
structure. The pregraphitic structure is initiated
by liquid crystals, the so-called mesophase, during
thermal degradation of pitches in the liquid state.
Such liquid crystals consist of aromatic molecules
with a molecular weight between 1000 and 2000, and
have a spherical morphology which is able to co-
alesce to larger areas during thermolysis between
650 and 800°K (Fig.2).

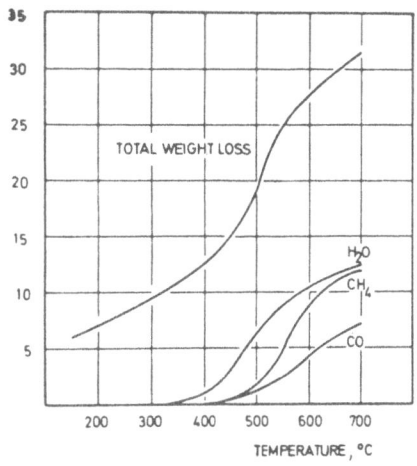
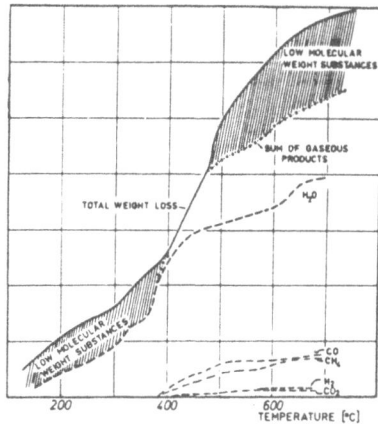

Fig. 3 : Volatile products during thermolysis of
        phenolics (left hand side $C_6H_5OH/HCOH$ 2:3
        right hand side: less crosslinked commer-
        cial type)

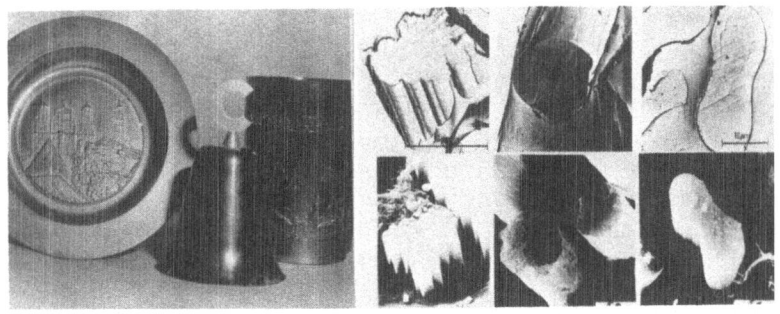

Fig. 4 : Conservation of precursor shape
        a) Articles of glasslike carbon (8,9,1o),
           which represent the exact shape of the
           precursor resin only with a linear
           shrinkage of at about 20 %;
        b) Precursor fibres (upper row),carbon
           fibres (lower row) made from them
           (Rayon, wet spun; PAN, dry spun)

If carbon is made of carbonaceous compounds which do not melt during thermal degradation, a pregraphitic structure can not be formed. This type of carbon is called "hard carbon", a carbon which never graphitizes by thermal treatment in contrary to the "soft" or graphitizing carbons mentioned before.

Polymer carbon, the subject of this paper, is a non graphitizing, hard carbon. The term polymer carbon describes firstly the origin, namely a carbon derived from polymers by carbonization. Also the term polymeric carbon was used for it (6). The term polymer carbon indicates also the molecular structure of such carbon, for which some similarities with that of polymers can be seen. If a thermosetting resin is slowly thermolyzed in non oxidizing atmosphere, the final product at temperatures of at about 1000 °C is a glass appearing black hard and brittle material, the so-called glasslike carbon. The volatile by products during pyrolysis of phenolics as typical precursor polymers for glass like carbon are shown in fig. 3 (7). In the best cases, no low molecular volatile products are formed thus enabling a solid carbon residue of more than 60 %.

The most important behaviour of such a residue is the conservation of the original shape of the crosslinked resin in spite of its shrinkage during carbonization due to the mass loss (fig. 4a) (8). Trade names for such commercially available impervious isotropic carbon products are "glassy carbon" (9) and "vitreous carbon" (10). This conservation of the polymer shape is also the basis for the carbon fibre formation from crosslinked polymer fibres (fig.4 b) i.e. rayon or oxidized polyacrylonitrile.

In fig. 5 the X-ray diffraction dagram of glasslike carbon is compared with that of a single crystal of natural graphite. The prismatic interferences 100 and 110 and the pyramidic ones 101 and 112 appear very weakly as so-called 2-dimensional reflexes 10 and 11 only. Also very low stacking heights of the polyaromatic layers can be seen from the line broadening of the 002 reflexes.

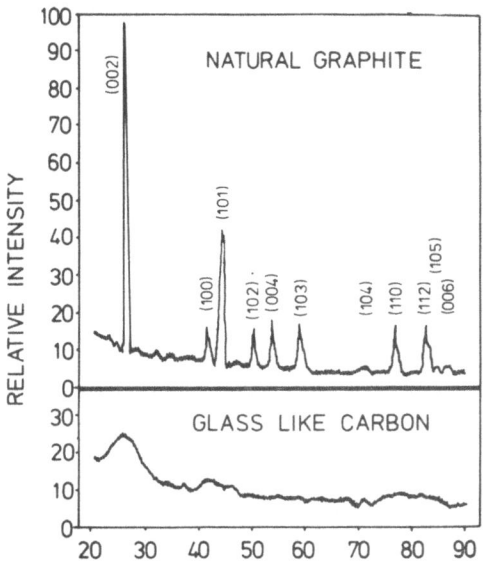

Fig. 5 : Wide angle diffraction of graphite single
         crystal and glasslike carbon (Cu Kɣ)

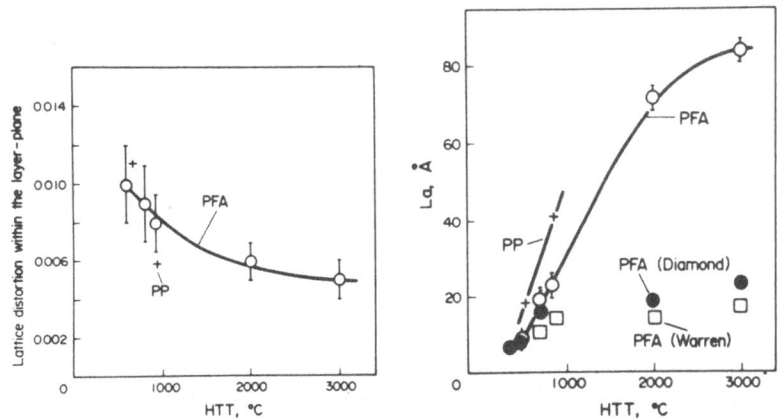

Fig. 6 : Lattice distortion in a-direction and
         L_a data of isotropic polymer carbon (15)

Since S. YAMADA (11) has first described his glassy carbon with high strength and hardness in the late fifties it remained an open question whether the structural arrangement of the carbon atoms has some similarities with that in diamond, that means whether sp$_3$ hybridization does exist. As S. ERGUN (12) and also we (13) were able to show by careful analysis of the whole diffraction profile, that the radial atomic distance distribution curves do not indicate any distances of 1.52 Å, characteristic for the diamond structure, one can conclude that no tetrahedral structural elements are present in glasslike carbon (sensitivity of the method at about 1 %). But there is also no indication (contrary to former assumptions) that singular isolated carbon atoms exist in glasslike carbon. It seems that all carbon atoms are organized in polyaromatic layers. Because of this two dimensional long range order, polymer carbon can not be classified as amorphous carbon. The term amorphous carbon is today reserved to a carbon material without long range crystalline order neither in a- nor in c-direction with reference to the lattice of hexagonal graphite, and with short range order with deviations of interatomic distances and of interbonding angles because of "dangling bonds".

If one uses the simple WARREN formula (14) for calculation of the layer sizes, or better $\bar{L}_a$ the mean size of defect free areas in carbon layer planes, maximum values of about 25 Å will result. If lattice distortion is taken into consideration (15), areas up to 80 Å can be found for carbon from polyfurfurylalcohol. For polyphenylene, even larger growth is observed, as shown in fig. 6.

The first inside view of the most probable structure of such glasslike carbon was obtained by absorption measurements with molecule probes on one hand and by high resolution transmission electron microscopy (TEM) on the other. Fig. 7, left hand side, gives results on microporosity of a polyfurfurylalcohol residue (16), right hand side that of a chlorinated polyphenylene (17). In both cases, slit pores or at least slit entrances of the pores are formed. These slit pores are narrowed by heat treatment. The temperature of final

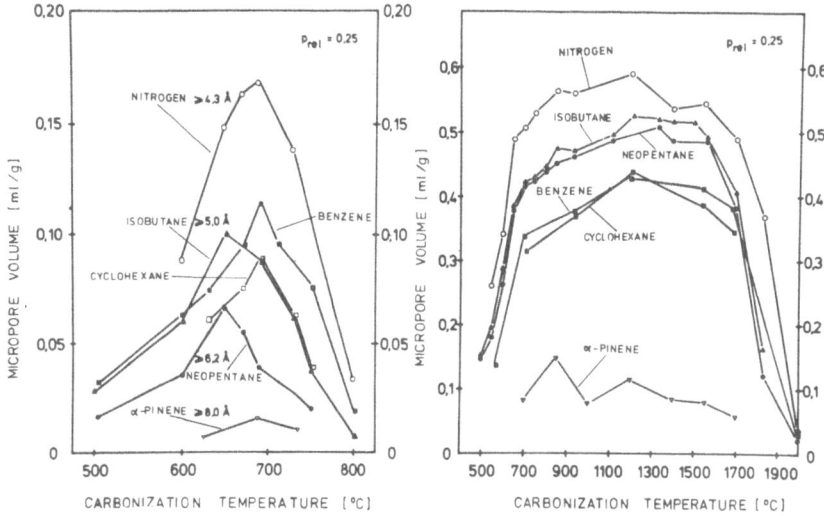

Fig. 7 : Micropore volume of residues from
         a) polyfurfurylalcohol (16)
         b) chlorinated polyphenylene (17)

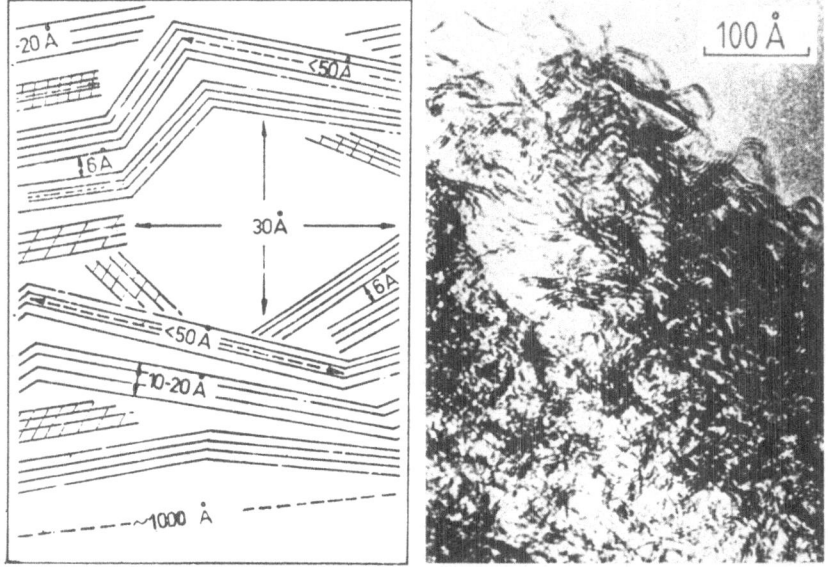

Fig. 8 : Two-dimensional model (18) and TEM (19)
         of glasslike carbon from PFA

Isotropic | Anisotrpic

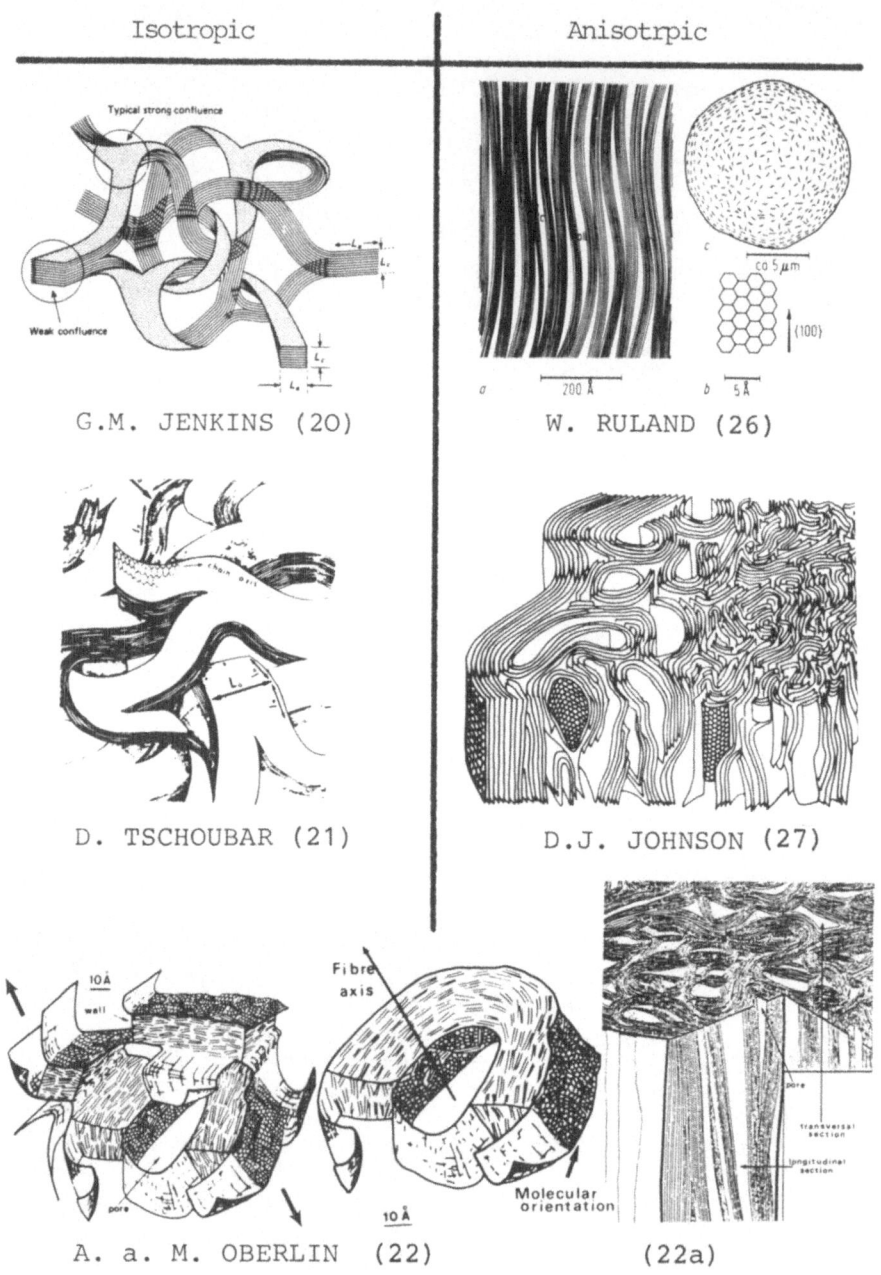

G.M. JENKINS (20)   W. RULAND (26)

D. TSCHOUBAR (21)   D.J. JOHNSON (27)

A. a. M. OBERLIN (22)   (22a)

Fig. 9 : Structural models for polymer carbon

shrinkage depends on the precursor type and thus
on the defect structure of the residue formed by
the volatile ligands.

Based on such results of microporosity, X-ray
diffraction and small angle scattering evaluation
as well as on density measurements, and finally
stimulated by RULAND's carbon fibre model (26) a
2-dimensional model for glasslike carbon was derived by us as shown in fig. 8, left hand side
(18). Simultaneously, by other research groups, the
first transmission electron microscopic pictures of
glass like carbons (19) confirmed this assumed
structure (fig.8, right hand side).

Other authors developed 3-dimensional models
like the ribbon model (fig.9) (20). The merit of
this ribbon model was that it has initiated the
imagination of a polymer structure with planar
macromolecules in glasslike carbon. This model,
however, can never explain the bulk density of glasslike carbon of about 1.6 (density of graphite
single crystal 2.26). More realistic is the bundle
model (21). Just recently, a tube model was developed by OBERLIN et al (22) which in principle can
explain similar X-ray diffraction, electron transmission and adsorption results. All these models
give an understanding for the high strength of the
glasslike carbon because of the strong σ -bonds in
sp_2 hybride structures, but also for the isotropy
of the mechanical and physical properties in bulk
samples, in spite of the anisotropic arrangement
of the atoms.

One can imagine, however, that anisotropic
arrangement of the planar macromolecules should result in higher mechanical properties of a bulk material. Such anisotropic polymer carbon has become
much more important for technical application in
form of the high strength and high modulus carbon
fibres.

Fig. 10 gives the stress/strain behaviour of
such anisotropic polymer carbon in comparison with
the isotropic one and with conventional granular
graphites. Some special types of high modulus carbon fibres can reach YOUNG's moduli up to 750 GN/m^2,

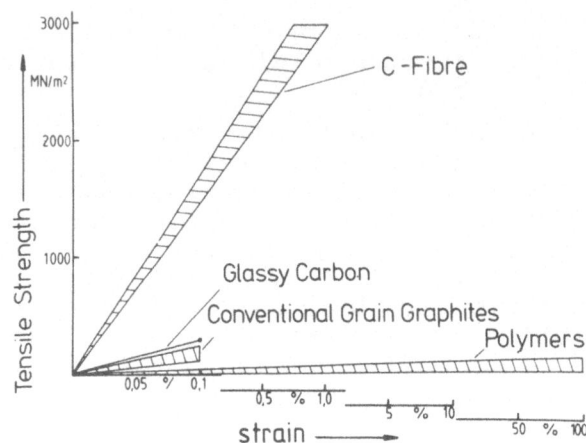

Fig.10:Stress-strain diagram of various C-materials

Table 1: Elastic constants of graphite single
         crystal (23)

ELASTIC CONSTANTS	C_{11}	C_{12}	C_{13}	C_{33}	C_{44}
GN/m^2	1060	180	16	36.5	4.5

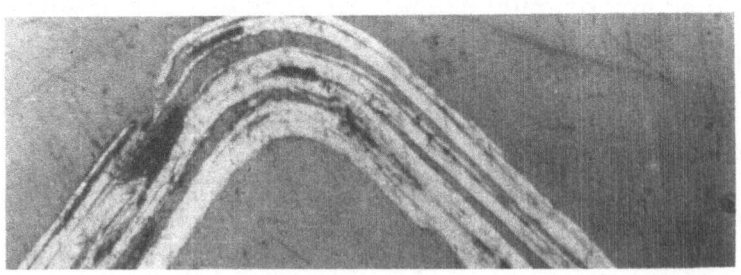

Fig.11: Natural graphite flake after bending (24)

corresponding to 75 % of the theoretical bonding
strength in the aromatic layer.

Table 1 shows the elastic constants for single
crystal graphite (23). The enormous high C 11 value
of 1060 GN/m^2 is caused by the strong σ -bonds in
the aromatic layers. The elastic constant C 33 in
perpendicular direction reflects the low bonding
strength between the layers. The shear modulus
C 44 between the layers, however, is extremely low
which limits the structural application of graphite
with high degree of 3-dimensional crystalline order.

Fig. 11 (24) shows the technical consequence
of this low shear modulus C 44. In spite of the
enormous strong σ -bonds within the layers, shear
stress reduces the resistance against bending to
a practically neglectible small value.

From these considerations on elastic constants
of the graphite single crystal we can derive the
structural precondition for this technically interesting anisotropic polymer carbon.

(1) Preferred orientation of the layers in the
direction parallel with the fibre axis for achievement of highest YOUNG's modulus.

(2) Avoiding of shear between the layers by
cross linkage between or by imperfect stacking of
the layers. By such means, complete delocalization of the π electrons like in the graphite single
crystal is avoided.

(3) Minimization of flaws or defects within
the layer with sizes smaller than the critical one
for strength increase of the brittle fibre material.

These demands, however, are partially contra-
dictory. The first demand for instance can be full-
filled by heat treatment at graphitization temperature. Fig. 12 (25) demonstrates the alignement of
parallel layers by graphitization treatment up to
2000 and 2700 °C which have been formed during pyrolysis with small areas and low degree of prefered orientation only. Such a graphitization
treatment, however, is disadvantageous from the
viewpoint of the second demand because imperfections within the layers are healed, and stacking
of the layers is improved.

The third demand must mainly be controlled by the type and structural perfectness of the precursor and by the thermal degradation step.

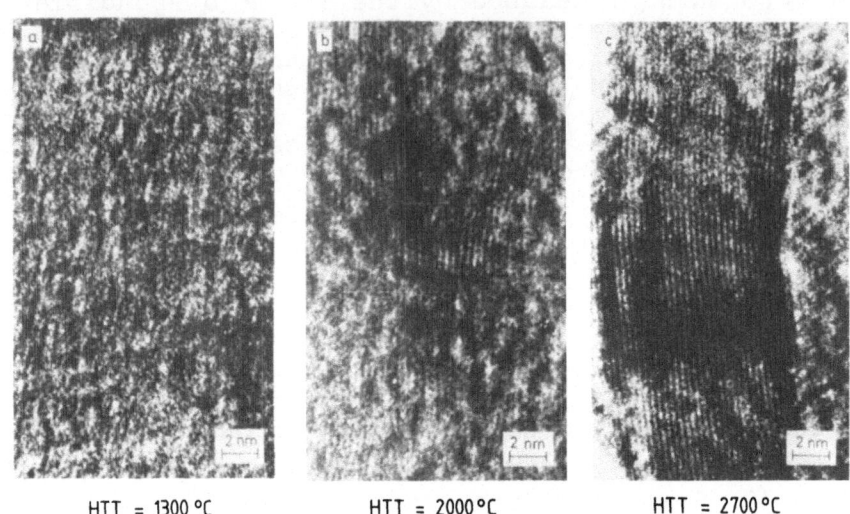

HTT = 1300 °C   HTT = 2000 °C   HTT = 2700 °C

Fig. 12 : TEM of PAN based carbon fibre in direc-
          tion parallel with the fibre axis after
          various final HTT's (1300, 2000, 2700 °C)
          (25)

CARBON FIBRES

The structural models for anisotropic polymeric carbon are compared in fig.9 with the isotropic types as discussed before. The first model
was developed by RULAND (26) for rayon based stress
graphitized fibres (the early THORNEL-type fibres
of UNION CARBIDE) from small angle scattering,
electron diffraction and TEM. The idea of ribbon
like polyaromatics was introduced by him.

The plates or bundle model of BENNETT et al.
(27) has been derived for PAN based carbon fibres.
It corresponds to the more realistic multilayer

sheet or bundle model of glasslike carbon of
TSCHOUBAR (20). The newest tube model of OBERLIN
(22) was developed for isotropic as well as for anisotropic polymer carbon.

As can be seen from fig. 13, the cross section
of PAN based carbon fibres appears in TEM similar
to the structure of isotropic polymer carbon (see
fig. 8). The preferred orientation which causes
the high strength and modulus of the carbon fibres
is achieved only in one direction that is parallel
with the fibre axis.

From a technological viewpoint it is interesting to compare the mechanical properties of the carbon fibre family with those of other reinforcing
fibres such as glass, polyaramide and boron. As can
be seen from fig. 14 the so-called type II-fibres
or high strength fibres are the strongest fibres
with current tensile strengths up to 4000 MN/m^2,
and a YOUNG's modulus still in the area of 250 to
300 GN/m^2. But most important is their high strain
to failure of 1,5 to 1,9 %. The high modulus of
type I carbon fibres as well as of the boron fibres
all have a higher YOUNG's modulus up to 550 or
even 600 GN/m^2. These exhibit a lower strength and
a very low strain to failure (below 1%), which is
most disadvantageous for many technical applications.

The polyaramide fibres have a middle position
between the high quality glass fibres and the high
strength - type carbon fibres. The glass fibres
have high strength and very high strain-to-failure
above 3 % but too low a YOUNG's modulus, which
means poor stiffness in spite of high strength.

CARBON FIBRES AS REINFORCEMENT IN POLYMERS

A revolution in material science, as described in popular literature recently (28), was
started by this family of carbon fibres. Fig. 15
shows carbon fibre textile as used in prepregs for
airplane structures. Such prepregs consist of epoxy
prepolymers reinforced by carbon fibres, which are
then combined by lay-up technique and subsequent
curing to C-fibre reinforced plastics (CFRP) the

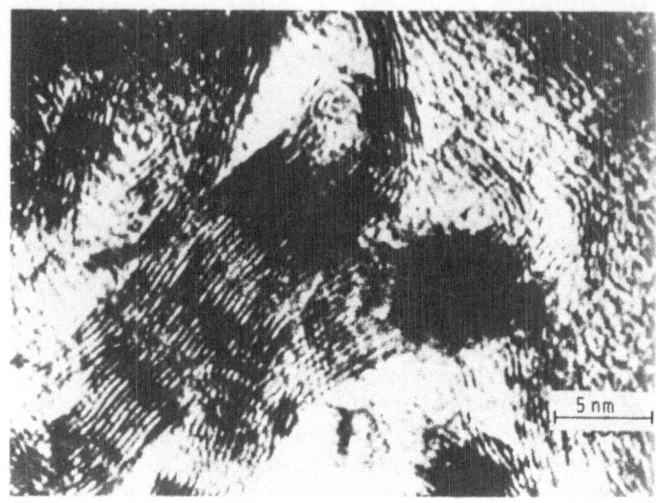

Fig.13: TEM of the cross section of a HM-C-fibre

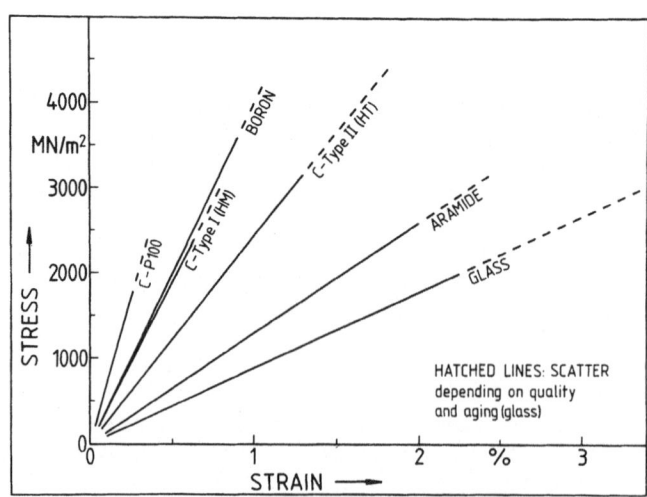

Fig.14: Stress-strain behaviour of fibres for
        reinforcement of polymers

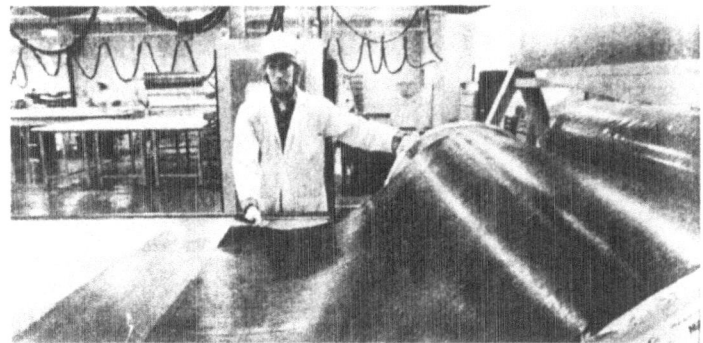

Fig. 15: Prepregs handling at NORTHROP AVIATION
         (28)

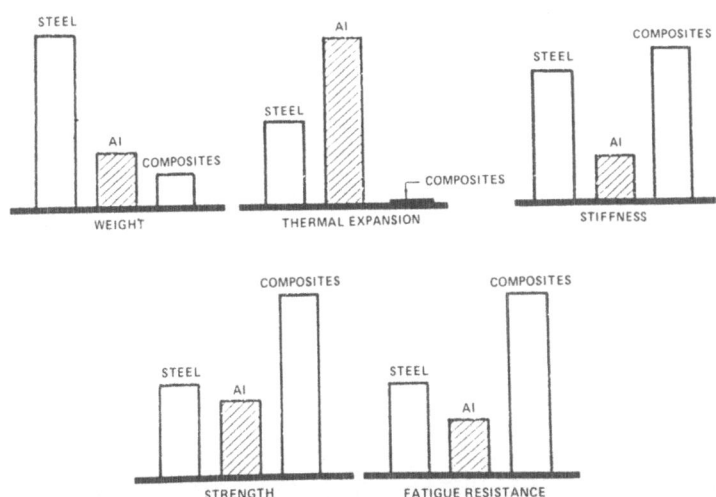

Fig. 16: Superior performance of composites
         compared with steel and aluminium (29)

the most promising advanced composites.

It has been known for years that such advanced composite materials have superior performance in weight, thermal expansion, stiffness, strength and fatigue resistance as compared with steel or aluminium (fig.16) (29). Indeed, it is the first time that polymers do not only replace metals but are developed for superior properties. One has to ask why public attention is focused to this material just recently only, and not years ago already when carbon fibre properties have been recognized first time and carbon fibres have been used for military applications but also for extravagant sporting goods.

During the last 30 years when many kinds of new materials were developed, propagated and disappeared again, material scientists described the situation by the bon mot: "The best what you can hear about a new material is when you hear first time about it". Therefore, the preconditions for a new structural material to be accepted as a candidate for revolutionizing the materials technology are not only superior extraordinary properties, but also:

(1) demonstration of performance,
(2) processibility, and
(3) availability at acceptable prices.

For non-military application the performance has been demonstrated in most spectacular way by the glorious flights and landings of the US space shuttle COLUMBIA. The large doors of the load room have been made from C-fibre reinforced epoxy because of the absolute need of weight saving. Fig.17 shows these doors during fabrication at ROCKETYNE INTERNATIONAL, in the late seventies.

Simultaneously with these tests for performance, new application fields have been explored, and new capacities for carbon fibre production have been set up. Table 2 gives a compilation of the demand of C-fibres during the last 2 years splitted into the fields of application. Also the price dropped down drastically from about 1000 Dollar/lb at the beginning of C fibre development to less

Fig. 17: Space shuttle door in fabrication

than 20 Dollar for some qualities. However, even
today, special qualities have an extremely high
price because of the complicated technology to
produce them (table 3). The technology will be discussed later.

On the basis of these prices one can calculate
the fibre costs per volume composite as shown in
fig. 18. The costs of the UD composites are based
on a volume fraction of C fibres of 60 %, and a
composite strength of 1400 MN/m^2. From the viewpoint of strength, glass fibres are the most economic ones followed by aramide fibres and finally
by PAN based carbon fibres. One must, however,
take into account the bulk density of the composite
material which is compiled in fig. 18 in the second line. There are applications such as in aerospace or even in engines for automotives, where
weight saving can easily compensate higher costs.

The decisive progress with carbon fibres,
however, is achieved by the YOUNG's modulus.
Fig. 19 gives some fibre costs per volume composite
as function of the maximum YOUNG's modulus of the
composite, which can be achieved by these fibre

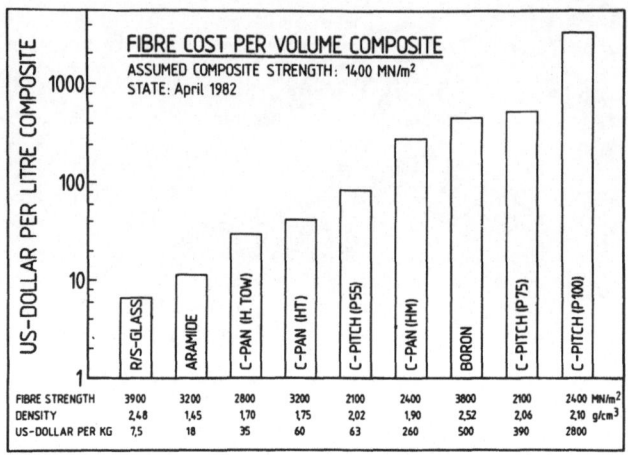

Fig. 18 : Fibre costs per volume UD-composite with a strength of 1400 MN/m²

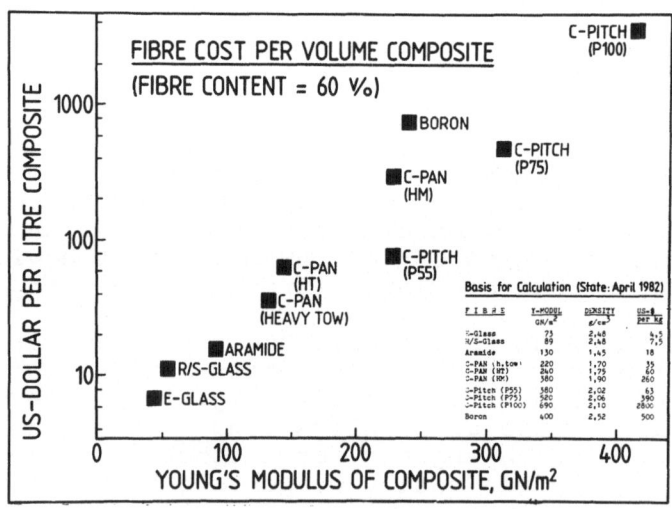

Fig. 19 : Fibre costs per volume composite as function of maximum YOUNG'S modulus of the composite achievable with a volume fraction of 60 % fibres of various types

Table 2 : Demand of C-fibres (t/yr)

	1981	1982
Aerospace	520 (42 %)	880 (47 %)
Sporting Goods	430 (35 %)	530 (28 %)
Other Industries	290 (23 %)	470 (25 %)
Total	1240	1880

Table 3 : Properties and prices of reinforcing fibres (1982)

F I B R E	TENSILE STRENGTH MN/m^2	Y-MODUL GN/m^2	DENSITY g/cm^3	US-\$ per kg
E-Glass	2500	73	2,48	4,5
R/S-Glass	3900	89	2,48	7,5
Aramide	3200	130	1,45	18
C-PAN (h.tow)	2800	220	1,70	35
C-PAN (HT)	3200	240	1,75	60
C-PAN (HM)	2400	380	1,90	260
C-Pitch (P55)	2100	380	2,02	63
C-Pitch (P75)	2100	520	2,06	390
C-Pitch (P100)	2400	690	2,10	2800
Boron	3800	400	2,52	500

types. Glass fibres and even aramide fibres have limitations in achievable bulk YOUNG's modulus of the composite above 100 GN/m^2. If YOUNG's moduli above 200 GN/m^2 are needed, pitch based carbon fibres promise to be more economic than high modulus PAN based fibres. Today, however, all high modulus fibres applied in critical structural elements such as in airplanes are made from PAN fibres.

FABRICATION METHODS FOR CARBON FIBRES

As far as the fabrication technology of such an isotropic polymer carbon is concerned, there exist in principle three types of processes (fig. 20). The main criteria is the method how the prefered orientation that means maximum anisotropy of the polymer carbon is achieved:

(1) by plastic deformation of isotropic polymer carbon at high temperatures, the so-called hot working or stretch graphitization process. This technique was applied by the first producer of high modulus carbon fibres (UNION CARBIDE) in the late sixties (30). Rayon fibres were used as precursor which results in an isotropic but also very porous and thus very easily deformable carbon fibre. In order to achieve high modulus, elongation up to 100 % is needed (fig.21). This technique has been left not only because of high costs of the rayon precursor, but also because of the difficulties with fracture during the extreme stretching at high temperatures. For rayon precursor this stretching technique was a conditio sine qua non, but it is not for PAN based or mesophase pitch based fibres. In principle, it can be applied for fabrication of ultra high modulus types. Obviously, the high prices for such types indicate application of this technique, although no open publication is available.

(2) Preferred orientation of the polymer carbon can be achieved by preorientation of the chain molecules in the polymer precursor fibre. Polyacrylonitrile has been proved as the most suitable and economic precursor. This thermoplastic polymer, however, needs a cross-linkage treatment before carbonization to preserve the fibre morphology. Simultaneously, cyclization of the side chains and dehydrogenation occur resulting in nuclei for the formation of the planar polyaromatic layers in polymer carbon. The cyclization can occur in non oxidizing as well as in oxidizing environement.

Fig. 22 indicates by DTA peaks, that the cyclization is strongly influenced by copolymers. In all technical processes the cyclization is per-

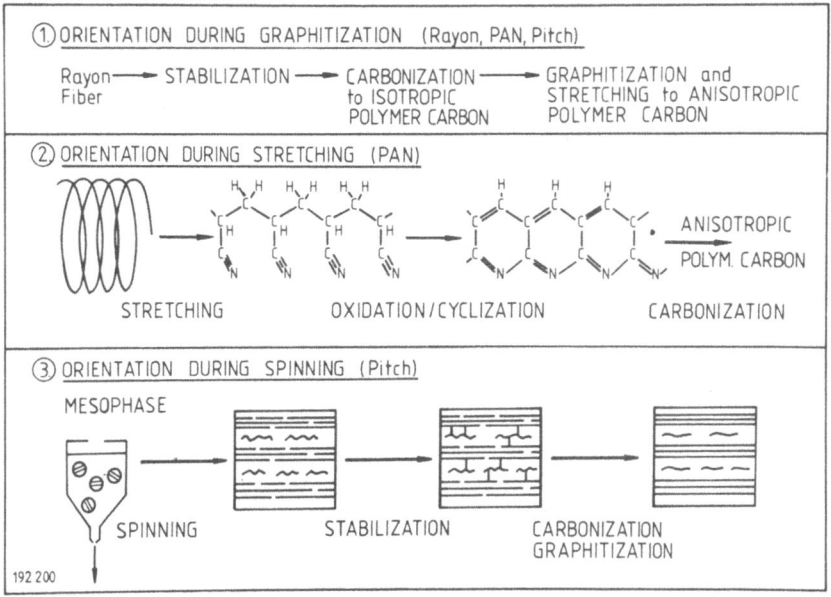

Fig. 20 : Three types of processes for fabrication
of anisotropic polymer carbon

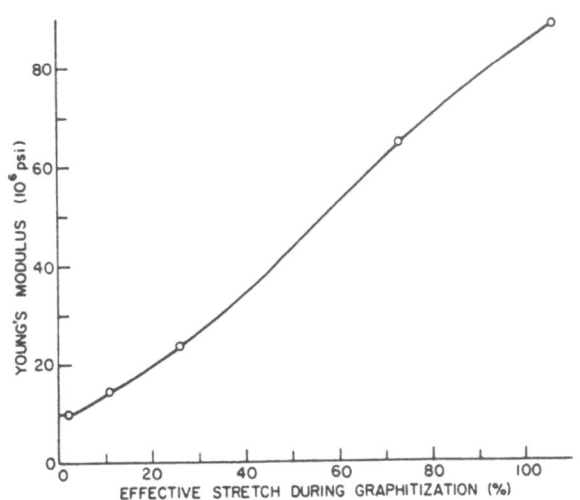

Fig. 21 : YOUNG's modulus of stretch graphitized
carbon fibres (20)

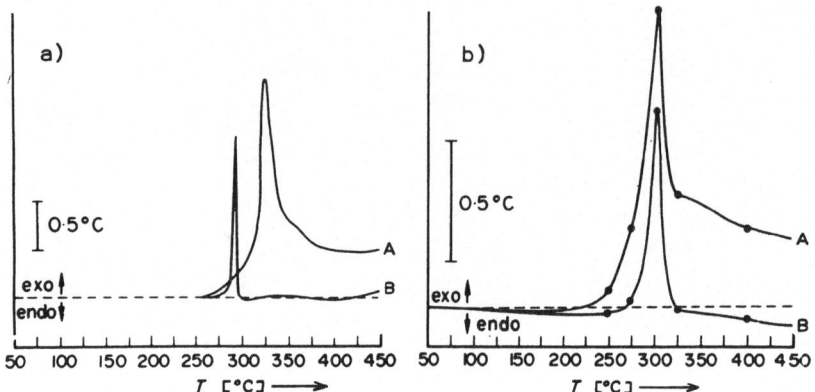

Fig. 22: DTA-curves of commercial PAN fibres in
air (A) and nitrogen (B) (heating rate
10 °C/min), a) Dry spun homopolymeric
acrylonitrile (DRALON T); b) wet spun copolymeric acrylonitrile (DOLAN 6%
methylacrylate) (31).

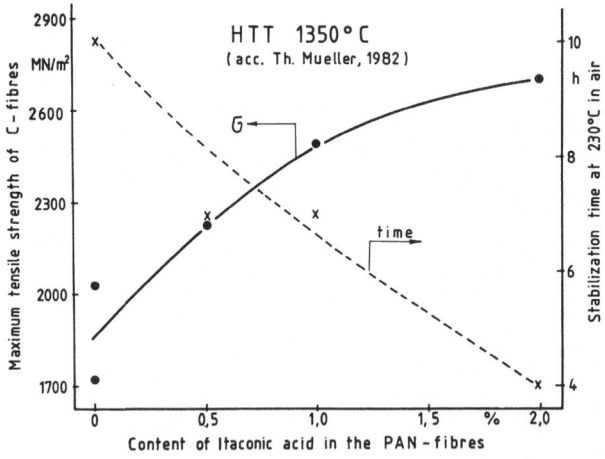

Fig. 23: Influence of Itaconic acid as copolymer
in PAN on tensile strength of C-fibres.

formed in oxidizing atmosphere for simultaneous
initiation of the crosslinkage reaction. Thus, the
rate of this heterogeneous reaction is diffusion
controlled, and long stabilization times in the
order of one hour are needed. Special interest is
therefore directed to catalysts for the overall
rate of the stabilization reaction.

Itaconic acid as copolymer has been found useful. Ionic cyclization mechanism is assumed in presence of this copolymer (32). Also final strength
was found to be increased by this copolymer
(fig.23).

There exist hundreds of publications on the
chemistry of the stabilization step, the volatile
byproducts and the probable structure of the stabilized fibres of at about 8 wt% is optimum to get
a maximum strength of the carbon fibres. It was published recently (33) that the oxygen will be present 80 % in acrydone type carbonyl and 20 % as
normal carbonyl or alcohol groups, whereas the nitrogen content is mostly present in acrydone structures and the rest as naphthyridine and hydro-
naphthyridine (fig. 24).

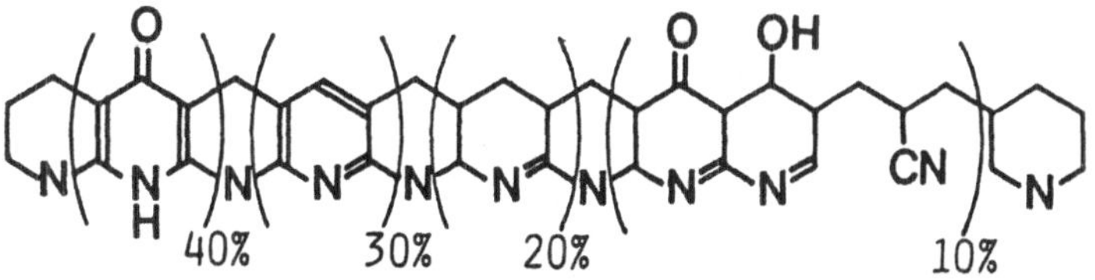

Fig. 24 : The chemical structure proposed for
          thermally stabilized PAN (33)

The keypoint in this process for formation of
anisotropic polymer carbon from polymer fibres
with a C/C-back bone which is preorientated in direction parallel with the fibre axis is the length
constancy or even elongation of the polymer fibre
during stabilization treatment by mechanical·
means (34). This process parameter is the same as
it has been applied for the not inflammable textile fibres (for instance black ORLON) before (35).
Also today, stabilized PAN is used as a non inflammable textile. For this purpose minimum oxygen
contents of 5-7 % are needed for inflammibility.

The final heat treatment of the carbon fibre
is a decisive process parameter for its mechanical
properties, besides of chemical composition of the
PAN precursor fibre (copolymers) and the process
measures to achieve maximum orientation of the polymer chains in the precursor and to maintain this
orientation during cyclization and carbonization.
It has been mentioned in the beginning, that polymer carbon is a non graphitizing so-called "hard"
carbon. By heat treatment at temperatures above
2500 K never the 3-dimensional order can be achieved. However, it has been shown in fig.12 that such
a heat treatment enables solid state reorganization
of the polyaromatic compounds to more perfect planarity and to improved preferred orientation in
direction parallel with the fibre axis. Also defects within the layers will be healed out, and
cross linkage between disordered layers will be
broken. In terms of mechanical properties, such a
reorganization will improve the YOUNG's modulus
of the fibre, but will decrease the strength. This
is confirmed in all cases with PAN based carbon
fibres. Table 4 gives some typical data. The
YOUNG's modulus increases from 250 to 400 GN/m^2
whereas the strength will decrease reciprocally
from 4000 to 2500 MN.

Commercially available PAN based carbon fibres
with final heat treatment temperatures of 1400 ºC
are so-called type II or high strength fibres and
those with heat treatment temperatures of 2700 ºC
are type I or high modulus fibres (compare fig. 14
and table 3).

	PAN-based			PITCH-based		
HTT ^{o}C	E GN/m^2	σ MN/m^2	ε %	E GN/m^2	σ MN/m^2	ε %
1400	250	4000	1.60	350	2000	0.57
2700	400	2500	0.63	540	2000	0.37

Table 4 : Influence of final heat treatment temperature on YOUNG'S modulus (E), tensile
strength (%), and
strain to failure (ε) of carbon fibers

There is another parameter mentioned in table 3
and fig. 18 and 19, namely the term "heavy tow"
leading to the lowest prices of PAN based carbon
fibres. This term is concerned to the wet spinning
technology of PAN, using spinnerets with a high
number of holes between 60 000 and 300 000. Thus,
the spinning process becomes extremely economic,
and fibre costs are reduced drastically if the
final PAN and the carbon can be handled as a heavy
tow with same numbers of monofilaments.

For carbon fibre production until now special tows were fabricated with a maximum number of
6000 monofilaments. It has been shown during the
last years, however, that such heavy tows can be
produced with improved quality at low costs
(fig. 25) (36).

(3) As shown in fig. 20, there exists a third
way to achieve orientation of the polymer carbon

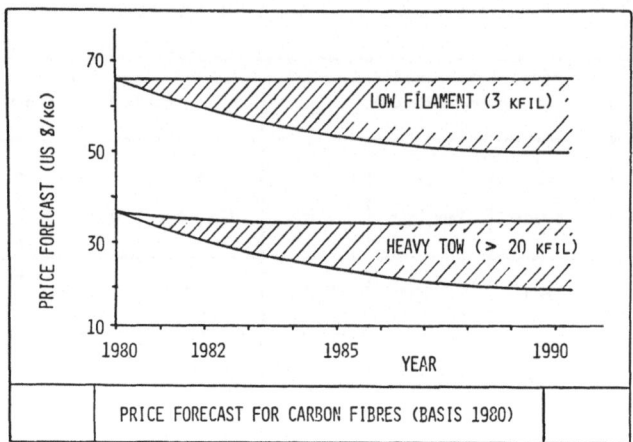

Fig. 25: Price forecast for carbon fibres from
         PAN (36)

in direction parallel with the fibre axis by spinning of a mesophase precursor. It has been shown
in fig.2 that pyrolysis of a pitch in the molten
stage will lead to the intermediate formation of
liquid crystals by the planar polyaromatic compounds formed during thermal degradation of the
precursor material. Such mesophase formation is a
precondition for achieving well graphitizing so-
called soft carbon. It has been discussed, however,
that soft carbon is unsuitable for carbon fibres
because of its graphitization tendency and thus
low shear modulus (table 1 and fig.11).

     At the first view, therefore, a carbon fibre
made from mesophase pitch seems to be a contra-
diction itself. The solution of this problem may
be seen in the stabilization treatment of the fibre
after spinning which is not only necessary for
conservation of the fibre morphology during subsequent heat treatment (the same as with the thermoplastic PAN) but also for crosslinkages between
the preformed polyaromatic compounds, by which formation of graphitizing carbon should be inhibited.
Thus again, the stabilization treatment in air

will turn into a heterogeneous diffusion rate controlled reaction. If one wants to preserve the
cost advantage of the inexpensive pitches compared
with the more expensive PAN polymer during the fabrication process, one has to try to shorten the
stabilization treatment. Incomplete stabilization,
mainly different degree of stabilization between
the surface areas and inner parts of the fibre,
however, will result in partial graphitizability
of the carbon fibre. This was described by the
only producer of this fibre type (37). Thus, the
mesophase pitch based carbon fibre will have completely different properties than that of PAN
based ones; the properties are compiled in table 4
in comparison with those of PAN based fibres, namely higher YOUNG's modulus and much larger increase
of YOUNG's modulus during graphitization treatment
but lower strength.

These properties of carbon fibres from mesophase pitch are very advantageous for carbon/carbon
composites (carbon fibre reinforced carbon materials
CFRC), which are mostly of military importance today in the missile technology. In spite of intensive studies on further advantageous applications
it can not yet be seen where these pitch based
fibres are superior for fibre reinforced plastics.
Even the present prices do not beat those of PAN
based fibres especially if heavy tows are con -
sidered.

Summarizing the three fabrication technologies
one can state that the hot working process starts
from isotropic polymer carbon and achieves preferred orientation by plastic deformation at graphitization temperatures, but does not introduce graphitizing carbon.

The PAN based process achieves preferred orientation by preorientation of the chain polymers and
transforms these chain polymers via ladder polymers into polyaromatic planar molecules ending
with polymer carbon with preferred orientation.

The third process starts with preformed polyaromatics in the form of liquid crystals and
achieves non graphitizing polymer carbon by cross

linkage via an oxidation process. Because this
cross linkage is less perfect, these C-fibres are
in the transition area between polymer carbon and
graphite.

TODAYS AND FUTURE APPLICATION OF CFRP's

        This paper should not end without giving some
information on present and future applications of
this anisotropic polymer carbon. In space technology, carbon fibre reinforced composites are the
unique structural material. In aircraft industry
all military planes make use of this revolutionary
material. Fig. 26 shows the model of a future
fighter which combines wings and stabilizers in one
structural element. The whole structure consists of
advanced composites. Only small parts, i.e. the engines are still made from metals. In today's commercial aircrafts only secondary and tertiary parts
are made from CFRP, as shown in fig. 27. There is

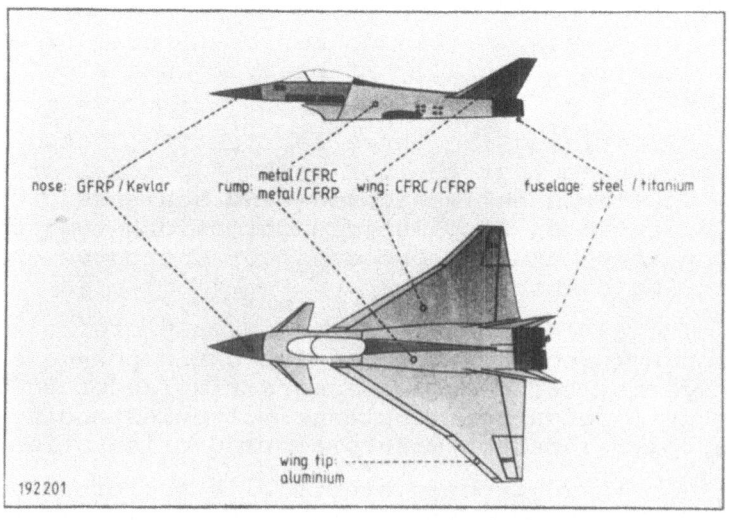

Fig. 26 : Model of a future fighter (38)

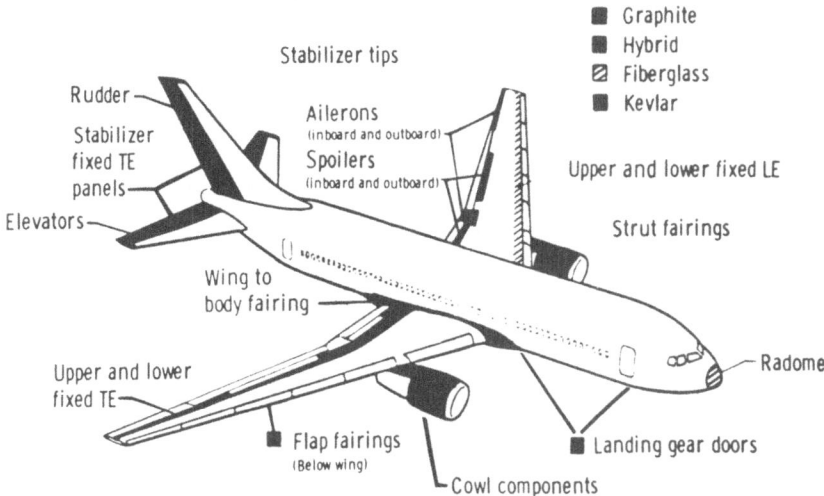

Fig. 27 : Composite applications in Boeing 767

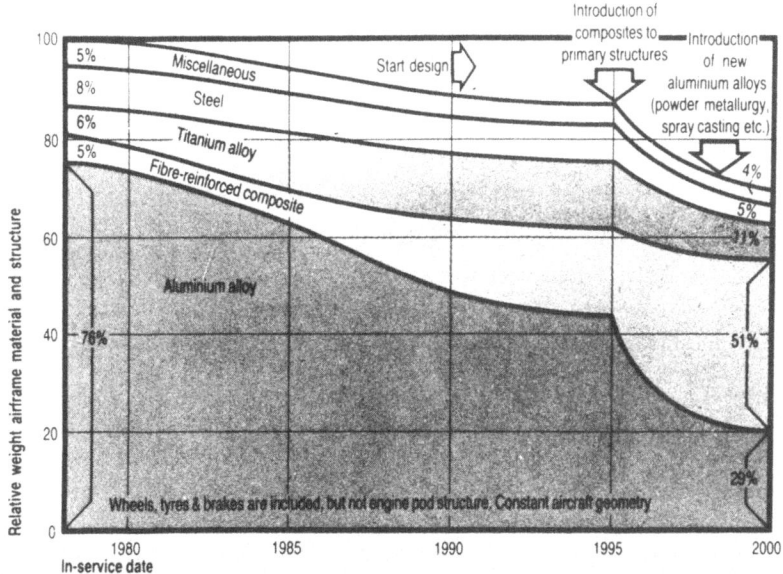

Fig. 28 : Structural materials for commercial
airplanes in future (39)

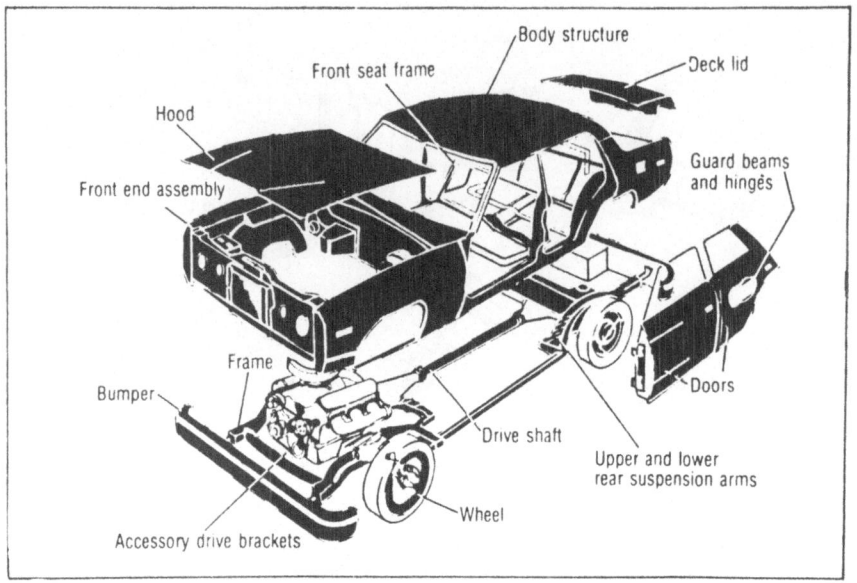

Fig. 29 : Possible applications of composites
          in automotive (40)

TRANSPORTABLE BRIDGE FROM CFRP + Al

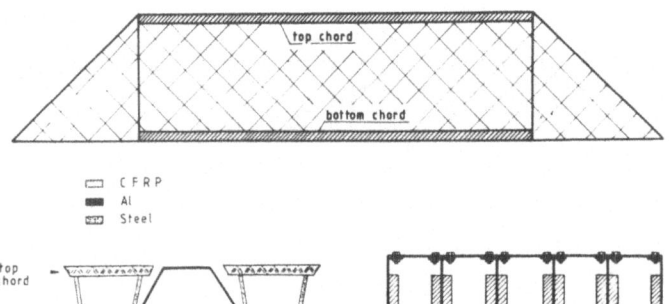

Fig. 30 : Transportable bridge with CFRP as
          load carrying material (41)

no question, that much more advanced composites
will be applied in future (fig.28). Gliders are
special examples for application where carbon
fibres have become unavoidable for high performance
types.

The enthusiasm to introduce carbon fibres into
the automotive industry (fig.29) (40) has been reduced because of the recession in this industrial
area. However, all leading companies in the automotive industry have started intensive studies on
this problem. The driveshaft and the leaf springs
are those parts which are introduced already or
will be introduced in the very near future.

Also a revolutionary transportable bridge has
been designed by DORNIER (41) with all load carrying parts made from carbon fibre reinforced cords.
The prototype of this bridge is 4¹ m long and has
a weight/m of 460 kg only. It is designed for a
maximum load of 60 tons for 20 000 crossings
(fig. 30).

Another revolutionary application is that of
submarine tubes for off-shore drilling (42). The
corrosion resistant tubes are lined by rubber and
can be used up to 2000 m depth.

Last not least the biocompatibility of carbon
fibres should be mentioned. Carbon fibres are used
for replacement of tendons and ligaments (43, 44)
and applications in composites such as with polymers or with carbon matrix are in preparation.

CONCLUSIONS

From the above report the following conclusions can be summarized:

(1) Polymer carbon is a non graphitizing
carbon material consisting of polyaromatic gra-
phite-like carbon layers in which the extremely
strong C/C- bonds are reflected in mechanical
bulk properties. The very low shear modulus as
known for crystalline graphite is avoided however,
by structural arrangement of the layer bundles and
by cross linkages between them. Chemists and ma-

terial scientists have developed methods for the
fabrication of polymer carbon from organic polymers with controlled morphology and structural preferred orientation.

(2) Depending on the degree of preferred
orientation polymer carbon can be used as glasslike carbon, as carbon fibres and textiles or in
composites first of all with polymers as matrix
(CFRP). Application as structural material is
starting now to replace metals because CFRP's can
be tailored according to the asked strength in the
various directions. Thus, the amount of material
needed for a special purpose is enormeously reduced. Additionally, the light weight of a CFRP reduces the total weight of structures by an order
of magnitude and makes polymer carbon to the future
material in motive application. The advantages because of the corrosion and fatigue resistance are
not yet fully explored for this purpose.

The polymer carbon in "all-carbon" materials
is the high-temperature material par excellence.
Finally, biocompatibility of polymer carbon promises a wide application as biomaterial in the
future.

(3) There remain some not yet solved problems
for the future. One is the polymer matrix itself.
For high performance composites epoxy resins on the
basis of bisphenol A are used. Some limitations are
given by the temperature sensitivity, long curing
times, inflammability, limited strain to failure
for special high performance qualities and relatively high water absorption.

Another problem not yet solved completely is
concerned with economic fabrication techniques for
mass production. Fully automatic wet winding has
been developed but is not yet applied. Pulltrusion
and sheet molding will be two of the methods to be
used in future for most economic fabrication.

Finally, a broad economic application of
CFRP's needs a redesign of structural components
made from metals so far. All such structures are
based on the isotropic properties of the metals.
A replacement of metals in such structural parts

f.i. in automotives which have been designed for metals is completely uneconomic. The redesign must be based on the sepcially tailorable properties of these new materials.

Nevertheless, if we generalize the chances of polymer carbon and composites in the future one could prognose with some imagination, that the metal age never has been replaced by a polymer age so far, but will be replaced in future by a carbon age with the support of polymers.

REFERENCES

1) BRAGG, W.H.·, BRAGG, W.L.,
   Proc. Roy. Soc. (London), 89, 277 ·(1913)

2) HULL, A.W.;
   Phys. Rev., 10, 661 (1917)

3) DEBYE, P. and SCHERRER, R.,
   Physik. Z., 18, p. 291 (1917)

4) HASSEL, D., and MARK, H.,
   Z. Physik, 25, p. 317 (1924)

5) BERNAL, J.D.,
   Proc. Roy. Soc. (London), A 106, p. 749 (1924)

6) JENKINS, G.M., KAWAMURA, K.,
   Polymeric Carbons - Carbon Fibre, Glass and Char
   (Cambridge University Press,
   Cambridge 1976)

7) FITZER, E., MÜLLER, K., SCHÄFER, W.,
   Chemistry of the Pyrolytic Conversion of
   Organic Compounds to Carbon, Chemistry and Physics of Carbon, vol.7, p. 237-283 (Marcel Dekker,
   New York, 1971)

8) Plate and Cup donated by Dr. H. BÖDER, SIGRI
   Elektrographit GmbH, Meitingen, (FRG),
   Bell donated by A.S. KOTOSONOV, Moscow (USSR)

9) Tokai Carbon C., Ltd., Tokyo, Japan

10) Plessey Co., Ltd.; Ilford-Essex, UK

11) YAMADA, S., SATO, W., ISHI, T.,
    CARBON 2, p. 253 (1964)

    compare: FITZER, E., SCHÄFER, W., YAMADA, S.,
    CARBON 7, p. 643 (1969)

12)   ERGUN, S.,
      X-ray Studies on Carbon
      in Chemistry and Physics of Carbon, Vol. 3,
      p. 211 (Marcel Dekker, New York, 1968)

      compare: ERGUN, S.,
      Phys. Rev. B 1, p. 3371 (1970)

13)   FITZER, E., BRAUN, W.,
      CARBON 16, pp. 81 - 83 (1978)

14)   WARREN, B.E.,
      Proc. of the 1. and 2. Conf. on Carbon, p. 49
      (Buffalo, New York: The University of Buffalo
      1956)

15)   BRAUN, W., FITZER, E.,
      Preprints of Carbon 72, Baden-Baden, p. 135
      (Deutsche Keramische Gesellschaft 1972)

16)   FITZER, E., KALKA, J.,
      CARBON 10, p. 173 (1972)

      compare: KALKA, J., Ph.D. Thesis,
      The University of Karlsruhe (1970)

17)   FITZER, E., LAUDENKLOS, P.,
      Preprints of the 12th Biennial Conf. on Carbon,
      Pittsburgh, p. 21
      (The American Carbon Society and the School of
      Engineering, University of Pittsburgh, 1975)

18)   FITZER, E., KALKA, J.,
      CARBON 10, p. 173-183 (1972)

19)   PHILIPS, V.A.,
      Report 71-C-167 GEC
      Research and Development Laboratory,
      (Schenectady, N.Y. 1971)

2o)   JENKINS, G.M.,
      Deformation Mechanism in Carbons   in
      Chemistry and Physics of Carbon 11, pp.189-242
      (Marcel Dekker, New York, 1973)

21)   ROUSSEAUX, F., TSCHOUBAR, D.,
      CARBON 15, p. 63 (1977)

22)   OBERLIN, A., OBERLIN, M.,
      paper 1807 in ext. abstr. of International
      Symp. on Carbon New Processing and New Appli-
      cations , Toyohashi, p. 77 (Carbon Soc. of
      Japan 1982)

22a) OBERLIN, A., OBERLIN, M.,
     paper FC 11 in ext.abstr. of the 15th Biennial
     Conf. on Carbon, Philadelphia, p. 288
     (The American Carbon Soc. and The University
     of Pennsylvania, 1981)

23) KELLY, B.T.,
     High Temp.- High Pressures 5, pp. 133-144 (1973)

24) COOPER, C.;
     Chemistry and Industry 18, p.678 (1982)

25) GRUNDLER, J.,
     Diplomarbeit, The University of Karlsruhe(1981)

26) RULAND, W.,
     J. Appl. Phys. 38, p.3585 (1967)

27) BENNETT, S.C., JOHNSON, D.J.,
     Proc.Fifth London Int.Carbon and Graphite in
     Conf. Vol. I, p. 377 (Soc. of Chem.Industry,
     London 1978)

28) Newsweek, February 1, p. 36 (1982)

29) DEUTSCH, G.C.,
     in 23th National SAMPE Symp. and Exhibition,
     Anaheim, p. 34 (Society for the Advancement
     of Material and Process Engineering 1978)

30) BACON, R.,
     Carbon Fibres from Rayon Precursors
     in Chemistry and Physics of Carbon 9,
     pp. 11 - 102, (Marcel Dekker, New York, 1973)

31) FITZER, E., HEYM, M.,
     Chemistry and Industry 21, pp. 663-676 (1976)

     compare: MÜLLER, D.J., Ph.D.Thesis,
     The University of Karlsruhe (1976)

32) FITZER, E., MÜLLER, Th., Ext. Abstr. of
     Carbon 82, London, p. 256 (Society of
     Chemical Ind. 1982)

33) TAKAHAGI, T., SHIMADA, I., FUKUHARA, M.,
     MORITA, K., ISHITANI, A.,
     in Ext. Abstr. of Internat. Symp. on Carbon
     New Processing and New Application , Toyohashi,
     paper 3A13 (Carbon Society of Japan, 1982)

34) WATT, W., JOHNSON, W., PHILIPPS, L.N.,
     Brit. Pat. 1 110 791 (1964/68)

35) JOHN-MANVILLE Corp., US Pat. 29 13 802 (1959)

36) BÖDER, H.,
     priv. communication, SIGRI Elektrographit GmbH,
     Meitingen, FRG (1982)

37) SINGER, L.S.,
     US-Patent 4,005,183 (1977)

38) MBB (Messerschmitt, Bölkow, Blohm)
     model exposed at the industrial world fair,
     Hannover 1981

39) Engineering, September 1982, p. 611

40) DHARAN, C.K.H., THOMS, J.,
     in 24. Nat. SAMPE Symp. and Exhibition,
     pp. 1550-1566, San Francisco, (Society for the
     Advancement of Material and Process Eng.,1979)

     BEARDMORE, P., HARWOOD, J.J., HORTON, E.J.,
     in Proc. of the 3rd Int. Conf. on Composite
     Materials, Paris, Vol. 1, p. 11
     (Pergamon Press 1980)

41) FÜSSINGER, R.,
     The First Aluminium CRP Bridge
     in Processing and Uses of Carbon Fibre Rein-
     forced Plastics, pp. 215-227, (VDI-Verlag
     Düsseldorf, 1981)

42) GRENIER, M.,
     private information, Aerospatiale,
     Médard on Jalles, France

43) BURRI, C., CLAES, L., FITZER, E., HÜTTNER,W.,
     KINZL, L.,
     in Proc. of Int. Fixation Congress, Ottawa,
     pp. 156 - 159 (Springer-Verlag, 1980)

44) HÜTTINGER, K.J.,
     Ext. Abstr. of International Symposium on
     Carbon New Processing and New Application
     Toyohashi, paper 1A08, p. 29 (Carbon Society
     of Japan 1982)

45) JENKINS, G.M.,
     Clin. Phys. Physiol. Meas., 1, No. 3,
     pp. 171-194 (1980)

RECENT DEVELOPMENTS IN THE SCIENCE AND TECHNOLOGY OF

ULTRA-HIGH MODULUS POLYOLEFINES

Ian M. Ward

Department of Physics
University of Leeds
Leeds LS2 9JT, UK

SYNOPSIS

   A comprehensive survey is presented of the development
of tensile drawing, hydrostatic extrusion and die drawing
processes for the manufacture of ultra high modulus polyethylene and polypropylene in the form of fibres, films and
extruded shapes. The properties of the highly drawn products
are described, including mechanical stiffness, creep, strength,
melting behaviour, shrinkage, thermal expansion, thermal
conductivity, barrier properties and chemical resistance.
Finally, there is an account of recent research into possible
applications of these materials with particular attention to
the reinforcement of brittle matrices, including polymer
resins, cement and concrete.

INTRODUCTION

   During the last ten years there have been several notable
advances in the development of ultra high modulus polyethylene
and polypropylene. This has been achieved most simply by
tensile drawing[1-4], but also by hydrostatic extrusion[5-7], ram
extrusion[8] and die drawing[9,10], all of which are solid phase
deformation processes. In polyethylene, an alternative
approach has been the production of fine ultra high modulus
fibres from dilute solution, either by crystallisation in an
elongational flow field[11] or by stretching fine fibres spun to
form a gel from dilute or reasonably dilute solution[12,13].

   In this paper the discussion will be confined to the
developments at Leeds University which relate to the prepar-

ation and properties of ultra high modulus polyethylene and
polypropylene by tensile drawing, hydrostatic extrusion and
die drawing.   In all cases small scale production units
have been constructed, in some cases followed by industrial
developments, so that substantial amounts of material could
be available for assessment of potential end uses.

TENSILE DRAWING

(1)   Polyethylene

The first guidelines for understanding the relationship
between modulus and tensile drawing behaviour came from the
studies of Andrews and Ward[14] on a series of polyethylenes
with different molecular weight and molecular weight
distribution.   Although there was no simple relationship,
it was concluded that the draw ratio obtained under
standard drawing conditions depended on the weight average
molecular weight $\bar{M}_w$, low $\bar{M}_w$ leading to high draw ratio.   A
very clear relationship was obtained between the Young's
modulus of the drawn samples and the draw ratio (Figure 1)
and this observation has proved to be a key observation
which unlies all attempts to produce high modulus oriented
polymers by solid phase deformation, and more recently by
gel drawing.

It was appreciated that there was likely to be much
advantage in controlling the initial morphology prior to
drawing, and this provided the key extra ingredient in the
researches of Capaccio and Ward[1,2,15], and later Capaccio,
Crompton and Ward[16] which led to the discovery of a number
of routes for the achievement of very high draw ratios and
consequently ultra-high modulus polyethylene.   It was
confirmed that $\bar{M}_w$ was the most important parameter in
determining the ultimate draw ratio, and by increasing the
draw temperature with increasing $\bar{M}_w$, conditions were
established for drawing polymers with $\bar{M}_w$ up to $\sim 8 \times 10^5$ to
give high modulus products.   The initial morphology was
most important for low molecular weight polymers, where
segregation of low molecular weight polymer during
crystallisation can also occur.   In all cases drawing
takes place at comparatively high temperatures to permit
adequate molecular mobility (probably associated with
similar mobility to that observed in the
α - relaxation process and the creep slip process to be
discussed) but below temperatures at which flow drawing

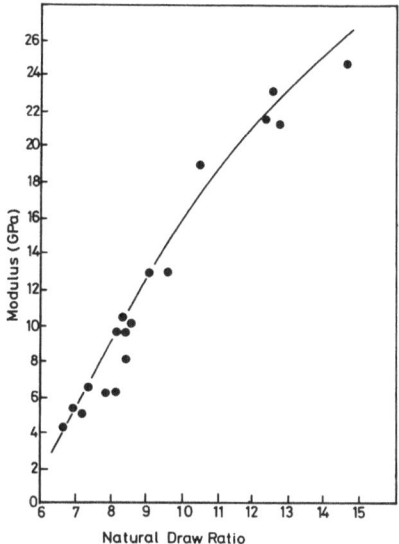

Fig.1.   Modulus vs draw ratio for cold drawn LPE.
Reproduced from J.Mater.Sci., 5, 411 (1970)
by permission of the publishers, Chapman &
Hall Ltd. (C).

occurs and there is no molecular alignment associated with
the macroscopic deformation.   The concept of a natural
draw ratio, familiar in polyester and nylon technology, is
therefore replaced by the concept of effective draw, draw
which produces molecular alignment and extension.   The
degree of effective draw is most conveniently monitored
by the modulus, and as shown in Figure 2, to a good
approximation this is determined only by the draw ratio
irrespective of molecular weight or initial morphology.
This geometric aspect to the drawing behaviour led to the
conclusion that the drawing process involves the
stretching of a molecular network.   For high molecular
weight polymers molecular entanglements form the network
junction points, and the situation is exactly analogous
to drawing an amorphous polymer such as polyethylene
terephthalate or polymethylmethacrylate.   For low
molecular weight polyethylene the crystalline regions
can also provide temporary junction points and the
morphology is therefore very important.

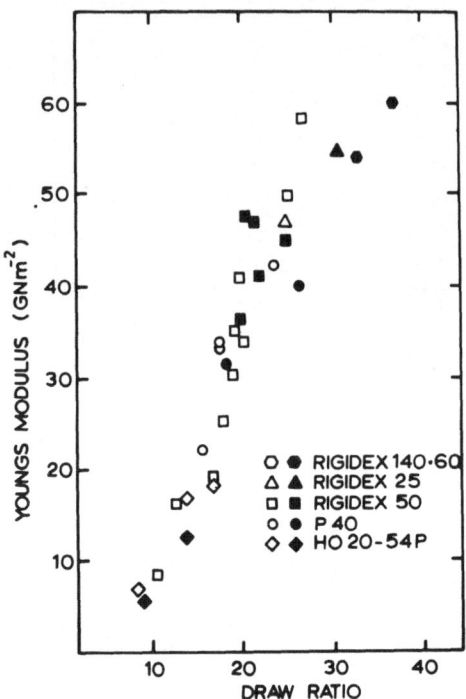

Fig.2.   Modulus vs draw ratio for a variety of quenched
          (open symbols) and slow-cooled (solid symbols)
          LPE samples drawn at 75°C.
          Reproduced from J. Polym. Sci., Polym. Phys.
          Ed. 14, 641 (1976), by permission of the publishers,
          John Wiley and Sons Ltd. (C)

(2)   Polypropylene

        In polypropylene it was found that the draw ratio
which could be achieved was greatly reduced with
increasing molecular weight, as in polyethylene, but
that initial morphology had no significant effect even
for low molecular weight samples[3,17].   The draw
temperature was a key variable, and an optimum draw
temperature of 110°C for effective drawing enabled draw
ratios ∿ 20 to be obtained.   This gave room temperature
moduli of ∿ 20 GPa, rising to ∿ 30 GPa measured at 5 Hz at
low temperatures.   This compares favourably with crystal
lattice modulus values ∿ 40 GPa, determined from X-ray
diffraction data.   We are therefore approaching a very

similar fraction of the theoretical modulus to that obtained
in polyethylene, but in this case the helical form of the
molecular chain reduces the ultimate modulus to a comparatively low absolute value.

(3)   Polyethylene Copolymers

Because of their improved creep behaviour (to be
discussed below) recent studies at Leeds University have
been concerned with the drawing behaviour of polyethylene
copolymers[18]. The introduction of a very small concentration of short branches ($\sim$ 1 butyl/1,000 carbon atoms)
produces a marked improvement in creep behaviour. It also
affects the strain hardening, and both the concentration and
nature of the side branches are important, butyl branches
being more effective than ethyl branches, and methyl
branches producing comparatively small changes in the
drawing behaviour.Figure 3 shows the load-extension curves
for a series of polyethylenes of very similar molecular
weight and molecular weight distribution comparing homopolymer (C) with a copolymer (B) containing about 1 butyl
branch/1,000 carbon atoms and a second copolymer (A)
containing about 4 butyl branches/1,000 carbon atoms. It
can be seen that there is a marked increase in the rate of
strain hardening and reduction in the final draw ratio with
increasing branch concentration. The relationship between
modulus and draw ratio is, however, very similar to that for
homopolymers, so that any limitation on the achievement of
high modulus appears to relate primarily to the reduction
in the maximum draw ratio attainable.

Further Developments

The research described above was concerned with the
drawing of dumbell specimens cut from compression moulded
sheets of polymer, and has been extended in two ways. First,
there has been the development of continuous drawing
processes for monofilaments and multifilament yarns, where
the drawing takes place between moving rollers, either
through a heated liquid bath (typically glycerol) or over
a heated plate. A major objective is to increase the
speed of drawing, which involves increasing the draw
temperature, taking care not to approach the flow drawing
state. By judicious choice of spinning and drawing
conditions processes have been developed for the production
of ultra high modulus polyethylene yarns on a pilot plant

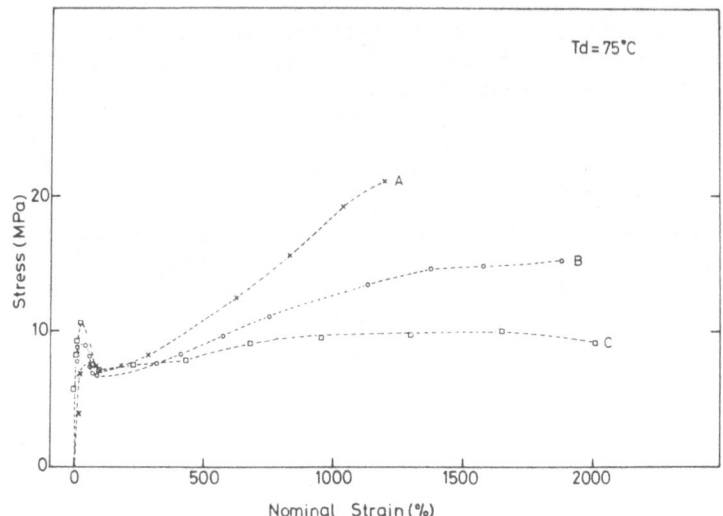

Fig.3.   Load-extension curves for Ethylene Hexene - 1
         copolymers (A   4 butyl/10³C; B   1 butyl/10³C)
         C   homopolymer.

scale[19,20].   Similar results have been obtained for polypropylene monofilament drawing at Leeds University[21].

        Secondly, it was appreciated that there was much to be
gained from more fundamental studies of the drawing process.
In the first instance the true yield or flow stress of linear
polyethylene (and polyoxymethylene) was determined as a
function of strain rate over a very wide range of plastic
strains (draw ratios up to 25).   The results were shown to
be compatible with the concept that for a given isotropic
polymer with specified structure in terms of molecular
weight and initial thermal treatment there is a true flow
stress-strain-strain rate relationship.   For a given
material the plastic strain level attained (i.e. the draw
ratio provided effective draw occurs) will define the true
stress-strain rate dependence and there is no strain rate
path dependence.   This enables us to define true stress-
strain-strain rate surfaces of the type shown in Figure 4
for Rigidex 50 linear polyethylene, and any tensile deformation process can be considered as defining a path across
this surface.

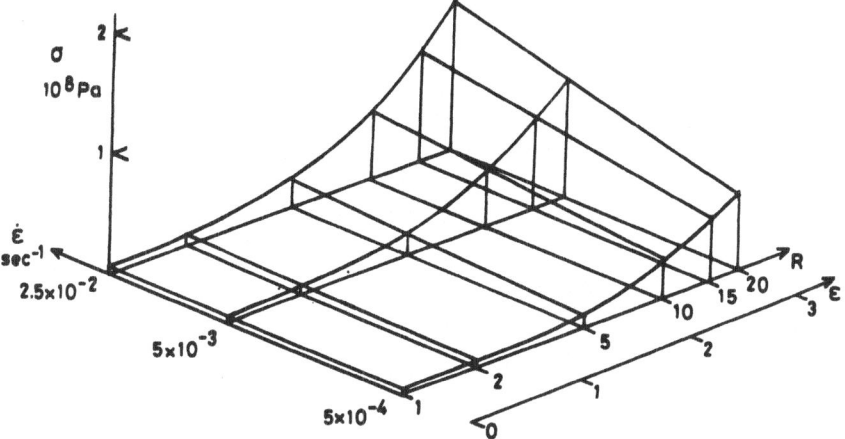

Fig.4.   True stress-strain-strain rate surfaces for
         Rigidex 50 LPE at 100°C.
         Reproduced from J.Mater.Sci., 13, 1957, (1978)
         by permission of the publishers, Chapman & Hall
         Leds. (C).

An important observation, which can be seen from the
shapes of the surfaces shown in Figure 4, is the increased
strain rate dependence of the flow stress with increasing
plastic strain ε (or equivalently draw ratio R or λ, where
$\varepsilon = \ell n R$).   We have taken as our starting point the Eyring
equation for a thermally activated rate process.   The
plastic strain rate $\dot{\varepsilon}$ is given at high stresses by

$$\dot{\varepsilon} = \dot{\varepsilon}_o \exp \frac{-\Delta U}{kT} \sinh \frac{\sigma_f v}{kT} \sim \dot{\varepsilon}_o \exp \left(\frac{-(\Delta U - \sigma_f v)}{kT} \right) \text{for large } \sigma_f$$
(1)

where $\dot{\varepsilon}_o$ is a constant (commonly termed the pre-exponential
factor, ΔU is the activation energy, σ_f is the flow stress,
v is the activation volume, k is Boltzmann's constant and T
is absolute temperature

and $\left(\dfrac{\partial \sigma_f}{\partial \ell n \dot{\varepsilon}} \right) = \dfrac{kT}{v}$

This implies that for a single, activated process at high
flow stresses, the stress-log (strain rate) relationship

should be linear, with a slope inversely proportional to the stress activation volume v.

The results for polyethylene show a non-linear relation-ship between the flow stress and strain rate, at given degree of plastic strain, as well as the increase in strain rate at high plastic strain already mentioned. This apparent change in stress activation volume with strain and strain rate can be formally represented either by two activated processes operating in parallel, each following the Eyring formulation[23], or by a single activated process in which the activation volume is allowed to vary with stress and/or strain[24]. The first formulation is of particular interest, because it appears to provide a link between tensile draw-ing and the creep behaviour of drawn LPE. The stress activation volume can be regarded as representing the volume swept out by the activated event. Very large activation volumes ($\sim 1000 \, \text{Å}^3$) are typical for isotropic polymers and have been taken to indicate the comparatively large-scale co-operative nature of the yield process. The decrease in activation volume with increasing draw can be attributed to the localisation of the yield process. More accurately, the behaviour of the drawn polymers (for $\lambda > 10$) can be represented by the superposition of a high stress, low activation volume process η_2 (operating at high strain rates) and a low stress, high activation volume process η_1 (operating over the whole range of strain rates, but dominant at low strain rates).

The total stress σ is given by

$$\sigma = \frac{2.3kT}{v_1} \left(\log \dot{\epsilon} - \log \left(\left[\dot{\epsilon}_o'\right]_1 /2\right) + \frac{kT}{v_2} \sinh^{-1}(\dot{\epsilon}/\left[\dot{\epsilon}_o'\right]_2)\right)$$

where the subscripts 1 and 2 refer to each process as illustrated in Figure 5, and we have retained the sinh representation for the low activation volume process.

In the drawing process, the polymer moves across the stress-strain-strain rate surfaces of Figure 4, and any incremental change in the flow stress $d\sigma$ with incremental

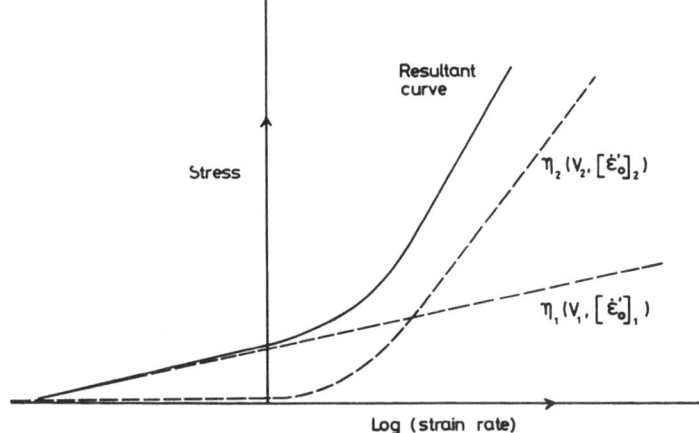

Fig.5.   Schematic representation of plastic flow for
the two process model: yield stress vs strain
rate.
Reproduced from Polymer, 22, 870 (1981)
by permission of the publishers, Butterworth
Scientific Ltd., (C).

change of draw ratio is given by

$$d\sigma = \left(\frac{\partial\sigma}{\partial\lambda}\right)_{\dot{\varepsilon}} d\lambda + \left(\frac{\partial\sigma}{\partial\dot{\varepsilon}}\right)_{\lambda} d\dot{\varepsilon} \qquad (2)$$

where $\left(\dfrac{\partial\sigma}{\partial\lambda}\right)_{\dot{\varepsilon}}$ is the strain hardening term and $\left(\dfrac{\partial\sigma}{\partial\dot{\varepsilon}}\right)_{\lambda}$ is

related to the strain rate sensitivity $\left(\dfrac{\partial\sigma}{\partial\ln\dot{\varepsilon}}\right)_{\lambda} = \dfrac{1}{\dot{\varepsilon}}\left(\dfrac{\partial\sigma}{\partial\dot{\varepsilon}}\right)_{\lambda}$

For a single activated process at high stress we have

$$\left(\frac{\partial\sigma}{\partial\lambda}\right)_{\dot{\varepsilon}} = -\frac{1}{\dot{\varepsilon}_0}\left(\frac{\partial\dot{\varepsilon}}{\partial\lambda}\right)_{\dot{\varepsilon}}\frac{kT}{v} \text{ and } \left(\frac{\partial\sigma}{\partial\ln\dot{\varepsilon}}\right)_{\lambda} = \frac{kT}{v}$$

and the strain hardening relates to a change in the pre-
exponential factor $\dot{\varepsilon}_0$ with draw ratio, whereas the strain
rate sensitivity relates to the activation volume for the

activated event.   In terms of our two process model the
exhaustion of all the flow processes as drawing proceeds is
represented by decreases in $\left[\varepsilon_o'\right]_1$ and $\left[\varepsilon_o'\right]_2$.   There may

however also be a change in the relative magnitudes of
$\left[\varepsilon_o'\right]_1$ and $\left[\varepsilon_o'\right]_2$ and there is evidence that v_2 (although

perhaps not v_1) decreases with draw ratio.   It is interesting to speculate that the large activation volume process
($v_1 \sim 500$ Å^3) is associated with the molecular network,
whereas the small activation volume process ($v_2 \sim 50\text{-}100$ Å^3)
is associated with the crystalline regions.

In a further investigation, Coates and Ward[25] determined
the neck profiles for a range of linear polyethylenes of
different molecular weight and morphology.   The strain rate
field can be defined by the change in strain rate for an
incremental change in draw ratio.   From equation (2)
this is given by

$$\frac{d\dot{\varepsilon}}{d\lambda} = \frac{\dfrac{d\sigma}{d\lambda} - \left(\dfrac{\partial\sigma}{\partial\lambda}\right)_{\dot{\varepsilon}}}{\left(\dfrac{\partial\sigma}{\partial\dot{\varepsilon}}\right)_{\lambda}} \qquad\qquad (3)$$

For a constant draw tension $\dfrac{d\sigma}{d\lambda}$ = constant, which is generally
known as the Considère relationship.   Hence the neck shape
relates to $\left(\dfrac{\partial\sigma}{\partial\lambda}\right)_{\dot{\varepsilon}}$ and $\left(\dfrac{\partial\sigma}{\partial\dot{\varepsilon}}\right)_{\lambda}$, and from inspection of their

results Coates and Ward concluded that the strain hardening
term $\left(\dfrac{\partial\sigma}{\partial\lambda}\right)_{\dot{\varepsilon}}$ was primarily affected by molecular weight

whereas $\left(\dfrac{\partial\sigma}{\partial\dot{\varepsilon}}\right)_{\lambda}$ the strain rate sensitivity term, was also

affected by the initial morphology of the samples.   It was
of particular interest that $\left(\dfrac{\partial\sigma}{\partial\lambda}\right)_{\dot{\varepsilon}}$ increased with increasing

molecular weight, and that $\left(\dfrac{\partial\sigma}{\partial\dot{\varepsilon}}\right)_{\lambda}$ was higher for slow cooled

than for quenched samples, which implied that the flow process
was less localised in the latter (i.e. a larger activation
volume).

HYDROSTATIC EXTRUSION

In hydrostatic extrusion, a billet of polymer is plast-
ically deformed in the solid phase by pushing it through a
converging die by application of pressure to a fluid which
surrounds the billet at the entrance to the die[26]. Provided
that the die cone angle is low (typically a semi-angle of
15° is used) the deformation imposed is tensile and the
products are identical to those obtained by tensile drawing.
Hydrostatic extrusion was shown by Gibson et al[5],[6] to be a
feasible route to the preparation of ultra-high modulus
polyethylene and substantial lengths of material with modulus
up to about 60 GPa were obtained.

It was originally thought that hydrostatic extrusion
would be advantagous because there is an entirely compressive
stress state overall, and the limitation of failure due to
tensile fracture which is the ultimate constraint on tensile
drawing, would therefore be removed. It was, however, found
that this intrinsic advantage is offset by the limitation that
there is a very rapid increase in the pressure required to
increase the extrudate velocity for high extrusion ratios,
leading rapidly to the situation where the extrudate velocity
cannot be increased by increasing pressure[27]. In Figure 6
the extrusion pressure is plotted as a function of the
extrusion ratio (equivalent to the draw ratio) for a constant
extrudate velocity. A series of LPE samples with a wide
range of molecular weights was selected, and it can be seen
that there is a limiting extrusion ratio which falls to
comparatively low values at high molecular weight[28]. These
results suggest that in general terms the molecular weight
effects observed in tensile drawing are also present in
hydrostatic extrusion. Furthermore, as Figure 7 shows,
there is an approximately identical relationship between
Young's modulus and extrusion ratio to that for tensile
drawing. A range of oriented products can be obtained,
albeit at rather slow rates, especially for the higher
molecular weight polymers and high extrusion ratios. A
similar relationship between Young's modulus and extrusion
ratio was reported by Williams[7] for polypropylene. The
maximum modulus obtained was about 15 GPa for an extrusion
ratio of about 15, which is very much in line with the
results reported above for the tensile drawing of polypropylene.

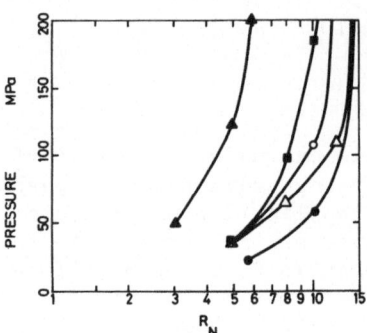

Fig.6.   Pressure/extrusion ratio characteristics (for
          extrusion velocity = 1cm/min) ▲ Ultra high mole-
          cular weight ■ MFI ∿ 0.2 O $\bar{M}_w = 10^5$, $\bar{M}_n = 6 \times 10^3$;
          Δ $\bar{M}_w = 10^5$, $\bar{M}_n = 13 \times 10^4$; ● $\bar{M}_w = 7 \times 10^4$, $\bar{M}_n = 1.3 \times 10^4$
          Reproduced from J. Polym. Sci., Polym. Phys. Ed.,
          16, 2015 (1978) by permission of the publishers
          John Wiley and Sons Ltd. (C).

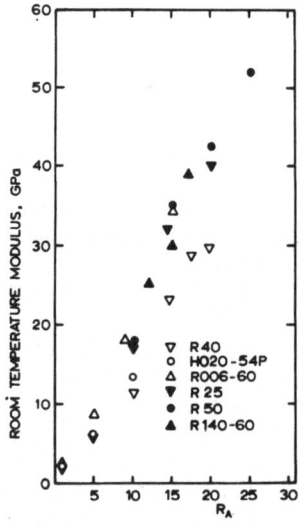

Fig.7.   Modulus vs extrusion ratio R_A for small diameter
          LPE extrudates.
          Reproduced from Plastics & Rubber Proc. & Applns.,
          2, 215 (1982) by permission of the publishers,
          Plastics & Rubber Institute (C).

The mechanics of the isothermal hydrostatic extrusion process has been analysed on the basis that because the deformation is essentially tensile, apart from the redundant work at the die entry and exit, it is only necessary to consider the axial equilibrium of thin disc-shaped elements within the die[29,30]. The Hoffman and Sachs treatment of flow in a conical die[31], developed for the extrusion of metals, was extended to include the dependence of the flow stress on strain rate (which has been discussed) and on pressure (which is a new factor).

The form of the pressure-dependent flow criterion was argued as a natural extension of the Eyring equation with the plastic strain rate $\dot{\varepsilon}$ given by[32]

$$\dot{\varepsilon} = \dot{\varepsilon}_o \exp - \left[\frac{\Delta U - \tau_f V + P\Omega}{kT} \right] \qquad (4)$$

where we have introduced a shear activation volume V and a pressure activation volume Ω. τ_f and P are the shear flow stress and hydrostatic pressure respectively and the other symbols are as in equation (1) above. For tensile deformation we then have

$$\dot{\varepsilon} = \dot{\varepsilon}_o \exp - \left[\frac{\Delta U - \sigma_f (V/2 + \Omega/3)}{kT} \right] \qquad (5)$$

Putting $(V/2 + \Omega/3) = v$, makes equation (5) exactly identical to equation (1) above. Equation (4) implies that the pressure dependence of the flow stress is given by $\sigma_f = \sigma_f^o + \gamma P$ where $\gamma = \Omega$, and for isotropic polymers γ is typically $\sim 0.1 - 0.2$. It is reasonable to anticipate that the smaller pressure activation volume Ω will change much less with plastic deformation than the shear activation volume V, hence the pressure dependence of the flow stress will increase with increasing plastic strain.

There is also the effect of friction between the billet and the die. Following the work of Briscoe and Tabor[33] on the shear and friction properties of thin polymer films, Hope and Ward[30] proposed that the coefficient of friction $\mu \sim \Omega/2V$. On this hypothesis, μ will also increase with increasing plastic strain i.e. towards the die exit.

Hope and Ward used the extended Hoffman-Sachs analysis

to predict best analytical fits to the experimental extrusion
pressure-extrusion ratio data. Tensile drawing data of the type
described above were used to determine the strain and strain
rate dependence of the flow stress. The equivalence of friction
and the pressure coefficient was assumed together with a
constant value of Ω.Typical results for three polymers, including LPE are shown in Figure 8. In POM the fit is excellent and
confirms that these assumptions are totally justified. In LPE
the fit can be improved by removing the assumption that Ω is
constant, but varies from $\sim 120 \text{Å}^3$ to $\sim 40 \text{Å}^3$ with increasing R_N.
In PMMA there is considerable scatter in the data due to
stick-slip behaviour associated with high friction, and again
it appears that Ω falls with increasing R_N. However, in all
cases it can be seen that the rapid upturn in pressure at high
extrusion ratios is predicted. The previous discussion shows
how this relates to (1) the unfavourable strain rate field
which means that the highest strain rates occur near the die
exit where the plastic strains are greatest (2) the increase
in the effect of hydrostatic pressure on the flow stress and
friction with increasing plastic strain.

These considerations led us to develop die-drawing as an
alternative process for large section oriented polymers.However,
there are two further aspects of hydrostatic extrusion which
are worthy of note. First, there is the hydrostatic extrusion
of short glass fibre reinforced polymers. It has been shown
that both the polymer matrix and the glass fibres become
oriented, so that the final product is an aligned glass fibre
composite with an oriented matrix. Research in this area on
short glass fibre reinforced polyethylenes is now in progress;
results for reinforced polyoxymethylene polymers have been
published[34,35]. Secondly, the analysis of the mechanics of the
hydrostatic extrusion process is only valid for small scale
extrusion where isothermal conditions are maintained. For
larger billets, the heat of deformation within the low thermal
conductivity polymer is less able to escape to the extrusion
tooling and there is a local increase in temperature. This can
be a considerable advantage. As shown in Figure 9, there are
two stable regimes of behaviour for the extrusion of large
billets. There is the low velocity isothermal regime where the
pressure is rising with increasing velocity. If the pressure
is reduced and the velocity increased, a second stable quasi-
adiabatic regime results. Here velocities ~ 50cm/min are reached,
which are very realistic in terms of productivity. Care must be
exercised in this adiabatic regime to ensure that there is
no major loss of properties due to annealing.

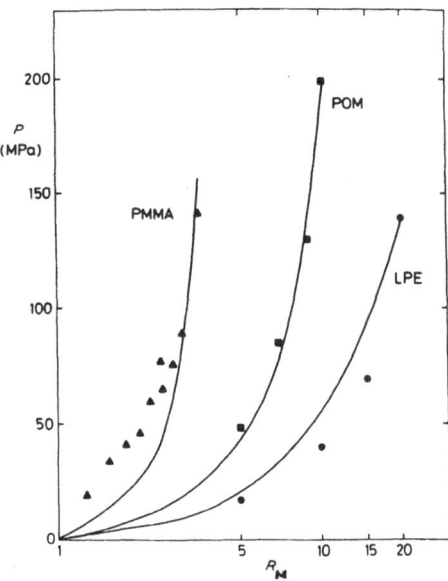

Fig.8.   Best analytical fits to experimental extrusion
          pressure-extrusion ratio R_N data.   Reproduced from
          J. Materials Science., 16, 1511 (1981) by
          permission of the publishers, Chapman & Hall Ltd.(C)

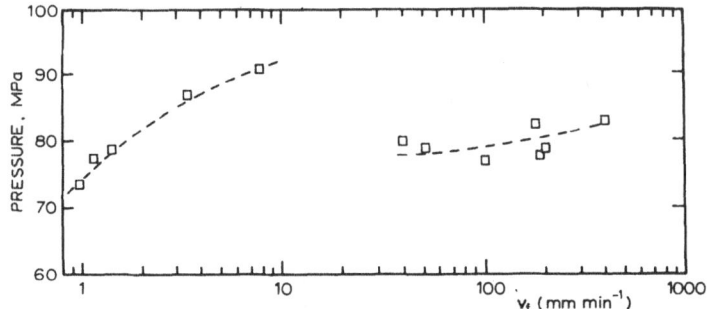

Fig.9.   Extrusion pressure vs extrudate velocity for
          LPE - large scale R40 extrusion ($T_N = 90°C$;
          product diameter 15.5mm). Reproduced from
          Plastics & Rubber Proc. & Applns., 2, 215 (1982)
          by permission of the publishers, Plastics and
          Rubber Institute (C).

Die Drawing

Figure 10 shows the drawing of a polymer rod through a heated conical die, where three regimes of deformation can be envisaged (I) conical die flow (II) isothermal tensile flow and (III) non-isothermal tensile flow. Depending on the strain hardening characteristics of the polymer, the material will leave the die wall at the die exit, or more usually at some point before the end of the conical region. A major virtue of die drawing is that the hydrostatic stress component only occurs in the die where in general there is a low deformation ratio. Furthermore the strain rate field is similar to a tensile drawing process where the highest strain rates are encountered at low levels of plastic deformation. In fact, for the die drawing of polypropylene[36] the maximum deformation ratio λ_{max} increases with draw speed, and in this case material with $\lambda = 20$ is produced at the comparatively fast rate of 50cm/min.

The properties of die-drawn polymers are related to the draw ratio in an identical manner to oriented materials

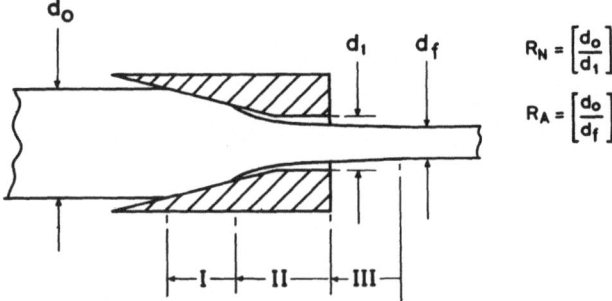

Fig. 10.   Schematic representation of the die-drawing process with heated die and stable post-die neck region.
Reproduced from Polym. Eng. Sci., 20, 1229 (1980) by permission of the publishers, Society of Plastics Engineers (C).

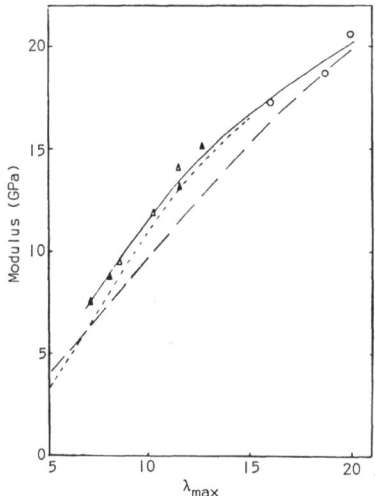

Fig.11.   Modulus vs maximum steady draw ratio λ_{max} for
           die drawn polypropylene ———, fibres data[3] — —
           and ———— hydrostatic extrusion data[7] shown for
           comparison.   Reproduced from Polymer, 20, 1553
           (1979) by permission of the publishers,
           Butterworth Scientific Ltd. (C).

produced by tensile drawing or hydrostatic extrusion
Figure 11 shows comparative results for polypropylene where
rods of up to 20.6 GPa were obtained.   In polyethylene[37],
very similar products have been obtained to those described
for hydrostatic extrusion, and there has been particular
interest in the manufacture of thin-walled tubes[38] and
other shapes, which are likely to be of value for practical
applications.

PROPERTIES OF ULTRA-HIGH MODULUS POLYOLEFINES

(1)   Mechanical Stiffness

     A central issue in the development of effective pro-
cesses has been the increase in axial modulus with deformation ratio (i.e. draw ratio or extrusion ratio).   It
has been shown that all the solid phase deformation processes give oriented materials which are equivalent to a
very good approximation.   In polyethylene, it is of some
interest to compare the measured mechanical anisotropy at
20°C and -175°C with that predicted theoretically (Table 1).

Table 1.   Comparison of measured mechanical
stiffness with theoretical predictions

	$-175^{\circ}C$	$20^{\circ}C$	Theoretical[f,g]
Axial Young's Modulus[a]			
λ = 30-35 20Hz	160	100	260-320
Transverse Young's Modulus[b]			
λ = 25 1-2 mins loading	–	1.35	5-10
Shear Modulus[c]			
λ = 25 1Hz	1.9	1.3	1-3
Poisson's Ratio[d,e]			
λ = 20-24 1 min loading	–	0.4-0.6	0.01-0.6

NOTES:

a   J.B. Smith, G.R.Davies, G.Capaccio & I.M.Ward,
        J. Polym. Sci., Polym.Phys. Edn. 13, 2331 (1975)

b   S.A. Jawad & I.M.Ward, J. Mater. Sci., 13, 1381 (1978)

c   A.G.Gibson, S.A. Jawad, G.R.Davies & I.M.Ward,
        Polymer, 23, 349 (1982)

d;e A.M. Zihlif, R.A.Duckett & I.M.Ward, J.Mater.Sci.,
        13, 1837 (1978); 17, 1125 (1982)

f   A. Odajima & M. Maeda, J.Polym.Sci., C15, 55,
        (1966)

g   K. Tashiro, M. Kobayashi & H. Tadokoro,
        Macromolecules, 11, 914 (1978)

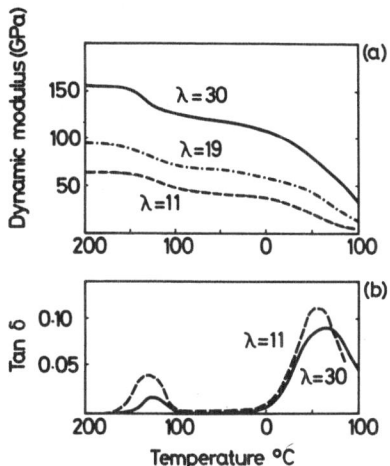

Fig.12.   Dynamic modulus (a) and tanδ (b) vs temperature
            for drawn LPE at indicated draw ratio λ.
            Reproduced from Phil. Trans. Roy. Soc., A294,
            473 (1980) by permission of the Royal Society (C).

It can be seen that the highly oriented polymer is very
anisotropic, and that the measured values are quite
similar to the predicted values.   The mechanical behaviour
(Figure 12) shows a very pronounced dependence on temperature and even at the highest draw ratios the α and γ
relaxations are clearly observed in both extension and
shear[39].   The axial modulus approaches the theoretical
value for the chain axis, yet there is strong temperature
dependence.   These observations have been explained along
the following lines.

     Although there is a very high degree of crystallite
orientation for draw ratios greater than 10, there is also
a two point small angle pattern corresponding to a long
period ∿ 200 Å which depends only on the deformation
temperature[40] (drawing or hydrostatic extrusion).   However,
the average crystal length in the chain axis direction,
as revealed by several techniques, including wide angle
X-ray diffraction[40] and dark field electron microscopy[41],
increases from ∿ 200 Å at λ = 10 to ∿ 500 Å at λ = 30.
The increase in axial modulus can therefore be most simply
explained on the basis of a Takayanagi model where there

is a degree of crystal continuity provided by crystalline
sequences which link two or more adjacent crystal blocks in
the drawn material. Such intercrystalline bridges are akin
to the taut tie molecules proposed by Peterlin[43] to explain
the increase in axial modulus on drawing. Assuming that the
intercrystalline bridges form randomly, their distribution
is determined by a single parameter p which defines the
probability of a crystalline sequence linking two adjacent
lamellae. It has been shown that the $-50^{\circ}C$ plateau moduli
in polyethylene (between the α and γ - relaxations) corres-
late well with this model where p is obtained from the
integral breadth of the (002) reflection.

To explain the temperature dependence of the moduli,
Gibson et al[39] have considered that the crystalline sequences
linking two or more crystal blocks act like short fibres in
an aligned short fibre composite. Quantitative predictions
of the temperature dependence of the axial modulus were
obtained, which were in good agreement with experimental data.
In physical terms, the α - relaxation is considered to be a
crystalline slip process and the β - relaxation relates
primarily to mobility of the non-crystalline regions.

The dynamic mechanical behaviour of ultra-high modulus
polypropylene[44,45] is shown in Figure 13. As in LPE, the
modulus is very temperature dependent, and in this case a
value of $\sim$ 25 GPa is observed at low temperatures, which is
to be compared with a crystal modulus value of 42 GPa. The
α and γ - relaxations are evident, but there is no detectable
β-relaxation in contrast to isotropic and low draw ratio
polypropylene. On annealing, the β-relaxation can be
observed and there is a drop in the modulus at high temperatures. These results suggest that the high draw produces
structural changes in the intercrystalline material, probably
by the introduction of taut tie molecules. On annealing
these are relaxed so that the conformational motions
associated with the β - relaxation are again permitted.

(2)   Creep and Recovery

Because of the possible use of ultra-high modulus LPE
for engineering applications, extensive studies of the creep
and recovery behaviour have been undertaken[46-48]. The creep
response is much reduced by increased orientation, but there
are also major effects of polymer molecular weight and
copolymerisation. It is very instructive to plot the creep

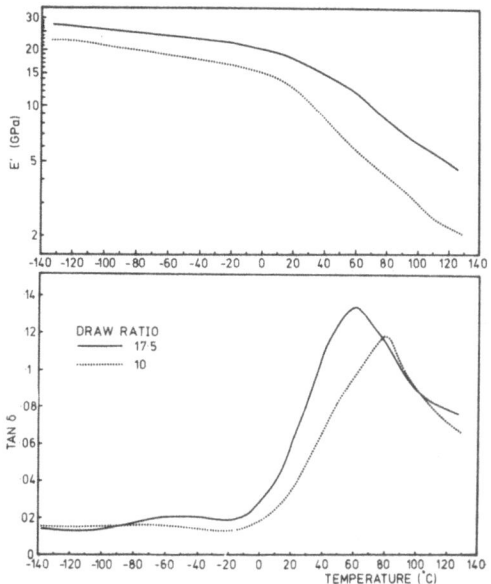

Fig.13. Dynamic modulus E and tanδ vs temperature for drawn polypropylene.

rate as a function of strain for a series of stress levels (Sherby-Dorn plots). Results for two grades of LPE are shown in Figure 14. The low molecular weight polymer shows creep rates which fall initially to a constant value, a plateau creep rate which depends on stress and in fact this is true to the lowest stress levels examined. Such behaviour is unacceptable if the material is to be in a load-bearing situation, because failure will be inevitable. However, the high molecular weight polymer shows a continuously decreasing creep rate at low stress levels, so that an equilibrium strain is reached and failure does not occur, provided that the stress levels do not exceed the critical stress, which can be quite high. A further instructive plot, shown in Figure 15, gives the strain rate as a function of stress for a range of oriented LPE samples of identical draw ratio (and hence, as discussed above, very comparable structure). A series of approximately parallel curves ensues, and it can be appreciated that there is a direct equivalence between these curves and the flow stress/strain rate curves described above in connection with tensile drawing. The high stress creep region can be represented by a single thermally activated process η_2, with an activation volume ~ 100 Å^3, and an

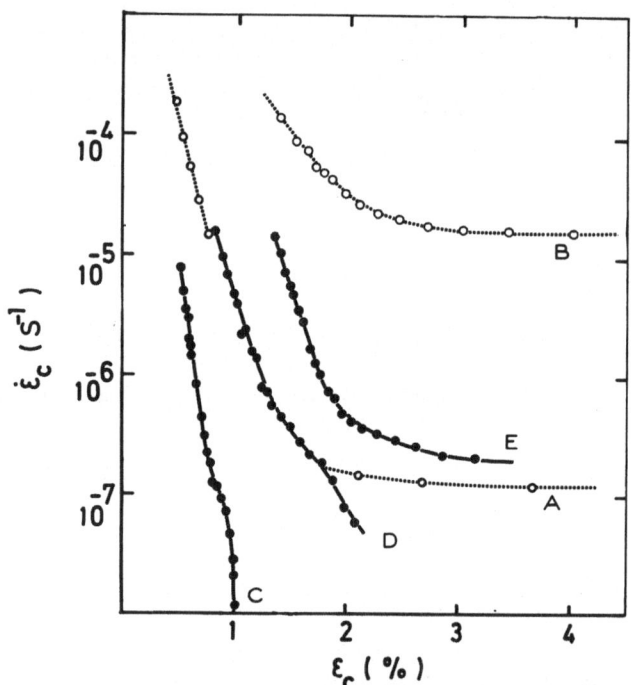

Fig.14.   Creep strain rate $\dot{\varepsilon}_c$ vs creep strain ε_c for low
           molecular weight ------ and high molecular weight
           ———— drawn LPE. Stress levels are A, C 0.1GPa;
           B, E 0.2 GPa; D 0.15GPa.
           Reproduced from Polymer, 19, 969 (1978)
           by permission of the publishers, Butterworth
           Scientific Ltd. (C).

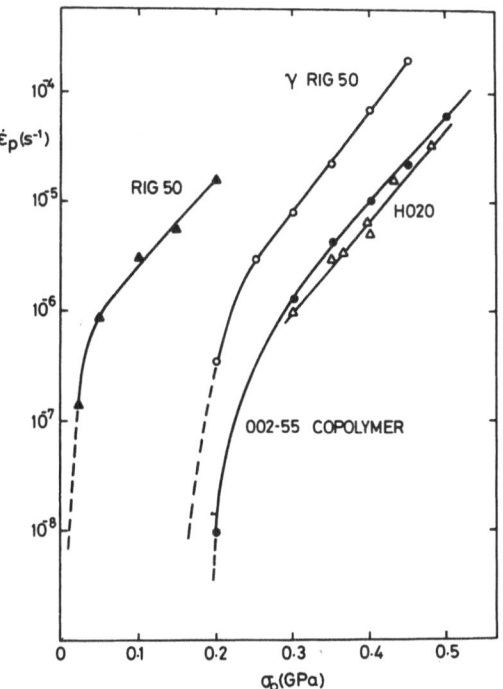

Fig. 15.   Steady plateau creep strain rate $\dot{\varepsilon}_p$ vs stress σ_o
for drawn LPE. ▲ R50 ($\bar{M}_w = 10^5$) O R50 γ-
irradiated before drawing ● 002.55 n-butyl
copolymer Δ HO2O ($\bar{M}_w = 3 \times 10^5$).
Reproduced from IUPAC Macromolecules ed. H.
Benoit & P. Rempp p.265 by permission of the
publishers, Pergamon Press (C).

activation energy of about 30 kcals/mole, similar to that
for the α - relaxation process.   It has been suggested that
this process may be associated with the movement of a defect
through the crystal lattice.   As we have proposed for the
flow stress, the behaviour over the whole range of strain
rates can be represented by a two process model, the second
process η_1 having an activation volume ~ 500 $\mathring{A}^3$ for the draw
ratio 20 data shown here.   It is extremely interesting that
this process makes a very small contribution to the behaviour
of low molecular weight polymer, but is very important in
high molecular weight polymer and polymer which has been
irradiated prior to drawing.   It has been proposed that this
high activation volume process is associated with a molecular
network.   It is particularly interesting to note that the
introduction of a very small concentration of short branches
is effective in reducing creep in a similar way, suggesting
that these also produce entanglements for a network.

At low stress levels in the high molecular weight and
similar polymers the stress is almost entirely taken by the
high activation volume process, so that there is very little
permanent flow.   The total viscoelastic and plastic behaviour
can be modelled to a first approximation by the mechanical
model shown in Figure 16, where springs have been placed in
series with the thermally activated Eyring dashpots.

The falling strain rate sections of the Sherby-Dorn
plots are associated with the extension of the springs to
reach constant extension at the plateau regions.   The
"critical stress" is merely seen as indicating a fall to a
strain rate level which is less than the sensitivity of the
measuring equipment, and below this stress, removal of the
load permits almost total recovery through flow in the
dashpot η_2.

As shown in Figure 17, the behaviour of ultra-high
modulus polypropylenes is very similar to that of LPE, and
these materials are now being studied in detail also[49].

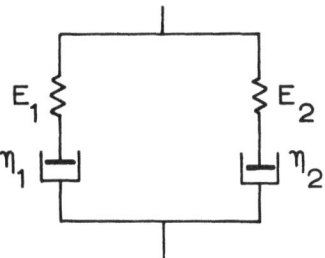

Fig.16.   Mechanical model for viscoelastic and plastic
           behaviour

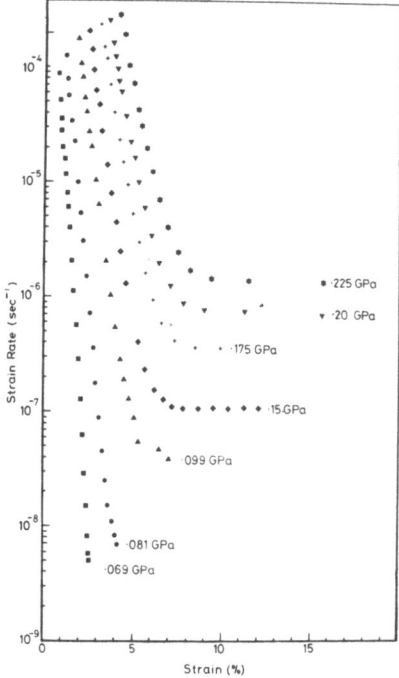

Fig.17.   Sherby-Dorn plots of creep strain rate vs creep
           strain for drawn polypropylene at 30°C, at
           indicated stress levels.

(3)   Tensile Strength

The breaking strengths of the ultra high modulus polyethylenes are very dependent on strain rate and temperature.
A typical sample undergoes a ductile-brittle transition with
increasing strain rate when tested at temperatures near to
ambient.   At low strain rates a yield point is observed, and
the yield stress/strain rate relationship shows the character-

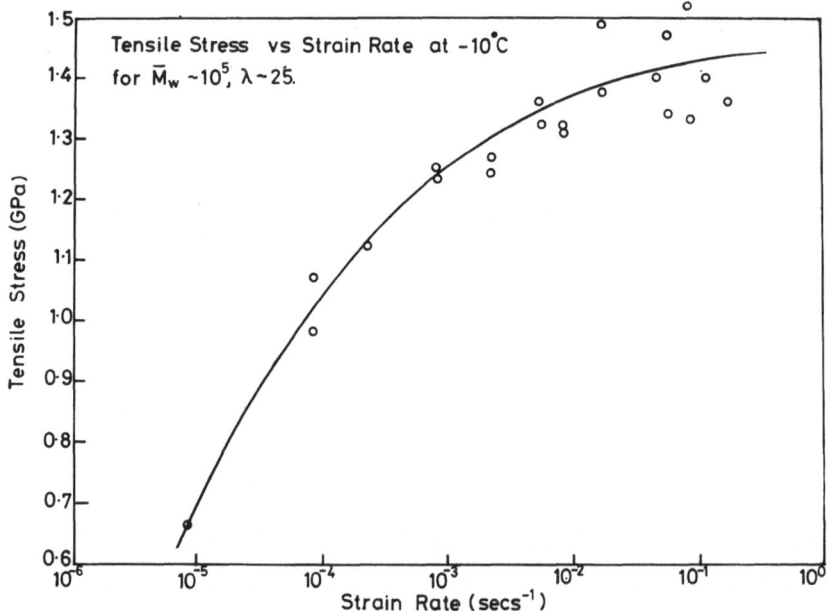

Fig. 18.   Tensile strength vs strain rate for drawn LPE at
            -10°C ($\bar{M}_w = 10^5$, $\lambda = 25$).

istics which have already been discussed.   With increasing
strain rate, there is a transition to brittle failure, the
failure stress rising with gradually decreasing ductility
until the failure stress reaches a constant value.   This
final brittle stress is dependent on both draw ratio and
molecular weight, and as shown in Figure 18 can reach a value
of 1.4 GPa even in low molecular weight samples[50].

(4)   Melting Behaviour

The melting temperatures of the drawn polyethylenes show
a marked dependence on the heating rate[51]. This is attributed to superheating[52], and the magnitude of the effect
increases markedly with increasing draw ratio and increasing
molecular weight.   It is also very greatly increased if the
sample is constrained[51,53], which may have useful practical
consequences.   The results are consistent with the view that
it is essential to retain orientation of the non-crystalline
component (i.e. so-called oriented amorphous material) so that
the change in entropy on melting is reduced and hence the
melting point increased.   The differences between samples of
different molecular weight are attributed to the more effective molecular network in the high molecular weight polymer.
At extremely low heating rates the melting temperatures fall
to 134-135°C.   This is significantly lower than the value of
138°C reported for extended chain LPE, and confirms that the
ultra high modulus drawn materials contain little if any
extended chain crystallisation.

(5)   Shrinkage, Shrinkage Force and Thermal Expansion
      Coefficient

There are several important features of the shrinkage
behaviour[54]. First, there is the remarkable stability of the
ultra drawn material, where the shrinkage is ∿ 5-7% at 135°C,
compared with ∿ 90% for the low draw sample.   This is shown
in Figure 19. Secondly, there are no major effects due to
molecular weight; the shrinkage depends primarily on draw
ratio.   Finally, if the drawn samples are heated to 150°C,
well above their melting point, total recovery is observed
to the initial length prior to drawing.   This last result
shows that there is a molecular network which retains its
identity both during drawing and during relaxation in the
total shrinkage test.   The reduction in shrinkage with
increasing draw ratio is attributed to the introduction of
increasing continuity in the crystalline phase due to the
crystalline bridges.   These crystalline sequences can be
considered to act in parallel with the network in a mechanical sense so that substantial shrinkage only occurs very
close to the melting point.

In several amorphous polymers, notably polyethylene
terephthalate[55] and polymethylmethacrylate[56] detailed
information on the nature of the molecular network has been

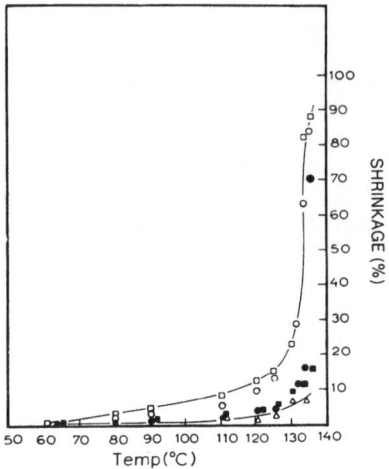

Fig.19.   Shrinkage as a function of temperature for LPE
           R50: (O) λ = 11, (●) λ = 20, (Δ) λ = 20-30;
           HO20: (◻) λ = 11, (◼) λ = 20.
           Reproduced from Colloid & Polymer Science,
           260, 46 (1982) by permission of the publisher,
           Dr. Dietrich Stemkopft Verlag. (C).

obtained from measurement of the shrinkage force, which is
best obtained by rapidly heating the oriented polymer to a
temperature just above its glass transition temperature.
Similar experiments on these drawn polyethylenes, heating
to temperatures in the range 75-120°C, show peak shrinkage
forces in the range 10-30 MPa, which decay rapidly with time
at the higher temperatures[54]. It has been noted that the
magnitude of the shrinkage force is high and comparable with
the drawing stress (cf. Figure 3).   It is also dependent on
molecular weight, increasing with increasing molecular weight,
and is larger for copolymers than for homopolymers. Although
there is a more complex situation than in amorphous polymers,
there is no doubt that the shrinkage forces are primarily
entropic in origin and associated with the extension of a
molecular network.

     The thermal expansion behaviour of ultra high modulus
polyethylene is very anisotropic.   Transverse to the draw
direction the thermal expansion coefficient is positive and
comparable to that for isotropic polymer.   In the draw
direction the coefficient is negative and very small
($\sim 10^{-5}$/°C). For low molecular weight polymers the value

is, in fact, identical with that of the crystalline regions, as determined by X-ray diffraction[57]. For high molecular weight polymers and copolymers, a higher negative value than that of the crystal has been observed[58]. It has been shown that this result can be explained on the basis of an internal stress acting on a structure whose stiffness changes with temperature. It is of particular interest that the magnitudes of the internal stresses estimated on this basis correspond very well with the measured shrinkage forces.

The behaviour of highly drawn polypropylenes is very analogous to that of ultra high modulus polyethylene. Typical shrinkage results are shown in Figure 20, and their similarity to the LPE results in Figure 19 is very striking. Studies at present in progress suggest that the behaviour with respect to shrinkage force and thermal expansion is also similar to that of LPE.

(6)   Thermal Conductivity

Ultra high modulus LPE shows a high anisotropy of thermal conductivity at temperatures above about 50K. The value in the draw direction $K_{||}$ = 90mV cm^{-1}deg^{-1} at 100K which is similar to that of stainless steel, and is about twenty times

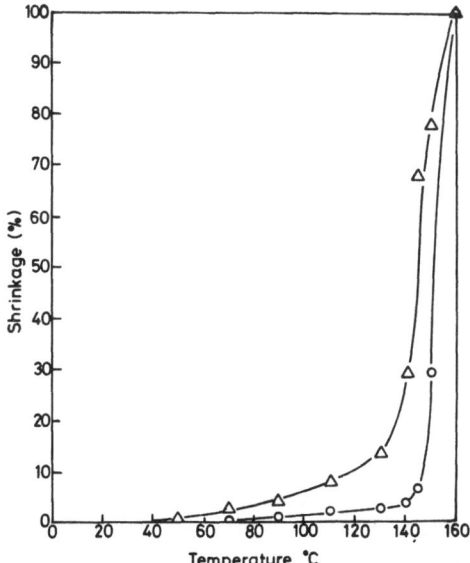

Fig.20.   Shrinkage as a function of temperature for drawn
          polypropylene (Δ) λ = 11 (0) λ = 20.

greater than the value in the transverse direction[59].   There
is an excellent correlation between K   at 100K and the -50°C
dynamic mechanical plateau modulus, and it has been shown that
the thermal conductivity can also be understood in terms of
the crystalline bridge model[59,60].

The thermal conductivity of oriented polypropylene
follows a similar pattern to that of LPE, although results
here so far have not been obtained on the most highly
oriented materials[61].

(7)   Barrier Properties and Chemical Resistance

Studies have recently been undertaken at Leeds University
on the barrier properties of drawn polyethylenes[62]. The
results are very dependent on the molecular size of the
permeant.   For example, the diffusion coefficient for helium
in homopolymer remains constant at $3.6 \times 10^{-6} cm^2 sec^{-1}$ up to draw
ratio 9, but falls first suddenly and then more gradually to
a value of 6.4×10^{-7} at draw ratio 20.   For a larger molecule,
methylene chloride, there is a dramatic fall in the diffusion
coefficient ca. draw ratio 9 and little change thereafter. The
results[63] show a fall in the diffusion coefficient from
$8.1 \times 10^{-8} cm^2 sec^{-1}$ for isotropic LPE to $4.9 \times 10^{-10} cm^2 sec^{-1}$ by
draw ratio 17.5, which compares well with earlier data of
Peterlin, Williams and Stannett[64].   It is considered these
observed differences in diffusion behaviour arise because
large molecules are unable to penetrate the microfibrils
present above draw ratio 9, whereas small molecules can do
so.   The slight reduction in the diffusion coefficient for
large molecules at high draw ratios could arise from the
closer packing of the fibrils.   The diffusion coefficient of
a small molecule such as helium reflects the changes in
microstructure within the fibrils, i.e. the development of
crystalline bridges and further increases in the orientation
of the non-crystalline material.

The changes in diffusion coefficient are accompanied by
changes in the solubility, and there is a similar difference
in behaviour between small and large molecules.   The solubility of a small molecule, helium, can be related very simply
to changes in density, confirming that the crystalline phase
is totally impermeable, but no other effects due to change
in structure are observed.   For a large molecule, methylene
chloride, the solubility falls sharply at draw ratio 9 and
remain approximately constant, suggesting that to a first

approximation the microfibrillar structure is impermeable
to the permeant.

These changes in diffusion coefficient and permeability
observed for gases and liquids, are accompanied by substantial
changes in chemical resistance. For example, highly drawn
LPE shows quite exceptional resistance to attack by concentrated nitric acid, and at λ=30 no measurable weight loss
occurs even after 10 day etching[65], compared with a weight
loss of about 30% for a sample with λ=11.

APPLICATIONS

(1)   Fibre-reinforced Composites

In Table 2 the principal mechanical properties of a
typical ultra high modulus polyethylene fibre are compared
with other high modulus fibres.   Although in absolute terms
the Young's modulus and tensile strength are less than those
of carbon fibre and Kevlar (the modulus is close to that of
glass) in terms of specific modulus the comparisons become
more favourable.   The ultra high modulus LPE fibres possess
a comparatively high elongation to break.   This, taken in
conjunction with that of high specific modulus and strength,
suggests that composites reinforced with LPE fibres would
combine acceptable stiffness and strength for a given
weight of material with a high energy absorption to failure.

Although there are limitations in the maximum working
temperature due to the comparatively low melting point, the
chemical inertness will be a positive advantage and in many
applications the relevant temperature range is close to
ambient.   A further positive advantage of ultra high
modulus polyethylene fibres is the comparative simplicity
of the fibre process which leads to a low projected cost
of manufacture.

Recent work at Leeds University has led to the
preparation of composites with continuous unidirectional
high modulus LPE fibres in either epoxy or polyester
matrices, the fibre concentration being about 50% by volume.
To obtain an adequate bond between the fibres and the matrix,
etching treatments with chromic acid and plasma were
examined[66].   It was found that plasma etching in the presence
of oxygen was most effective, the treatment giving a well
developed cellular surface structure.   The resin penetrates

Table 2.   Properties of Reinforcing Fibres (Room Temperature)

Property Fibre	Tensile Modulus (GPa)	Tensile Strength (GPa)	Elong- ation at Break %	Density ρ g/cm^3	Specific Modulus GPa/ρ	Specific Strength GPa/ρ	Maximum Working Temperature $^\circ$C
Carbon	230	1.5	1.5	1.8	128	0.8	>1500
Glass	75	2.0	2.5	2.5	30	0.8	250
Kevlar 49	125	3.0	3.0	1.45	86	2.1	$\approx$ 180
Polyethy- lene	50	1.0	18	0.96	52	1	120

Table 3.   Characteristics of Unidirectional Fibre
            Reinforced Composites

Flexural Modulus	23	GPa
Tensile Strength	0.35	GPa
Compressive Strength	85	MPa
Charpy Impact Energy	125	kJ/m^2

these small pits to produce mechanical keying between the
fibres and the matrix.   Failure in fibre pull-out tests or
in interlaminar shear tests is then due to rupture within
the fibres. The flexural modulus and tensile strength of the
composites are in the range expected, giving the expected
benefit of the fibre reinforcement.   The compressive
strength of these composites is low, but the impact energies
obtained in an unnotched Charpy test are in the same range as
those for Kevlar and carbon fibre reinforced composites.
Typical results[67] are summarised in Table 3.

(2)   Cement and Concrete Reinforcement

The excellent chemical resistance cf ultra high modulus
polyethylene, especially in respect of alkali, makes it an
attractive candidate for the reinforcement of cement and
concrete.   Here there are three distinct possibilities, all
of which are now being actively explored either at Leeds
University or in collaboration with other organisations.

One interesting possibility is the incorporation of
short fibres in cement, as an alternative to glass fibres.
Preliminary results are encouraging in suggesting that there
are significant improvements in the toughness of the cement
when only a very small volume fraction (∿ a few%) of fibres
are incorporated.

A second application stems from the pioneering research
of Dr. D. Hannant and his colleagues at Surrey University[68],
who have shown that cement can be very effectively toughened
by the incorporation of a low modulus polypropylene net,
again in a very small volume fraction. In this instance the

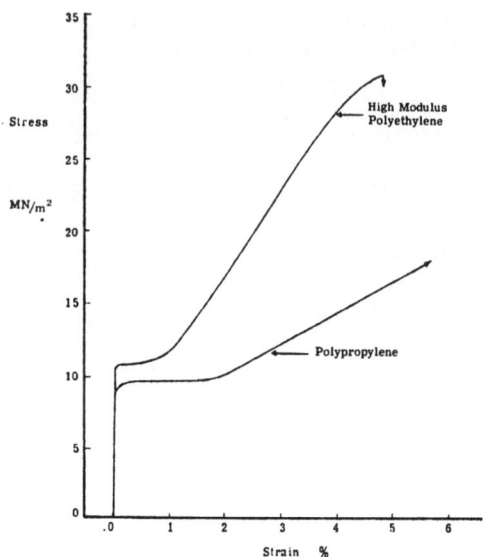

Fig.21.   Tensile stress–strain curves for cement composites
           containing 7% volume of aligned fibrillates
           film networks.
           Reproduced from Polymers in Civil Engineering
           (1980) by permission of the publishers, Plastics
           & Rubber Institute (C).

primary aim has been the replacement of asbestos fibres for
applications in roofing and similar situations. Recent
research by Dr. Hannant[69], using a high modulus net, made
from a highly drawn fibrilled polyethylene film, has produced
reinforced cement with two advantages over the low modulus
polypropylene nets. The results are illustrated in Figure
21, from which is is clear that the high modulus net provides
much greater post-cracking strength to the cement, as would
be anticipated. It also increases the initial strength of
the cement, i.e. it raises the level at which cracking
initiates.

    A third potential application of ultra high modulus
LPE in this area, involves using modified polyethylene rods
to reinforce concrete structures in place of the traditional
steel reinforcement.   In exposed conditions steel tends to
corrode and its concrete covering spalls away. Thus in
coastal structures and in certain semi-structural applications such as cladding units polyethylene could prove to be
an acceptable alternative to steel. A feasibility study is

Table 4.   Sorption and Diffusion Data at 50°C

	Equilibrium Sorption (weight %)	Diffusion Coefficient $(cm^2 s^{-1})$
Isotropic pipe		
Toluene	12	1.3×10^{-7}
THF	10	1.1×10^{-7}
Oriented pipe		
Toluene	1.5	1.6×10^{-9}
THF	1.0	1.9×10^{-9}

at present underway in the Department of Civil Engineering at Leeds University[70] with the aims of examining the bonding of polyethylene and concrete and on the general structural behaviour of columns reinforced with modified polyethylene.

(3)   Miscellaneous Applications

The excellent chemical resistance has suggested applications for ultra highly drawn LPE in the form of insulating tapes for electrical cables, or as pipes produced by hydrostatic extrusion or die-drawing.   Table 4 summarises the behaviour of oriented pipes[71] with regard to toluene and tetrahydrofuran (THF) at 50°C.   There are clearly dramatic improvements in sorption and diffusion resistance, and this has been confirmed by limited studies of other solvents, including methanol, 4-methyl pentan-2-one and xylene.

The high energy absorption of the composites reinforced with ultra high modulus PE, suggests that there may be more direct applications of the comparatively high energy to break, for example, using woven fabrics as protective clothing. The production of new variants, with higher creep resistance, should also widen the possible applications to include ropes and nets.

CONCLUSIONS

    The principal guidelines have now been established for
the production of ultra high modulus polyethylene and polypropylene by tensile drawing to give fibres and tapes, and
by hydrostatic extrusion and die drawing to give a variety
of shapes including rod, sheet and tube.

    In addition to high modulus, there is considerable
enhancement in other properties, including strength, chemical
resistance and thermal stability. It is now a considerable
challenge to identify the technological applications which
can stem from the portfolio of enhanced properties which
these new materials possess.

REFERENCES

1.   G. Capaccio & I.M.Ward, Nature, Physical Sci.,
          243, 143 (1973)
2.   Idem, Brit. Pat. Appl. 10746/73 (filed 6 March 1973)
3.   D.L.M. Cansfield, G. Capaccio & I.M.Ward, Polym. Engng.
          Sci., 16, 721 (1976)
4.   W.N. Taylor & E.S. Clark, Polym. Eng. Sci.,
          18, 518 (1978)
5.   A.G.Gibson & I.M.Ward, Brit. Pat. Appl. 30823/73
          (filed 28 June 1973)
6.   A.G.Gibson, I.M.Ward, B.N.Cole & B. Parsons, J.Mater.
          Sci., 9, 1193 (1974)
7.   T. Williams, J. Mater. Sci., 8, 59 (1973)
8.   N. Capiati, S. Kojima, W. Perkins & R.S.Porter,
          J. Mater. Sci., 12, 334 (1977)
9.   A.G.Gibson & I.M.Ward, J. Mater.Sci., 15, 979 (1980)
10.   P.D.Coates & I.M.Ward, Polymer 20, 1439 (1979)
11.   A. Zwijnenberg & A.Pennings, J.Polym.Sci., Polym.
          Lett. Edn., 14, 339 (1976)
12.   P.Smith & P.J. Lemstra, J.Mater.Sci., 15, 505 (1980)
13.   B. Kalb & A.J.Pennings, Polymer 21, 3 (1980)
14.   J.M.Andrews & I.M.Ward, J.Mater.Sci., 5, 411 (1970)
15.   G.Capaccio & I.M.Ward, Polymer 15, 233 (1974)
          16, 239 (1955) Polym.Eng.Sci., 15, 219 (1975)
16.   G.Capaccio, T.A. Crompton & I.M.Ward, J.Polym.Sci.,
          Polym. Phys. Edn., 14, 1641 (1976);
          18, 301 (1980)
17.   A.J. Wills, G.Capaccio & I.M.Ward, J.Polym.Sci.,
          Polym. Phys. Edn., 18, 493 (1980)
18.   G.Capaccio & I.M.Ward, to be published.

19. G.Capaccio, F.S. Smith & I.M.Ward, British Patent
    No. 1506565 (filed 5 March 1974)
20. W.Wu & W.B. Black, Polym.Eng. Sci., 19, 1163 (1979)
21. B.Brew, D.L.M.Cansfield & I.M.Ward, unpublished work.
22. P.D.Coates & I.M.Ward, J.Mater.Sci., 13, 1957 (1978)
23. C.Bauwens-Crowet, J.C. Bauwens & G.Homès, J.Mater.
    Sci., 7, 176 (1972)
24. L.A.Davis & C.A.Pampillo, J.Appl. Phys., 42,
    4659 (1971)
25. P.D.Coates & I.M.Ward, J.Mater.Sci., 15, 2897 (1980)
26. B.Parsons & I.M.Ward, Plastics & Rubber Processing
    & Appl. 2, 215 (1982)
27. P.S.Hope, A.G.Gibson & I.M.Ward, J. Polym.Sci.,
    Polym. Phys. Edn., 18, 1243 (1980)
28. A.G.Gibson & I.M.Ward, J.Polym.Sci., Polym. Phys.
    Edn., 16, 2031 (1978)
29. P.D. Coates, A.G.Gibson & I.M.Ward, J.Mater.Sci.,
    15, 359 (1980)
30. P.S.Hope & I.M.Ward, J.Mater.Sci., 16, 1511 (1981)
31. O. Hoffman & G.Sachs "Introduction to the Theory of
    Plasticity for Engineers" (McGraw-Hill, New
    York, 1953).
32. I.M.Ward, J.Mater.Sci., 6, 1397 (1971)
33. B.J.Briscoe & D.Tabor in "Polymer Surfaces" edited
    by D.T. Clark & W.J. Feast (Wiley Interscience,
    New York 1978) Ch.1.
34. P.S.Hope, A. Richardson & I.M.Ward, Polym.Eng. &
    Sci., 22, 307 (1982)
35. A.C.Curtis, P.S.Hope & I.M.Ward, Polymer Composites
    3, 138 (1982)
36. P.D. Coates & I.M.Ward, Polymer 20, 1553 (1979)
37. A.G.Gibson & I.M.Ward, J.Mater.Sci., 15, 979 (1980)
38. A.G.Gibson & I.M.Ward, Polym.Eng. & Sci., 20, 1229
    (1980)
39. A.G.Gibson, S.A. Jawad, G.R.Davies & I.M.Ward, Polymer
    23, 349 (1982)
40. J. Clements, R.Jakeways & I.M.Ward, Polymer,
    19, 639 (1978)
41. C.J. Frye, I.M.Ward, M.G.Dobb & D.J.Johnson,
    Polymer 20, 1310 (1979)
42. A.G.Gibson, G.R.Davies & I.M.Ward, Polymer 19,
    683 (1978)
43. A. Peterlin, Ultra-High Modulus Polymers ed.
    A. Ciferri & I.M.Ward, Appl. Science
    Publishers, London 1979, Ch.10.

44. A.J.Wills, G.Capaccio & I.M.Ward, J. Polym. Sci.,
    Polym. Phys. Edn., 18, 493 (1980)
45. J. Duxbury, unpublished work.
46. M.A.Wilding & I.M.Ward, Polymer, 19, 969 (1978)
47. M.A.Wilding & I.M.Ward, Polymer, 22, 870 (1981)
48. M.A.Wilding & I.M.Ward, Plastics & Rubber Proc. and
    Appls., 1, 167 (1981)
49. J. Duxbury & I.M.Ward, to be published.
50. D.L.M. Cansfield, D.W. Woods & I.M.Ward, to be
    published.
51. J. Clements, G.Capaccio & I.M.Ward, J.Polym. Sci.,
    Polym. Phys. Edn., 7, 693 (1979)
52. H.G. Zachmann, Kolloid Z.Z. Polym. 206, 25 (1965)
    ibid 216-217, 180 (1967)
53. J. Clements & I.M.Ward, Polymer 23, 935 (1965)
54. G.Capaccio & I.M.Ward, Colloid & Polymer Sci.,
    260, 46 (1982)
55. P.R. Pinnock & I.M.Ward, Trans. Farad.Soc. 62,
    1308 (1966)
56. N. Kahar, R.A.Duckett & I.M.Ward, Polymer 19, 136 (1978)
57. A.G Gibson & I.M.Ward, J.Mater.Sci., 14, 1838 (1979)
58. G.A.J. Orchard, unpublished work.
59. A.G.Gibson, D. Greig, M. Sahota, I.M.Ward & C.L.Choy,
    J. Polym. Sci., Polym. Lett.Edn., 15, 183 (1977)
60. A.G.Gibson, D. Greig & I.M.Ward, J. Polym. Sci.,
    Polym. Lett. Edn., 18, 1481 (1980)
61. C.L.Choy & D. Greig, J.Phys. C: Solid State Physics,
    10, 169 (1977)
62. P.S.Holden, unpublished work.
63. P.S.Holden, G.Capaccio & I.M.Ward, unpublished work.
64. A. Peterlin, J.L.Williams & V. Stannet, J.Polym.Sci.,
    A1, 5, 957 (1967)
65. G.Capaccio & I.M.Ward, J. Polym.Sci., Polym. Phys.
    Edn., 19, 667 (1981); ibid 20, 1107 (1982)
66. N.H. Ladizesky & I.M.Ward, J.Mater.Sci., (in press)
67. N.H. Ladizesky & I.M.Ward, unpublished work.
68. D.J. Hannant, J.J. Zonsveld & D.C.Hughes, Composites,
    9, 83 (1978)
69. D.J. Hannant, Symposium 'Polymers in Civil Engineering'
    Plastics & Rubber Institute, London 20th May,1980.
70. E.W. Bennett, A.R. Cusens & M.M.Kamal, unpublished
    work.
71. J.M.Marshall, P.S.Hope & I.M.Ward, Polymer 23,
    142 (1982)

LIQUID CRYSTAL POLYMERS: VI. LIQUID CRYSTALLINE

POLYESTERS OF SUBSTITUTED HYDROQUINONES

W. Jerome Jackson, Jr.

Research Laboratories, Eastman Chemicals
Division, Eastman Kodak Company
Kingsport, Tennessee   37662

SYNOPSIS

The use of monosubstituted hydroquinones enables the
preparation of liquid crystalline polyesters with lower
melting points than can be obtained with hydroquinone.
When the substituent is chloro, methyl, tert-butyl, or
1,1-dimethylhexyl, the polyterephthalate homopolyesters
still melt at too high a temperature to be
injection-molded or melt spun without thermal degradation,
but additional modification with a flexible aliphatic
component, a kinked rigid group, or
2,6-naphthalenedicarboxylic acid reduces the melting point
further and enables the preparation of melt processable
liquid crystalline polyesters. The stability of
substituted hydroquinone polyesters in air at 150°C
decreases with various substituents in the order of
phenyl, tert-butyl, chloro, methyl. At 300°C in air, the
hydroquinone polyesters containing phenyl or tert-butyl
substituents are the most stable. The effects of
composition on liquid crystallinity and on the properties
of injection-molded plastics and melt spun fibers also are
discussed.

INTRODUCTION

In an earlier paper[1a] in this series, an overview
was given of the effect of structure on the melting points
and polymer properties of thermotropic liquid crystalline
aromatic polyesters. The basic structure in these

polyesters is benzene rings interlinked at para positions
through ester groups. The high melting point of the basic
structure, i.e., poly(1,4-phenylene terephthalate) of
about 600°C, can be decreased below 400°C to a temperature
at which the polyester is sufficiently stable to be melt
spun or injection molded by (1) introduction of flexible
aliphatic units into the basic, rod-like aromatic
structure, (2) substitution of the aromatic rings,
(3) modification of the polymers with certain rod-like
comonomers, or (4) introduction of rigid kinks into the
straight polymer chains. Often, it is necessary to use a
combination of two of these approaches to reduce the
melting point of the basic structure to a desirable range
and to achieve high mechanical properties on melt
processing. The objective of this paper is to show the
effect of composition on the melting points and properties
of thermotropic liquid crystalline polyesters prepared
with substituted hydroquinones.

Goodman, McIntyre, and coworkers at ICI prepared
polyesters by heating diacid chlorides with various
substituted hydroquinones in an inert solvent at 200 to
300°C.[2,3] Because of what we know now about the effect
of composition on liquid crystallinity,[1a] the
all-aromatic compositions prepared with symmetric aromatic
dicarboxylic acid chlorides, such as terephthaloyl
chloride or substituted terephthaloyl chlorides, and
substituted hydroquinones (substituents were CH_3,
CH_3O, Cl, Br) would be expected to exhibit liquid
crystalline properties if they could be melted or
dissolved in a solvent. The all-aromatic compositions
prepared, however, melted above 350°C, too high for them
to be melt spun or injection molded without decomposition,
and they were insoluble in all the solvents tested. Lower
melting compositions were obtained when aliphatic groups
also were present, but since these polymers were not spun
or molded, it was not discovered whether they would
exhibit enhanced mechanical properties.

The first disclosures of polyesters having very high
mechanical properties after melt processing are in two of
our patents,[4,5] which describe the preparation and
properties of copolyesters of poly(ethylene terephthalate)
modified with units derived from p-hydroxybenzoic acid.
Our first two papers in this series[6,7] give additional
details and properties of fibers and bars and evidence

that these polymers are indeed thermotropic liquid
crystalline polyesters.

Schaefgen and coworkers disclosed the use of
substituted hydroquinones for the preparation of
thermotropic liquid crystalline all-aromatic polyesters
and demonstrated that fibers, films, and plastics having
high mechanical properties could be obtained on extrusion
or molding.[8,9]

EXPERIMENTAL

The aromatic polyesters were prepared by melt
condensation from aromatic dicarboxylic acids, the
diacetates or dipropionates of aromatic diols, and
acetoxybenzoic acids. Specific preparations for the
different types of polyesters are given in the references,
especially references 5, 11, and 20.

Inherent viscosities (I.V.'s) were measured at 25°C
in a 25/40/35 wt % mixture of phenol/p-chlorophenol/
tetrachloroethane at a polymer concentration of
0.1 g/100 mL of solution.

A differential scanning calorimeter (DSC)
(Perkin-Elmer DSC-2) was used for determining melting
endotherms (first cycle) and glass-transition temperatures
(second cycle). The recorded melting point is the peak
temperature of the melting endotherm.

Softening points were determined on pressed films
with a Du Pont 941 Thermomechanical Analyzer having a 10-g
weight on a tipped probe (0.025-in. diameter) and a scan
rate of 10°C/min.

Relaxation times were determined with a Rheometrics
Mechanical Spectrometer Model RMS-7200 by the use of the
eccentric-rotating-disk mode.

Melt viscosities were determined with an Instron
capillary rheometer.

Molecular weights were determined with a Chromatix
KMX-6 low-angle light scattering photometer in m-cresol or

in an isorefractive mixed solvent consisting of
o-dichlorobenzene and p-chlorophenol (1:1 by weight).

The polymers were injected into unheated molds in a
1-oz Watson-Stillman injection-molding machine to produce
2 1/2- x 3/8- x 1/16-in. bars for tensile measurements and
5- x 1/2- x 1/8-in. bars for determination of flexural
properties, notched and unnotched Izod impact strength,
heat-deflection temperature, and mold shrinkage. ASTM
procedures were used for measuring tensile strength and
elongation at break (ASTM D1708), flexural modulus and
flexural strength (ASTM D790), notched Izod impact
strength (ASTM D256 Method A), heat-deflection temperature
(determined at 264 psi, ASTM D648), and mold shrinkage
(ASTM D955).

The polyesters were melt spun through a one-hole
capillary on an Instron rheometer and taken up on a roll
at about 2000 to 4000 ft/min. The fibers were then
wrapped around a wire frame that was placed in a glass
tube contained in an aluminum block, dried by heating
under a nitrogen or argon flow at 125°C for one hr, and
then heated at the indicated elevated temperatures and
times to increase the polyester molecular weights. The
tensile properties of the as-spun and heat-treated fibers
were determined by averaging 10 breaks of 1-in. gage
length single filaments. The use of longer gage lengths
or plied yarn gives appreciably lower tenacity and modulus
values.

RESULTS AND DISCUSSION

We attempted to prepare polyesters from terephthalic
acid and the diacetates of several substituted
hydroquinones.

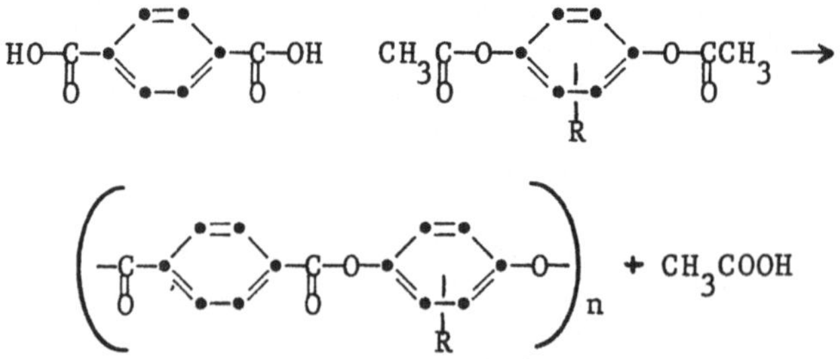

I

When the substituent R was chloro, methyl, <u>tert</u>-butyl, or
1,1-dimethylhexyl, the polymers solidified <u>during</u>
preparation (temperature of the Wood's metal heating bath
was 350°C or higher). Therefore, these polymers melt at
too high a temperature to be melt spun or injection molded
without thermally degrading. When the substituent was
phenyl, however, the polyester melted at 340°C and could
be spun into high tenacity fibers[1a,10] or injection
molded to give plastics having very high tensile strength
and modulus.[1a] When unsubstituted hydroquinone is used

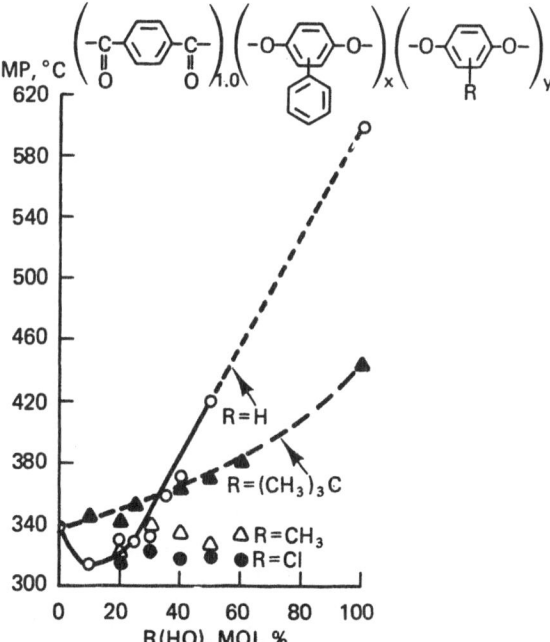

Fig. 1.   Melting points of poly(phenyl-1,4-phenylene
            terephthalate) copolyesters.

TABLE 1.   EFFECT OF HYDROQUINONE ON PROPERTIES OF INJECTION—
MOLDED POLY(PHENYL—1,4—PHENYLENE TEREPHTHALATE)

Hydroquinone, mol %	0	10	20	30
Tensile strength, psi	26,000	27,000	25,700	28,900
Elongation to break, %	6	6	6	7
Flexural modulus, 10^5 psi	19.3	23.7	26.0	29.2
Flexural strength, 10^3 psi	30,000	25,500	25,400	35,200
Notched Izod impact strength, ft—lb/in.	1.6	1.2	0.8	3.4
Heat—deflection temp, °C	258	>230	>230	>230
Rockwell hardness, L	111	108	108	113

(I, R=H), the polyester melts with decomposition at about
600°C.[1] (Korshak[1b] reported the melting point to be
above 500°C, and Eareckson[1c] reported the polymer to be
infusible.)

Figure 1 shows the effect of hydroquinone content on
the melting points of copolyesters of terephthalic acid,
phenylhydroquinone, and hydroquinone (open circles).[11]
Except for the polymers melting above 400°C, which were
increased in molecular weight by solid—state
polymerization, the compositions were prepared in the melt
at bath temperatures of 360 to 380°C. The copolyesters
prepared with up to 30 mol % hydroquinone were easily melt
processable at 360°C, and Table 1 lists the properties of
test bars, injection molded at 360°C, of some of these
polyesters having I.V.'s of about 2.5. (The tensile
properties shown in all tables were determined with
1/16-in. thick bars [lower strengths are obtained with
thicker bars because of lower orientation[6]], and other
properties were obtained with the standard 1/8-in. thick
bars.) The smooth bars were internally fibrous, as is
typical of liquid crystalline polyesters, because of the

orientation that occurred during molding. The
compositions are characterized by high tensile and
flexural strengths, and it is noteworthy that the flexural
modulus increased 50% to almost 3 x 10^6 psi when the
hydroquinone content was increased to 30 mol %. These
high tensile and flexural properties apply only in the
direction of orientation. As shown earlier,[6] liquid
crystal polyesters do not have these enhanced properties
perpendicular to the direction of flow of the melt. The
exact values of the heat-deflection temperatures of the
copolyesters in Table 1 were not determined because the
thermal limit of the oil bath used for the test at that
time was 230°C.

It is of interest to compare the properties of the
liquid crystalline polyester plastics in Table 1 with the
properties of polyesters which do not exhibit liquid
crystalline melts. Table 2 lists the properties of
injection-molded test bars of the polyterephthalate of
phenylhydroquinone (Table 1), the polyisophthalate of
phenylhydroquinone (I.V. 0.5), a polyester prepared from
bisphenol A diacetate and a 65/35 ratio of terephthalic
and isophthalic acids (I.V. 0.6), and a widely used
engineering thermoplastic, bisphenol A polycarbonate. The
polyisophthalate (T_g 150°C) is not liquid crystalline
because of the kinks introduced into the rigid polymer
chain by the meta-oriented isophthalic acid derived unit.
The bisphenol A polyester and polycarbonate also have
highly kinked structures because of the bisphenol A (BPA)
derived unit and, therefore, are not liquid crystalline.
Consequently, these polymers have not shown the high
tensile and flexural properties that are characteristic of
liquid crystalline polyesters, and the properties are
isotropic (do not vary with direction of measurement in
'the plastic or direction of flow during molding). These
polymers also do not have the low mold shrinkage values in
the flow direction (usually zero) of liquid crystalline
polyesters. The 0.16% mold shrinkage of the
polyterephthalate of phenylhydroquinone is probably the
result of the high crystallinity of this polymer
(crystalline polymers shrink more on cooling than
noncrystalline polymers). The crystallinity also
increases the heat-deflection temperature, as shown by the
different values for the polyterephthalate and
polyisophthalate of phenylhydroquinone in Table 2. The

TABLE 2. PROPERTIES OF INJECTION–MOLDED PLASTICS

Plastic	T(ϕHQ)	I(ϕHQ)	65T35I(BPA)	BPA Poly–Carbonate
Tensilé strength, psi	26,000	12,600	10,000	8,500
Elongation to break, %	6	35	50	75
Flexural modulus, 10^5 psi	19.3	4.8	3.1	3.4
Flexural strength, 10^3 psi	30,000	9,800	14,600	13,500
Notched Izod impact strength, ft–lb/in.	1.6	0.5	4.5	16
Heat–deflection temp, °C	258	114	168	132
Mold shrinkage, %	0.16	0.5	0.6	0.6

injection-molded bars of the three nonliquid crystalline polymers in Table 2 are amorphous.

Figure 1 also shows the effects of several substituted hydroquinones on the melting points of copolyesters prepared in the melt state from terephthalic acid, phenylhydroquinone diacetate, and the substituted hydroquinone diacetates (polymer I.V.'s about 1.0 to 3.0). The copolyesters prepared with up to 60 mol % chlorohydroquinone or methylhydroquinone diacetate had slightly lower melting points than the unmodified polyterephthalate of phenylhydroquinone, but further polymerization of these copolyesters in the solid state at about 280°C gave polymers with an additional melting endotherm at about 370°C. In spite of the greater bulk of the tert-butyl group, the melt-prepared copolyesters had higher melting points than those prepared with chloro- or methylhydroquinone diacetate.[12] The DSC melting endotherm of the homopolyester of tert-butylhydroquinone (I, R=(CH$_3$)$_3$C–) peaked at 444°C. Because of

decomposition, the melting points of the chloro- and
methylhydroquinone polyterephthalates could not be
determined.

The softening points of the polyesters in Figure 1,
determined by thermomechanical analysis, were within 10 to
20°C of the crystalline melting points, determined by
differential scanning calorimetry. This is an indication
that the crystallinity is sufficient for fibers of these
polymers to be heat-treated near their softening points
without fusing together. Heat treatment increases the
polymer molecular weight and order and thereby increases
the fiber tensile properties. (A discussion of fiber
heat-treatment procedures is given in a Du Pont
patent.[13])

Figure 2 shows the correlation between the fiber
tenacity of poly(phenyl-1,4-phenylene terephthalate) and
melt viscosity, which is a measure of the polymer
molecular weight. Melt-spun fibers were heat-treated in
an oven at 300°C and at 320°C for various times, removed
and chopped, and then compression molded into sample disks
for use in determining the melt viscosity of the remelted
fibers with a Rheometrics Mechanical Spectrometer in the

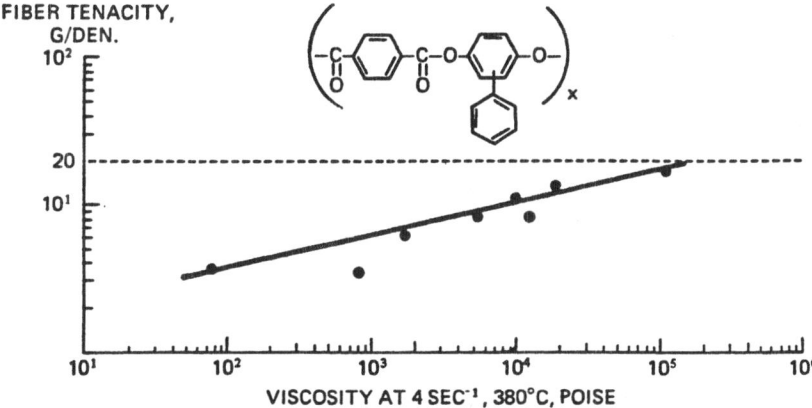

Fig. 2.   Correlation of fiber tenacity and melt viscosity
           of poly(phenyl-1,4-phenylene terephthalate).

eccentric-rotating-disk mode. The increased tenacities
obtained on increasing the fiber heat-treatment time are
due to the increased molecular weight, which caused the
increase in melt viscosity. It is noteworthy that when
the tenacity is 20 g/den. (fiber I.V. 12.0), the melt
viscosity at 380°C is over 10^5 poise, about an order of
magnitude too high for melt spinning (even considering
that the higher shear rate during spinning gives a lower
melt viscosity).

Copolyesters of terephthalic acid and
phenylhydroquinone modified with 50 mol % of chloro-,
methyl-, and tert-butylhydroquinone (I.V.'s about 2.0)
were melt spun at 380°C to give fibers having as-spun
tenacities of 3 to 4 g/den. Fibers heat-treated for 0.5
to 1.0 hour at 330 to 340°C had tenacities of 15 to
20 g/den. (Table 3).

Polyesters of terephthalic acid and
phenylhydroquinone containing alkyl or halogen
substituents attached to the phenyl group are disclosed in
one of our patents.[14]

TABLE 3.   FIBER PROPERTIES OF COPOLYESTERS OF POLY(PHENYL-
1,4-PHENYLENE TEREPHTHALATE)

x/y	1.0/0	0.5/0.5	0.5/0.5	0.5/0.5
R	—	Cl	CH_3	$(CH_3)_3C$
Den./fil	1.4	2.3	2.6	1.6
Tenacity, g/den.	28	20	20	15
Elongation, %	3.7	3.0	3.1	2.8
Elastic modulus, g/den.	810	710	780	660

Modification With Flexible Units

Instead of attaching a large substituent, phenyl, to
the hydroquinone molecule to reduce the high melting
points of hydroquinone terephthalate polyesters, flexible
aliphatic components can be incorporated in the polymer
chain. In our earlier paper,[1a] we noted that the
polyester of 4,4'-(ethylenedioxy)dibenzoic acid and
hydroquinone (II, n=2, R=H) melted at

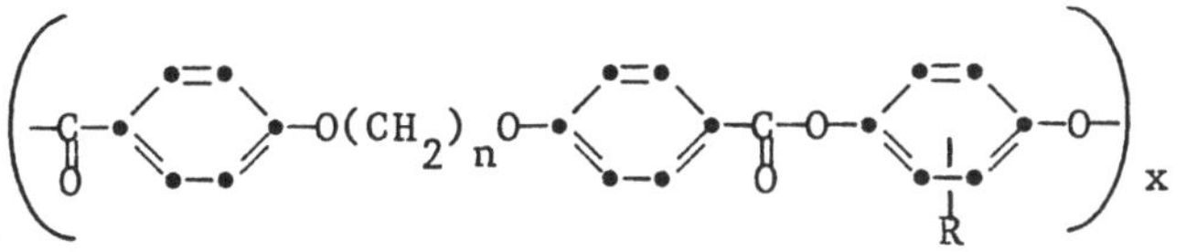

II

362°C, and the similar methylhydroquinone polyester
(R=CH$_3$) melted at 316°C. In the series of
methylhydroquinone polyesters, we showed that the strength
and stiffness of injection-molded test bars decreased as
the polyester flexibility increased (n=2 to 4). Lenz and
coworkers have prepared similar polyesters in which n=10
and R=H, CH$_3$, Cl, or Br and studied the effect of
structure on the thermotropic mesomorphic
behavior.[15,16]

Another method of introducing flexibility in the
polyesters is to use 4-carboxybenzenepropionic acid as a
monomer with the substituted hydroquinones:

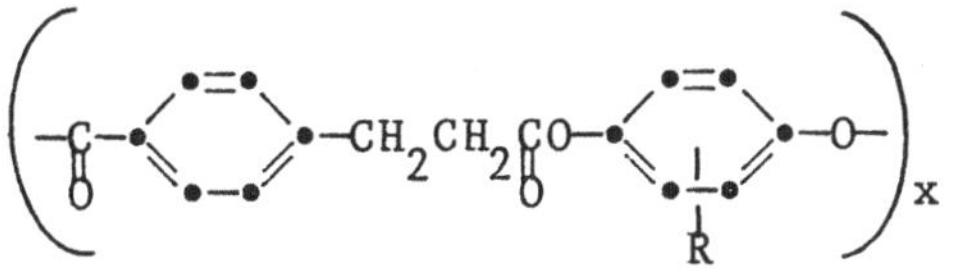

III

The following melting points were obtained:

R	MP, °C
H	425
CH$_3$	387
Cl	372

Even though the methyl and chloro substituents gave lower
melting liquid crystalline polyesters, their melting
points are still too high for the polymers to be thermally
processed without thermal degradation, because of the
presence of the aliphatic component. When R was phenyl,
however, the melting point was greatly reduced, apparently
to 186°C according to a thermogram (Figure 3), which was
determined with a clear, pressed, quenched film. The
thermogram also showed a T_g at 109°C and an exotherm at
126°C. The exotherm (3.2 cal/g) is not due to
crystallization in the normal sense, however, but was
caused by the formation of liquid crystalline order. When
viewed on the hot stage of a Fisher–Johns melting point
apparatus, the polymer was molten, not solid, and the melt
was hazy, not clear. When heated to the region of the DSC

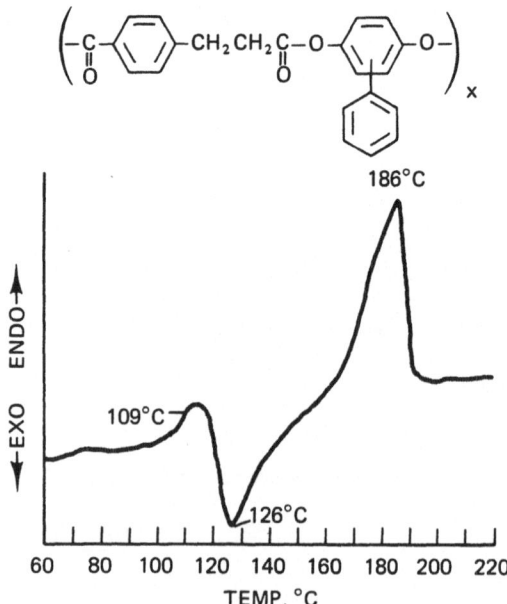

Fig. 3.   DSC thermogram of polyester of 4-carboxybenzene-
propionic acid and phenylhydroquinone.

endotherm (3.7 cal/g), the melt became clear. When heated
on the hot stage of a polarizing microscope, a pressed,
quenched film became birefringent at 110°C and remained
birefringent until the region of the DSC endotherm was
reached. No birefringence was present at 195°C. The DSC
endotherm at 186°C, therefore, is due to the transition
from the liquid crystalline to the isotropic state. The
degree of liquid crystallinity in the polymer apparently
was reduced by the bulky phenyl substituent. (The similar
polymers with the smaller methyl and chloro substituents
had turbid, liquid crystalline melts until they solidified
during preparation at 365°C.) The DSC thermogram obtained
when the phenylhydroquinone polyester was cooled showed
only an exotherm (4.1 cal/g) at 156°C corresponding to
the isotropic-to-mesophase transition. Crystallization of
the polymer apparently did not contribute to the DSC
exotherm obtained at 126°C when the phenylhydroquinone
polyester was initially heated; a wide-angle X-ray
diffraction pattern of a birefringent sample, which had
been heated at 150°C and then cooled, did not show any
multiple sharp reflections characteristic of a crystalline
polymer but contained only a single ring that was sharper
than the diffuse halo which is characteristic of an
amorphous polymer. It is of interest that an X-ray
diffraction pattern of the initial pressed, quenched film,
which was clear, was similar to that of the opaque, white,
birefringent film obtained after heating.

When R in III is methyl or chloro, the high melting
points can be reduced to a more processable range by
replacing part of the carboxybenzenepropionic acid with
terephthalic acid. Figure 4 compares the melting points
of these copolyesters of the substituted and unsubstituted
hydroquinones. The eutectic occurred at about 315°C when
R was hydrogen. When R was CH_3 or Cl, the DSC
endotherms were broad and small at 50 mol %
carboxybenzenepropionic acid content. No endotherms were
observed in the copolyesters containing 60 or 70 mol %.
It is apparent that the carboxybenzenepropionic acid was
very effective in reducing the high melting points of the
polyterephthalate homopolyesters. The polyester prepared
from chlorohydroquinone diacetate, 70 mol % terephthalic
acid, and 30 mol % 4-carboxybenzenepropionic acid was melt
spun at 350°C to give fibers having a tenacity of 6 g/den.
as-spun and 20 g/den. after heat treatment in a nitrogen

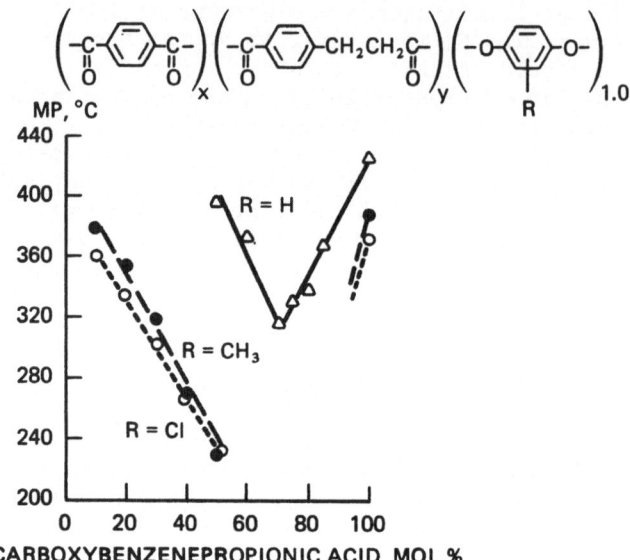

Fig. 4. Melting points of 4-carboxybenzenepropionic acid
         copolyesters.

atmosphere (consecutively for one hour each at 260, 270,
and 280°C).

Another method of reducing the melting points of the
higher-melting polyterephthalates of substituted
hydroquinone is to introduce flexibility into the polymer
chain by preparing them as copolyesters with poly(ethylene
terephthalate). The properties of injection-molded test
bars are described in an earlier paper in this series[7],
and the properties of melt-spun fibers are disclosed in a
patent.[17]

Modification With p–Hydroxybenzoic Acid

'We attempted to reduce the high melting points of the polyterephthalates of methylhydroquinone and chlorohydroquinone by modification with p–hydroxybenzoic acid (PHB):

IV

When y in structure IV was 0.3 to 0.5 (x + y = 1.0), the copolyesters prepared from terephthalic acid, methylhydroquinone or chlorohydroquinone diacetate, and p–acetoxybenzoic acid solidified during melt preparations at 360°C or higher. When phenylhydroquinone diacetate (R=C_6H_5) was used, however, copolyesters[18] containing up to 50 mol % PHB (y=0.5) had melting points in the range of 300 to 360°C. The melting points, determined by differential scanning calorimetry, and the softening points, determined by thermomechanical analysis, are compared in Figure 5 (open circles and triangles). The relatively low softening points of these copolyesters containing from 30 to 70 mol % PHB make it possible to thermally fabricate shaped articles at temperatures appreciably below the crystalline melting points without loss of the high strength and stiffness that are characteristic of liquid crystalline polyesters. A 1/8-in. thick injection–molded bar of a composition containing 50 mol % PHB (y=0.5) was formed (bent) easily with a Wabash press at 210°C[18] even though the DSC melting peak was 358°C, and an injection–molded bar of a composition containing 60 mol % PHB (DSC melting endotherm at 373°C) was shaped at 225°C. In contrast to the low softening points of these copolyesters, highly crystalline liquid crystal copolyesters, such as those in Figure 1, have softening points only about 10 to 20°C below their melting points. When 2,6–naphthalenedicarboxylic acid was substituted for the terephthalic acid component in Figure 5, the compositions had similar low softening points, and the crystallinity was so low that no crystalline melting points were detected by differential scanning calorimetry at 50 and 80 mol % PHB

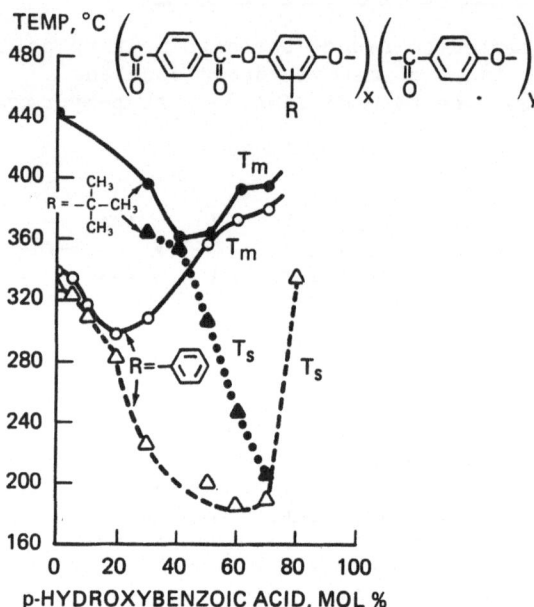

Fig. 5. Melting and softening points of copolyesters of
terephthalic acid, p-hydroxybenzoic acid, and
phenyl- or t-butylhydroquinone.

contents.[18,19] The similar PHB copolyesters prepared
with hydroquinone diacetate instead of phenylhydroquinone
diacetate had melting and softening points only about 5 to
10°C apart.[19]

Figure 5 also shows the effect of PHB content on the
melting and softening points of copolyesters when R in IV
is tert-butyl. The melting and softening points of these
copolyesters are higher than those of the
phenylhydroquinone copolyesters in Figure 5, and the
softening points of both compositions range up to almost
200°C lower than the melting points of the same
compositions.

Some of the properties of injection-molded test bars of IV when R is phenyl and y is 0 to 0.5 are listed in Table 4. The polyester I.V.'s were 2.5 to 3.5, and the injection-molding temperatures were 350 to 380°C, depending upon the I.V. In general, the mechanical properties are similar to those in Table 1, including the increase in flexural modulus when the homopolymer of phenylhydroquinone and terephthalic acid was modified. The heat-deflection temperatures decreased as the PHB content increased, however, because of the decrease in crystallinity (cf. Figure 5).

The relaxation times of several polymers are compared in Figure 6. Poly(ethylene terephthalate) (PET, I.V. 0.60) has low relaxation times which were not affected by the shear rates (rad/sec) at which they could be measured whereas the less flexible PET copolyester modified with 60 mol % PHB[6] (I.V. 0.62) had appreciably higher relaxation times, particularly at the lower shear rates. Replacement of the flexible PET portion of this copolyester with the more rigid phenylhydroquinone (ϕHQ) terephthalate (T) gave 40T(ϕHQ)60(PHB), and at the typical 360°C

TABLE 4.   EFFECT OF PHB ON PROPERTIES OF INJECTION-MOLDED POLY(PHENYL-1,4-PHENYLENE TEREPHTHALATE)

x/y	1.0/0	0.9/0.1	0.7/0.3	0.5/0.5
Tensile strength, psi	26,000	22,600	20,500	24,400
Elongation to break, %	6	3	6	11
Flexural modulus, 10^5 psi	19.3	22.3	30.5	28.1
Flexural strength, 10^3 psi	30,000	37,000	35,100	28,400
Notched Izod impact strength, ft-lb/in.	1.6	2.8	1.6	1.1
Heat-deflection temp, °C	258	222	193	179
Rockwell hardness, L	111	113	109	105

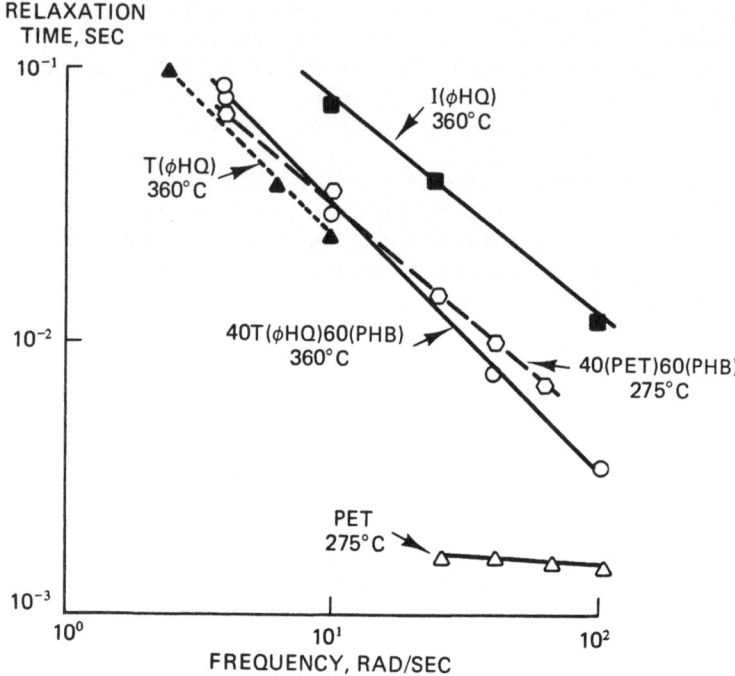

Fig. 6.   Relaxation times of polyesters.

processing temperature of this polymer (I.V. 2.4), the
relaxation times at the various shear rates were almost
identical to those of 40(PET)60(PHB) at its 275°C
processing temperature. The homopolyester of
phenylhydroquinone and terephthalic acid (I.V. 2.0) has
similar relaxation times at its 360°C processing
temperature. The relaxation of a polymer during cooling
in an injection-molded bar is a low-rate process, and
these three liquid crystalline polyesters have
sufficiently long relaxation times at low shear rates to
retain their extended chain orientation when injection
molded so that the test bars have very high strength and
stiffness properties (cf. Table 4). The polyisophthalate
of phenylhydroquinone, on the other hand, also has long
relaxation times (Figure 6, I.V. 0.5), about three times

longer than those of the polyterephthalate because the
Maxwell viscosities of the polyisophthalate are about
three times higher than those of the polyterephthalate.
(The Maxwell moduli are very similar, and relaxation time
is equal to the Maxwell viscosity divided by the Maxwell
modulus.) The polymer chains in the polyisophthalate are
highly kinked because of the meta-oriented dicarboxylic
acid component and, consequently, the polymer is not
liquid crystalline (properties in Table 2). For high
property values to be obtained, the polymer chains must be
properly oriented and the relaxation time must be long
enough for the polymer to cool below its crystallization
temperature (or T_g, if not crystalline) and thereby lock
in the orientation.

Modification With 2,6-Naphthalenedicarboxylic Acid

The high melting points of the polyterephthalates of
chloro- methyl-, and ethylhydroquinone (I, R=Cl, CH_3,
C_2H_5) have been reduced to a melt-processable range by
Schaefgen[8,9] by modification with
2,6-naphthalenedicarboxylic acid (N). Flow points of 292
to 302°C were reported for these three polyesters modified
with 30 mol % of N. The polyterephthalate of
phenylhydroquinone (structure I, R=C_6H_5) melts low
enough to be melt processable, and modification with 30 to
50 mol % N reduces the softening points to about
200°C.[19] This depression of the softening points
limited the fiber compositions that could be heat-treated
without the individual filaments fusing together. The low
softening points and low crystallinity in these
compositions do permit, however, the thermal forming of
extruded sheets and molded specimens to give various
shapes while retaining high strength and stiffness. For
instance, injection-molded bars of the 70/30 T/N
copolyester (softening point 205°C) were thermally shaped
(bent) at 160°C to give sharp angles.

Table 5 shows the effect of N content on the plastic
properties of the polyterephthalate of phenylhydroquinone
(I.V.'s 2.4 to 2.6) injection molded at 350 to 380°C. It
is of interest that the temperature at which the above
70/30 T/N copolyester could be thermally formed is close
to its heat-deflection temperature of 158°C. The
copolyesters had somewhat higher tensile and flexural
strengths and lower stiffness values (flexural moduli)

TABLE 5. EFFECT OF 2,6-NAPHTHALENEDICARBOXYLIC ACID ON
PROPERTIES OF INJECTION—MOLDED POLY(PHENYL—1,4—PHENYLENE
TEREPHTHALATE)

Naphthalene diacid, mol %	0	30	50	80
Tensile strength, psi	26,000	28,000	28,000	26,200
Elongation to break, %	6	9	8	7
Flexural modulus, 10^5 psi	19.3	17.2	16.9	16.0
Flexural strength, 10^3 psi	30,000	35,200	35,400	31,700
Notched Izod impact strength, ft—lb/in.	1.6	5.0	4.7	3.0
Heat—deflection temp, °C	258	158	164	189
Rockwell hardness, L	111	114	116	118

than the homopolyester. The higher notched Izod impact
strengths of the copolyesters indicate that the
copolyester plastics have higher toughness than the
homopolyester, which is also true for the copolyester
fibers.[19] The heat-deflection temperatures of the
copolyesters are appreciably lower than that of the
homopolyester because of the reduced crystallinity. These
data suggest, therefore, that if toughness is important
for an application and if a heat-deflection temperature of
about 160°C is adequate, modification of the homopolyester
with 30 to 50 mol % of N should be of benefit.

Modification by Addition of Kinks to Chains

Our earlier paper[1a] discussed polymer intermediates
that can be used to introduce rigid kinks in the straight,
rod-like aromatic polyester chains to reduce the polyester
melting points sufficiently for the polymers to be melt
processed without thermal degradation. About 20 mol % of

resorcinol[1a,20] or m-hydroxybenzoic acid[1a,21] or 30
mol % of isophthalic acid is required to reduce the
melting points sufficiently for the polyesters of chloro-
or methylhydroquinone and terephthalic acid (I, R=Cl or
CH₃) to be melt processed. As the the meta-component is
increased, the tensile and flexural properties of
injection-molded plastics decrease because of a reduction
of the apparent level (which we do not yet know how to
measure) of liquid crystallinity. The heat-deflection
temperatures of the plastics also decrease (to about 120°C
with these compositions) because of a reduction in
crystallinity. As the level of crystallinity decreases,
the temperature at which yarn filaments will stick
together during heat treatment also decreases, and a
longer time will be required to increase the tensile
properties by build-up of the molecular weight because of
the required lower temperature. Crystallinity is less
important in plastics than in fibers since a heat
treatment to increase the properties is not required, and
plastics containing up to about 60 mol % (acids and diols
each total 100 mol %) of the meta-oriented component can
have tensile and flexural properties that are almost twice
as high as those of similar polyesters that are not liquid
crystalline.[1a]

Although it is not necessary to reduce the melting
point of the polyterephthalate of phenylhydroquinone to
achieve melt processability, it is of interest that the
level of liquid crystallinity apparently is greatly
decreased if the homopolyester is modified with only 25
mol % isophthalic acid (I) (diacids total 100 mol %). At
340°C the melt was clear and optically isotropic whereas
the melt of the polyterephthalate of methylhydroquinone
modified with 60 mol % of the isophthalic acid kinking
component exhibited the stir opalescence that is typical
of liquid crystalline polyesters. A pressed, quenched
film of the phenylhydroquinone copolyester (I.V. 1.1) was
transparent, and when heated above its T_g (150°C) to
about 175°C, it became hazy. At 200°C the film was opaque
and white. A wide range X-ray diffraction pattern of the
opaque film at room temperature showed only a single
diffuse ring, similar to that discussed earlier with the
phenylhydroquinone polyester of 4-carboxybenzenepropionic
acid (III, R=phenyl). When a pressed, transparent film of
the 75T25I copolyester was viewed under a polarizing
microscope, the sample was birefringent because of strains

from pressing and remained birefringent up to the maximum
temperature capability (300°C) of the instrument, with
colors of yellow, blue, green, and magenta simultaneously
present. (On the hot stage of a Fisher-Johns melting
point apparatus, the melt became isotropic at about
340°C.) The haze which appeared in the film when it was
heated above the polymer T_g is caused by the formation
of liquid crystalline order. The haze also occurred in
film made from III when R was phenyl. With additional
heating, both polymers became isotropic (clear). In
contrast, melts of similar polyesters prepared with
methyl- or chlorohydroquinone instead of
phenylhydroquinone do not become isotropic below their
decomposition temperatures. The bulky phenyl group in
phenylhydroquinone reduces the thermal stability of the
liquid crystalline mesophase and the amount of a flexible
spacer or kinked component that can be tolerated in the
polyester without loss of liquid crystallinity, perhaps
because the steric effect of the phenyl group in the
phenylhydroquinone unit reduces the attractive forces
between the polymer chains.

In a series of liquid crystalline polyesters
containing a decamethylene flexible spacer between the
mesogenic units in the polymer backbone, Lenz and
coworkers observed that a methyl, chloro, or bromo
substituent on a hydroquinone moiety (II, n=10, R=CH$_3$,
Cl, Br) decreased the temperature at which the nematic
(liquid crystalline) melt became isotropic, and two
chlorine atoms para to each other on the hydroquinone
derived unit decreased this temperature twice as much as
one chloro substituent.[15] The single phenyl substituent
on the hydroquinone moiety in our polyesters appears to be
exerting a similar effect in reducing the temperature at
which the polymers become isotropic.

It is instructive to compare the melt viscosities of
an all-aromatic liquid crystalline polyester with those of
a similar polyester that is not liquid crystalline because
of kinks in the polymer chain. Figure 7 compares the melt
viscosities at 360°C of the polyterephthalate and
polyisophthalate of phenylhydroquinone at the shear rates
which are employed in melt processing (about 1000 sec^{-1}
for melt spinning and about 10,000 sec^{-1} for injection
molding). The polyterephthalate has the low melt
viscosities that are typical of liquid crystalline

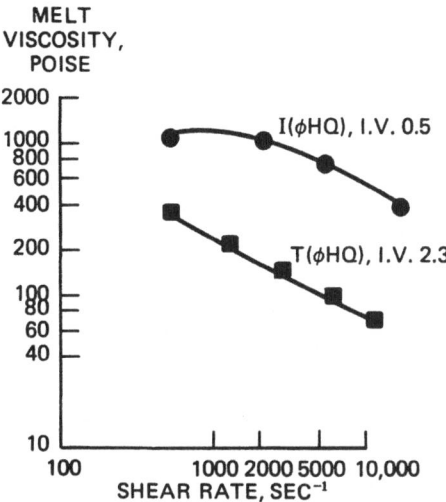

Fig. 7. Effect of shear rate and liquid crystallinity on
         melt viscosities at 360°C of the polyisophthalate
         and polyterephthalate of phenylhydroquinone.

polyesters, and the polyisophthalate has significantly
higher melt viscosities. (Table 2 lists the plastics
properties of these two polyesters.) Even though the
I.V.'s of these two polymers are quite different (0.5 for
the polyisophthalate and 2.3 for the polyterephthalate),
the molecular weights are similar. For the
polyisophthalate an absolute value of 34,200 was obtained
in m-cresol solution by low angle laser light scattering
(LALLS) (30,500 by gel permeation chromatography [GPC]
using a universal calibration); and for the
polyterephthalate a value of 27,000 was obtained (LALLS)
by using an isorefractive mixed solvent consisting of
o-dichlorobenzene and p-chlorophenol (1:1 by weight).
(Our earlier estimates of the molecular weights of the
polyterephthalate by GPC were too low because they were
derived from GPC data alone and we had assumed that the
universal calibration is applicable to rigid rod
molecules.[1a])

In a study of the effects of temperature on the melt viscosities of copolyesters of terephthalic acid, isophthalic acid, and methylhydroquinone, McFarlane and Davis[22] observed that minima in the melt viscosity versus temperature curves occurred at about 340 to 360°C in compositions containing 40 to 60 mol % isophthalic acid (total diacids equal 100 mol %). The increase in the melt viscosities with increasing temperature after the minima presumably is due to the increase in the isotropic content of the polymers and the decrease in the degree of liquid crystallinity. We did not observe this phenomenon in the rigid-rod, all-aromatic, liquid crystalline polyesters that did not contain any meta component because of the high temperatures involved (above the decomposition temperatures of the polyesters). A poly(terephthalate-isophthalate) of methylhydroquinone containing 70 mol % isophthalic acid was not liquid crystalline and, therefore, did not exhibit a minimum in a plot of melt viscosity versus temperature.[22]

When a liquid crystalline polyester is modified by a component that imparts rigid kinks or flexibility to the polymer chain, the degree of liquid crystallinity is decreased. Substituents attached to an aromatic ring also have an effect. The degree of liquid crystallinity of a thermotropic liquid crystalline polyester is also decreased by increasing the temperature of the molten polymer or by decreasing the molecular weight below some limiting value. Work now being conducted by Lenz may lead to a quantitative determination of the degree of liquid crystallinity of a liquid crystalline polyester by small-angle laser light-scattering measurements.[16,23]

Thermal-Oxidative Stability

The most thermally and oxidatively stable polyesters are those containing only aromatic rings, attached through ester groups, with no halogen, alkyl, or alkoxy substituents attached to the aromatic rings. A measure of the stability can be obtained by heating films of the polymers in air and determining the time required for the solubility of the films to be decreased. Increasing levels of crosslinking are present when the initially soluble film changes in the order: partially soluble, highly swollen, slightly swollen, insoluble, brittle. The

chemical resistance of the polymers can also be increased
in this manner, but it is desirable not to heat the film
(or shaped object) until it is crosslinked so much that it
is brittle. In our test, pressed 10-mil films of the
polyterephthalate of phenylhydroquinone modified with
chloro- and methylhydroquinone were placed in glass tubes
which were then heated in an aluminum block at 300°C while
ambient air was passed through the tubes at a rate of two
standard cubic feet per hour. The solubilities were
determined by stirring the films in a 65/35 (by volume)
solution of methylene chloride/trifluoroacetic acid at
room temperature for one hour and then allowing the films
to stand in the solution for several hours. The air
treatment time required for the films to become insoluble
(no indication of swelling) decreased as the level of the
chlorine or methyl substituent increased (Table 6), and at
40 mol % chloro- or methylhydroquinone (diol components
total 100 mol %), the films were insoluble after being
heated in air for 30 minutes at 300°C (about four hours
were required when heated at 275°C). The crosslinking

TABLE 6. CHEMICAL RESISTANCE OF POLYESTERS

R	x/y	Treatment time in 300°C air	
		30 min	40 min
—	1.0/0	Soluble	Mostly soluble
Cl	0.8/0.2	Soluble	Highly swollen
Cl	0.7/0.3	"	Slightly swollen
Cl	0.6/0.4	Insoluble	Insoluble
CH_3	0.8/0.2	Soluble	Highly swollen
CH_3	0.7/0.3	"	Slightly swollen
CH_3	0.6/0.4	Insoluble	Insoluble

reaction appears to be primarily a surface reaction, because films with a sufficiently high molecular weight to be creasable before being heated were still tough and creasable after being heated in air long enough to become insoluble. On additional heating, the films became brittle.

When a film of the polyterephthalate of phenylhydroquinone modified with 40 mol % methylhydroquinone was heated in an argon atmosphere at 300°C for one hour, the film remained soluble; therefore, the mechanism of the crosslinking reaction involves oxidation of the methyl group. A tert-butyl group is more resistant to oxidation than a methyl group, and a film of the polymer containing even 50 mol % tert-butyl-hydroquinone was still soluble after 30 minutes in air at 300°C. On the other hand, a film of the polyterephthalate of phenylhydroquinone modified with 40 mol % chlorohydroquinone was insoluble after being heated under argon for one hour at 300°C. Crosslinking probably involved loss of some chlorine atoms, free radical formation, and coupling of free radicals. The presence of air facilitates the crosslinking reaction, presumably after loss of chlorine, because the film was insoluble after 30 minutes in air at 300°C whereas it was almost completely soluble after 30 minutes under argon at 300°C. A copolyester containing 50 mol % chlorohydroquinone was insoluble after 30 minutes under argon at 300°C. Additional examples of polyesters that become insoluble after being heated in air are given in our patent.[24]

As indicated above, when the films were heated for extended times in air, the films became brittle (broke on creasing). In liquid crystalline aromatic polyesters containing substituted hydroquinones, the effect of the substituent on stability in air is as follows (decreasing stability, left to right, based on time required for 10-mil thick films suspended in a forced-air oven at 150°C to become brittle):

phenyl >> tert-butyl > chlorine > methyl

A film of the polyterephthalate of phenylhydroquinone was still creasable after 30 weeks, when the test was stopped, and our experience with other liquid crystalline all-aromatic polyesters containing no alkyl or halogen

substituent suggests that the film would have remained
tough (creasable) over two years. On the other hand,
films of chlorohydroquinone polyesters became brittle
within a few weeks in the 150°C oven, and films of
polyesters containing high levels of methylhydroquinone
generally became brittle in less than one week. The
methoxy group also contributed greatly to instability.

Table 7 shows the effect of air at 260°C on the
tensile strength and the Izod impact strength of a
polyester of terephthalic acid, 60 mol %
phenylhydroquinone, and 40 mol % hydroquinone. The
heat–deflection temperature (264 psi load) was 269°C.
Even after test bars suspended in a forced–air oven at
260°C were heated for six weeks, they retained 60% of
their initial tensile strength and 50% of their initial
unnotched Izod impact strength. A 10–mil film of this
composition was still creasable after six weeks, but broke
after eight weeks in the oven at 260°C. When
methylhydroquinone was used instead of hydroquinone in a
similar composition, however, the film was brittle and
black after one day in the oven.

TABLE 7. EFFECT ON PLASTIC PROPERTIES OF AGEING AT 260°C IN
AIR

Time, wk	Tensile strength, psi	Izod impact strength, ft–lb/in.	
		Notched	Unnotched
0	28,300	1.7	12.3
2	23,400	0.9	6.8
4	22,300	0.6	7.4
6	17,500	0.6	6.3

Although these polyesters containing methyl or chloro substituents have limited stability when exposed to air at high temperatures for long periods of time, they are sufficiently stable to be melt spun at 360 to 380°C into air and then heat treated in an inert atmosphere to increase their molecular weights and give fibers having very high tenacities (20 g/den.). These polyesters which contain only aromatic rings, however, are sufficiently stable to be heat-treated, even in air, to give fibers with high tenacities. Table 8 shows the results of experiments with fibers of the polyterephthalate of phenylhydroquinone. A tenacity of 22 g/den. was obtained when the fibers were heated in an argon atmosphere for 0.5 hour at 300°C (13 g/den. in air). The tenacity increased to 18 g/den. as the treatment time in air at 300°C was increased to 1.5 hours, but the tenacity decreased somewhat during the one-hour treatments when the temperature was increased from 300 to 340°C, probably because more branching and crosslinking took place in the fibers as the temperature was increased. Since a film of this polymer was highly swollen by solvent after the film had been heated in air at 300°C for one hour, a significant amount of crosslinking had occurred. Similar crosslinking taking place in the fibers limits the tenacities that can be attained during the heat-treatment step.

In the absence of air, higher temperatures can be tolerated, particularly when the polyesters contain only aromatic rings. When the polyterephthalate of phenylhydroquinone was heated in the melt state in a Tinius-Olsen Plastometer at 360°C for 45 minutes, no change in flow rate (13.2 g/10 min) or I.V. (2.3) occurred. According to thermogravimetric analysis

TABLE 8. EFFECT OF HEAT TREATMENT IN AIR ON TENACITY OF POLY(PHENYL-1,4-PHENYLENE TEREPHTHALATE) FIBERS

Temp, °C	Tenacity, g/den.		
	0.5 hr	1.0 hr	1.5 hr
280	—	9	—
300	13	17	18
320	14	16	—
340	—	15	—

(heating rate at 20°C/min), the initial weight loss of
this polymer occurred at about 400°C in air and 430°C in
nitrogen.

CONCLUSIONS

Polymers with lower melting points can be obtained
when monosubstituted hydroquinones are used in the
preparation of liquid crystalline aromatic polyesters from
symmetrically oriented aromatic dicarboxylic acids and
aromatic diols than can be obtained when hydroquinone is
used. When the substituent is chloro, methyl, tert-butyl,
or 1,1-dimethylhexyl, the polyterephthalate homopolyesters
still melt too high (appreciably above 350°C) to be
injection molded or melt spun without thermal
degradation. Additional modification by incorporation of
flexible aliphatic components or kinked rigid groups into
the polymer backbone reduces the polymer melting point
further and thereby allows the preparation of
melt-processable liquid crystalline polyesters. Very high
plastic and fiber tensile properties can be attained if
the level of modification is low (20 mol %), but these
properties decrease as the flexible or kinked component
increases. In addition to reducing the liquid
crystallinity in the melt and, therefore, the
orientability of the polyesters from the melt, the
introduction of kinked groups, such as meta-oriented
aromatic modifiers, also reduces the crystallinity in the
polymers; consequently, heat-deflection temperatures of
the injection-molded plastics are reduced, and the
temperatures to which fibers can be heat treated (to
increase the polymer molecular weights and fiber tensile
properties) are reduced because of fusion of filaments.

The high melting points of the polyterephthalates of
substituted hydroquinones can also be reduced to a
melt-processable level by modification of the polyesters
with 2,6-naphthalenedicarboxylic acid. Since this is a
symmetrical aromatic dicarboxylic acid, the polyester
melts have a high degree of liquid crystallinity and,
therefore, plastics and fibers with high tensile
properties can be produced from them.

If the substituent on the hydroquinone is phenyl, the
polyterephthalate homopolyester is liquid crystalline and
melt processable. Modification of this polyester with

p-hydroxybenzoic acid or with 2,6-naphthalenedicarboxylic
acid reduces the melting point and, particularly, the flow
point and crystallinity (but not the liquid
crystallinity). Modification with
2,6-naphthalenedicarboxylic acid also increases the
toughness of the fibers and the notched Izod impact
strength of the plastics. Modification of the
polyterephthalate of phenylhydroquinone with up to about
30 mol % hydroquinone or tert-butylhydroquinone or with up
to about 60 mol % chloro- or methylhydroquinone has very
little effect on the melting or softening points of the
melt-prepared polyesters, and plastics and fibers with
very high tensile properties can be obtained from them.

A methyl or chloro substituent attached to a
hydroquinone moiety in a liquid crystalline aromatic
polyester decreases the thermal-oxidative stability of the
polyester, but the substituent does permit greater
chemical resistance of a shaped object to be obtained by
heat treatment in air at 300°C. The stability of
substituted hydroquinone polyesters in air at 150°C
decreases with various substituents in the order of
phenyl, tert-butyl, chloro, methyl.

To be able to achieve maximum tensile and stiffness
properties in melt-processed polyesters, the molten
polyesters must have both a high degree of liquid
crystallinity, which enables high orientation to be
attained on injection molding or melt spinning, and
sufficiently long relaxation times so that this
orientation is maintained as the melt cools. Kinked
components in the polymer chains reduce the amount of
polymer orientation that can be achieved, and flexible
components reduce the relaxation time.

ACKNOWLEDGEMENTS

I would like to acknowledge the numerous
contributions of H. F. Kuhfuss and J. C. Morris, who also
prepared the polyesters. I am also indebted to R. M.
Schulken and R. H. Cox for determination of the relaxation
times and melt viscosities, B. L. Neff for determination
of the molecular weights, P. D. Griswold for determination
of the correlation of fiber tenacity and melt viscosity,
Karen J. Watkins for the X-ray analyses, and J. B. Davis
for the microscopy data.

REFERENCES

1. (a) W. J. Jackson, Jr., Br. Polym. J., 12, No. 4, 153 (1980). (b) V. V. Korshak, Russ. Chem. Rev., 29, 269 (1960). (c) W. M. Eareckson, J. Polym. Sci., 40, 399 (1959).

2. I. Goodman, J. E. McIntyre, and J. W. Stimpson (to ICI), British Patent 989,552 (1965), U.S. Patent 3,321,437 (1967).

3. I. Goodman, J. E. McIntyre, and D. H. Aldred (to ICI), British Patent 993,272 (1965), U.S. Patent 3,368,998 (1967).

4. H. F. Kuhfuss and W. J. Jackson, Jr. (to Eastman Kodak Co.), U.S. Patent 3,778,410 (1973).

5. H. F. Kuhfuss and W. J. Jackson, Jr. (to Eastman Kodak Co.), U.S. Patent 3,804,805 (1974).

6. W. J. Jackson, Jr., and H. F. Kuhfuss, J. Polym. Sci., Polym. Chem. Ed., 14, 2043 (1976).

7. F. E. McFarlane, V. A. Nicely, and T. G. Davis, "Contemporary Topics in Polymer Science," Vol. 2, E. M. Pierce and J. R. Schaefgen, Eds., Plenum, New York, 1977, p 109.

8. J. J. Kleinschuster, T. C. Pletcher, and J. R. Schaefgen (to Du Pont), Belg. Patent 828,935 (1975).

9. J. R. Schaefgen (to Du Pont), U.S. Patent 4,118,372 (1978).

10. C. R. Payet (to Du Pont), U.S. Patent 4,159,365 (1979).

11. W. J. Jackson, Jr., and H. F. Kuhfuss (to Eastman Kodak Co.), U.S. Patent 4,360,658 (1982).

12. W. J. Jackson, Jr., and H. F. Kuhfuss (to Eastman Kodak Co.), U.S. Patent 4,238,600 (1980).

13.   R. R. Luise (to Du Pont), U.S. Patent 4,183,895
      (1980).

14.   W. J. Jackson, Jr., G. G. Gebeau, and H. F. Kuhfuss
      (to Eastman Kodak Co.), U.S. Patent 4,153,779
      (1979).

15.   J.-I. Jin, S. Antoun, C. Ober, and R. W. Lenz, Br.
      Polym. J., 12, No. 4, 132 (1980).

36.   S. Antoun, R. W. Lenz, and J.-I. Jin, J. Polym.
      Sci., Polym. Chem. Ed., 19, 1901 (1981).

17.   J. R. Schaefgen (to Du Pont), U.S. Patent 4,075,262
      (1978).

18.   W. J. Jackson, Jr., G. G. Gebeau, and H. F. Kuhfuss
      (to Eastman Kodak Co.), U.S. Patent 4,242,496
      (1980).

19.   W. J. Jackson, Jr., Macromolecules, in press.

20.   W. J. Jackson, Jr., and H. F. Kuhfuss (to Eastman
      Kodak Co.), U.S. Patent 4,156,070 (1979).

21.   J. C. Morris and W. J. Jackson, Jr. (to Eastman
      Kodak Co.), U.S. Patent 4,146,702 (1979).

22.   F. E. McFarlane and T. G. Davis (to Eastman Kodak
      Co.), U.S. Patent 4,011,199 (1977).

23.   R. W. Lenz, private communication.

24.   W. J. Jackson, Jr., and H. F. Kuhfuss (to Eastman
      Kodak Co.), U.S. Patent 4,287,332 (1981).

NEW DEVELOPMENTS IN HIGH TEMPERATURE AND

HIGH STRENGTH POLYQUINOLINES

J.K. Stille

Department of Chemistry
Colorado State University
Fort Collins, Colorado   80523

SYNOPSIS

The relationships between the structures and the morphology, solution properties and mechanical properties of polyquinolines depend upon the manner in which the quinoline units are linked together. A wide variety of structures account for polyquinolines that either are amorphous, soluble in common organic solvents and have moderately high glass transition temperatures (~250°C) or are crystalline, are soluble only in acidic solvents, forming anisotropic solutions, and have high crystalline transition temperatures (>500°C). The conducting properties of certain doped polyquinolines containing either total conjugation or sulfur linkages between quinoline units are described. Polyquinolines containing biphenylene units either in the main chain or at the chain ends undergo crosslinking and chain extension reactions. The use of biphenylene end-capped oligomeric polyquinolines as matrix resins in graphite composites is discussed.

INTRODUCTION

Polymers that are synthesized for use as high performance materials generally contain a polyaromatic main chain. As a result, if the polymers are amorphous, they may be melt processed or processed from solution, but their ultimate use temperature is then limited to the glass transition temperature, which is well below the temperature at which these polymers undergo thermal degradation or thermooxidative decomposition. If, on the other hand, the polymers are crystalline, then they usually possess very high crystalline transition temperatures, and cannot be fabricated by melt techniques. Because of the limited solubility of such crystalline materials, their fabrication from solution usually involves corrosive solvents of low volatility.

Polyquinolines have been synthesized by a polymerization reaction that allows the introduction of a wide range of structural features. As a result, the linearity and rigidity of the polymer can be adjusted to provide materials that are amorphous and have good solubility in common organic solvents and lower glass transition temperatures (250°C) or materials that are highly crystalline, insoluble in the common solvents, and have crystalline transition temperatures greater than 550°C. In addition, inert or reactive groups can be incorporated into the main chain of the polymer, appended from the main chain or placed at the chain ends. The synthesis and properties of such polymers are discussed herein. Specifically, the structure-property relationships of semirigid and rigid-rod polyquinolines, the effect of large pendent groups on the solubility of such polymers and the electrical properties of certain polyquinolines are considered. Finally, the crosslinking reaction of thermally stable polymers, particularly polyquinolines that contain biphenylene groups, and the use of such materials as matrix resins in composites are described.

RESULTS AND DISCUSSION

Polyquinoline-Forming Polymerization Reactions

The step-reaction polymerization of an aromatic bis-o-aminoketone with an aromatic bis ketomethylene monomer yields high molecular weight polyquinolines by acid catalysis. Alternatively, a single monomer containing both the aminoketone and ketomethylene functions undergoes a self polymerization. The reaction is first order in aminoketone and first order in ketomethylene functions. The polymerization is carried out conveniently in m-cresol at 135°C. The reaction can be carried out in the presence of poly(phosphoric acid)[2-6]

R = nil, p-C₆H₄

or the products of the reaction of phosphorous pentoxide with m-cresol,[7,8] although with either of these catalysts, the reaction stopped after achieving a moderate molecular weight, ($\overline{DP}$ = 320). The products of the reaction of phosphorous pentoxide with m-cresol were shown to be an equimolar mixture of mono-(1) and di-m-cresol (2) esters of phosphoric acid, as the stoichiometry suggests.[10]

$$P_2O_5 + 3 \underline{m}\text{-}CH_3C_6H_4OH \longrightarrow$$

$$\underline{m}\text{-}CH_3C_6H_4O\text{-}\overset{O}{\overset{\|}{P}}(OH)_2 + (\underline{m}\text{-}CH_3C_6H_4O)_2\overset{O}{\overset{\|}{P}}OH$$

$$\mathbf{1} \qquad\qquad\qquad \mathbf{2}$$

Phosphate ester 1 was only marginally effective as a polymerization catalyst, but 2 effected the polymerization at the same rate as the m-cresol/phosphorous pentoxide reaction mixture. An advantage to using pure 2 is that much higher molecular weights could be obtained (DP > 550 in 14h).

Polyquinoline Structure vs Properties

As a result of the availability of a wide variety of aminoketone and ketomethylene monomers that could be synthesized, and the ability of the Friedlander reaction to generate high molecular weight polymers under relatively mild conditions, a series of polyquinolines containing a variety of structural differences could be obtained. The chain stiffness could be altered from a semirigid main chain to a linear, rigid-rod main chain. The monomers were designed to yield only thermally stable materials, such that the resulting polymers had excellent thermal stability, with the initial weight loss occurring between 500 and 600°C in air (TGA),[1] the exception being that polyquinolines having perfluoromethylenes in the main chain started oxidative degradation at somewhat lower temperatures (~400°C).[11] Isothermal ageing studies of certain polyquinolines showed them to be among the most thermally stable organic polymers known.

High molecular weight polyquinolines (Mn=325,000) containing flexible oxygen linkages between the aromatic units, for example, structure 3, are largely amorphous and are soluble in common organic solvents.[7,8,12] Clear, transparent films can be cast from chloroform or tetrachloroethane, and although the materials are largely amorphous, crystalline transitions can be detected. The semirigid polymers have glass transitions in the 255-300°C range.[13,14] None of these semirigid polyquinolines were crystalline enough such that they maintained good mechanical properties above the glass transition temperature.[12,14] (Figure 1).

Unoriented films of these polyquinolines (eg. 3) have high moduli (10^6 psi), moderate tensile strength (3.5×10^3 psi), and low elongation (0.6%) combined with good electrical properties (Table 1).[10] Oxygen and water vapor permeabilities are relatively high [O_2, cm^3 (100 in^2-24h-atm)$^{-1}$=278; H_2O vapor, g (100 in^2-24h-44mm)$^{-1}$mil^{-1}=9.4]. The thermal stability of 3

TABLE 1

Electrical Properties of Polyquinoline 3

Dielectric Strength, kV/mil	7.8
Volume Resistivity, Ω-cm/mil	2.5 x 10^{15}
Dielectric Constant	2.6/2.5 ($10^2/10^5$ Hz)
Dissipation Factor	0.017/0.005 ($10^2/10^5$ Hz)

Film thickness = 0.3-0.4 mil; $\overline{Mn}$ = 81,700; density = 1.174.

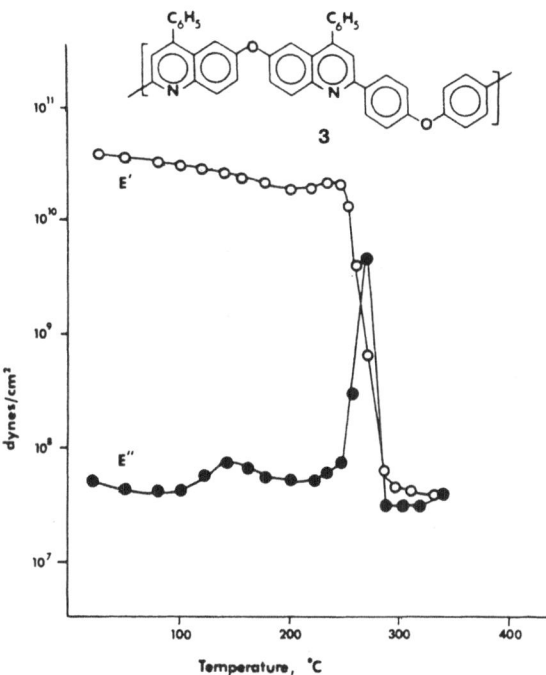

Figure 1. Dynamic storage and loss moduli vs. temperature: heating rate 5°C/min, He atmosphere, frequency 35 Hz.

is excellent as determined by isothermal ageing (Table 2). A life in air (crease test) of about 10 years at 200°C can be obtained by extrapolation of the Arrhenius data for isothermal ageing.

TABLE 2

Thermal Stability of Polyquinoline 2[a] Film Failure

Time	Temperature °C
2 h	450
20 h	350
11 days	300
6 months	250
10 years	200

[a]See footnote a, Table 1. TGA breaks 525°C (air); 630°C (N_2).

Because the use temperature of the amorphous semirigid polyquinolines is limited to their glass transition temperatures, an effort was made to increase the glass transition temperature by synthesizing a more rigid chain, but still maintaining the amorphous nature to enable solution processability. Polyquinolines containing pendent loops attached to the polymer main chain (eg. 4) give the polymer enhanced rigidity.[15-18] These loops are perpendicular to the main chain, and have the effect of raising the glass transition temperature, while maintaining the good solubility in common organic solvents that is characteristic of the amorphous polyquinolines. Replacing the oxygen linkages in the semirigid polyquinoline (3) by pendent loops (4) has the effect of raising the glass transition temperature by about 120°C, raising its use temperature nearly to 400°C.

Polyquinolines having directly bonded quinoline units in the 6-positions and connected in the 2-positions by p-phenylene units (eg. 5) are rigid rods that are highly crystalline, having crystalline transition temperatures at 500°C or above.[8,19] These polymers are soluble only in the polymerization medium

(<u>m</u>-cresol/di-(<u>m</u>-cresyl phosphate) or strong acids (eg. trifluoromethanesulfonic acid), in which they form anisotropic solutions at about 9 wt %. Even solutions of very low concentrations (~1%) will give anisotropic solutions on shearing. These dilute solutions become isotropic on heating, but form anisotropic solutions again at ambient temperatures on shearing.

The rigid-rod polyquinolines are characterized by high glass transition temperatures, but not any higher than the polyquinolines containing the pendent loops. In some cases the Tg was not apparent due to the relatively high degree of crystallinity, and in some cases the crystalline transition temperatures could not be observed because it was apparently above the decomposition temperature. Thus, for these rigid-rod polyquinolines, the decomposition temperature and the crystalline transition temperature nearly coincide. The comparision of the mechanical properties of the semirigid and rigid-rod polyquinolines is exemplified by a comparison of the dynamic storage moduli of the two types, 3 and 5 (x=1) (Figure 2).

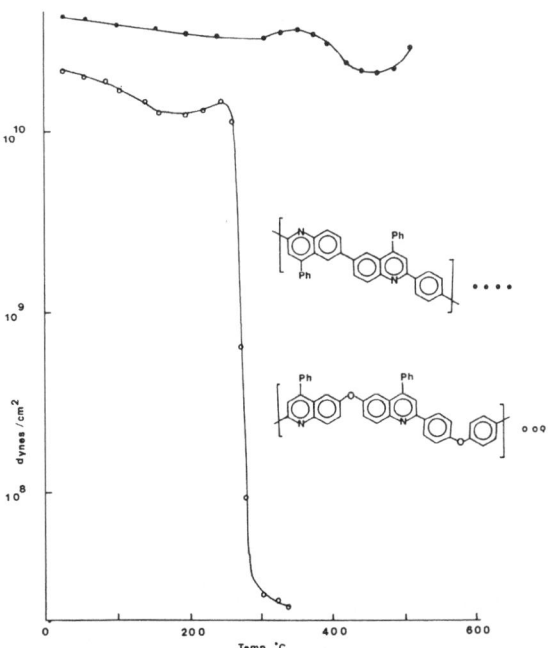

Figure 2.   Thermomechanical properties of polyquinolines.

Rigid-rod polyquinolines could be wet spun in some preliminary trials (8-18.5 wt % solids) from the polymerization dope to give fibers with good tensile strengths (9 gpd) and high moduli. The x-ray structure of a fiber of poly[2,2'-(p,p'-biphenyl)-6,6'-bis-(4-phenyl-quinoline)], (5, x=2 idealized) shows that the parallel chains oriented along the direction of the fiber axis tend to stack in nearly coplanar sheets.[21.]

Part of the difficulty in studying the solution properties of such rigid-rod polymers is that the solubility is limited to strongly acidic media. In a number of cases, the solubility of polyaromatics has been improved by the attachment of phenyl groups on the main chain. Polyquinoline 5 (x=2) already has two pendent groups per recurring unit, but does not have the desired solubility in the more usual organic solvents.

In order to disrupt the crystal structure and improve solubility, longer pendent arms were placed in the 4-positions of the quinoline units. With an arm at least two phenylene units long, the crystal packing observed for 5 (x=2) could not be achieved, and a third phenyl group would necessitate rotation of the quinoline units out of the plane. In addition, it was hoped that the aromatic arms would keep adjacent chains at a distance that would tend not to allow a dense packing.

The synthesis of such polyquinolines has been carried out to give structures of the type 6.

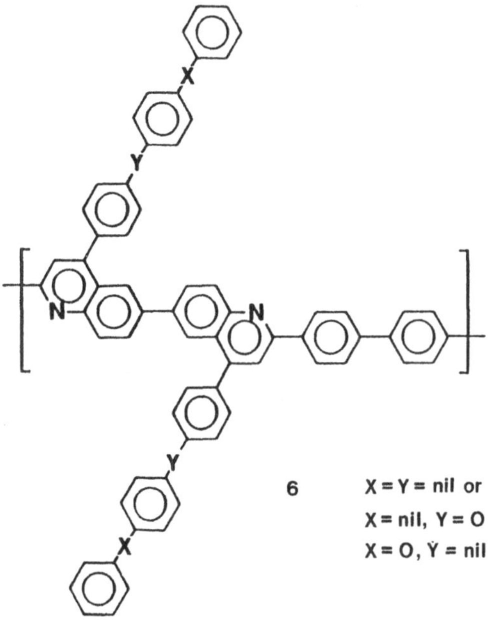

6   X = Y = nil or O
           X = nil, Y = O
           X = O, Y = nil

Electrical Properties of Polyquinolines

The wide range of structural features that can be incorporated into polyquinolines, plus the fact that these materials are readily processable and possess good mechanical properties, made certain polyquinolines prime candidates for semiconducting and conducting materials. Although polyquinoline 3 is an excellent high temperature insulating material, (Table 1) certain other polyquinolines should possess conducting properties on doping. Two types of polyquinolines were utilized for this purpose, those containing conjugated stacked planar structures (eg. 5, x=2) and those which are non-planar, but contain sulfur linkages in the main chain (7,8).

TCNQ forms charge-transfer complexes and simple complex salts with quinoline that have high conductivity.[21] The structure of 5 (x=2), having stacked coplanar aromatic sheets, resembles, in some respect, the structure of graphite, for which very high conductivity is observed for the arsenic pentafluoride intercalate.[22,23] Preliminary conductivity measurements on oriented fibers of 5 (x=2) doped with arsenic pentafluoride (400 torr) show that the conductivity rises rapidly from $\sigma = 10^{-11}$ to 10^{-8}.

Although poly(p-phenylene sulfide) contains phenyl rings nearly at right angles to each other, the doped polymer yields a material that exhibits conductivity.[23] Films of polyquinolines containing sulfur connecting groups as exemplified by 7 and 8 are insulators. Preliminary experiments show that doping raises the conductivity from 10^{-12} to 10^{-5}.

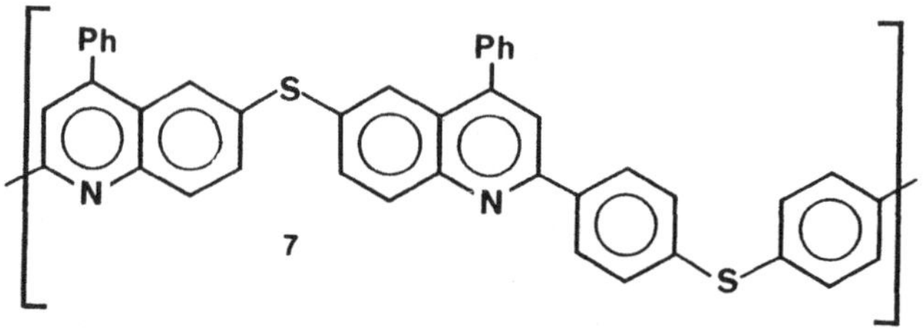

7

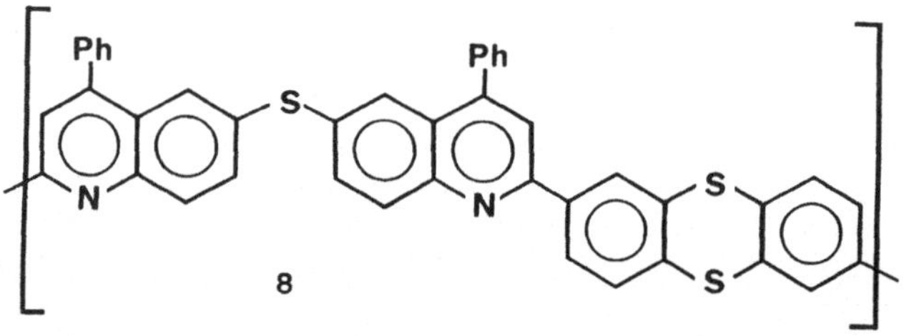

8

Crosslinking and Chain Extention Reactions of Polyaromatics Containing Biphenylene Groups

In an effort to produce thermally stable polymers that could be fabricated readily and then converted to a material with high use temperatures, the crosslinking reactions of aromatic polymers have been explored. Unfortunately, because of the chemically inert structure and reduced chain mobility of polyaromatics, there are few good crosslinking reactions for them. Ideally, such a crosslinking reaction should take place without the

evolution of volatiles to produce a thermally stable
product by the generation of a new thermally stable link.
The temperature of the crosslinking reaction should be
above Tg, but below the polymer decomposition temperature.

Such a reaction that is applicable to crosslinking and
chain extension is the ring-opening reaction of
biphenylene (9). At about 350°C, biphenylene
undergoes a ring opening to yield mainly tetrabenzocyclooctatetraene (10) and some polymer. When the reaction
is carried out in the presence of 2,4-diphenylquinoline
(11), small amounts of biphenyl-substituted quinolines
in addition to 10 are obtained. The catalytic
thermolysis of biphenylene containing small amounts of

norbornadiene rhodium chloride dimer or bis(triphenylphosphine)dicarbonylnickel, takes place nearly 150°C
lower, at 200°C. Thus, this reaction appeared to meet the
criteria for crosslinking aromatic polymers.

Polyaromatic structures such as polyquinolines 12,
that have the biphenylene units incorporated into the main
chain, undergo crosslinking reactions.[9,24-27]
Polyquinolines, polyamides, polyquinoxalines, and
polybenzimidazoles, all containing biphenylene units,
crosslink thermally, with or without the aid of transition
metal catalysts.

The crosslinked polymers were insoluble in all solvents, had higher storage moduli at ambient temperature, higher storage moduli above the Tg, and higher Tg's than the uncured samples. Several features of this curing reaction became evident: 1. Crosslinking takes place only above the Tg, even though the thermal or transition metal catalyzed reaction was known to take place at a temperature below the Tg. 2. The onset of the ring opening, as observed by the exotherm, was above, but paralleled the Tg of the sample. 3. In the presence of a transition metal catalyst, lower biphenylene reaction temperatures and shorter reaction times could be used to effect the same crosslinking reaction. 4. Relatively low incorporations (2.5 mole %) of biphenylene into the main chain were sufficient to raise significantly the modulus (at temperatures above Tg) of the cured sample. 5. The thermal stability of the crosslinked polymers as measured by TGA appeared to be as good as the same polymers not containing the biphenylene units. These crosslinking reactions undoubtedly are taking place primarily by tetrabenzocyclooctatetraene formation between chains.

The use of biphenylene monomers for incorporation into a polymer main chain requires the synthesis of a difunctional biphenylene in monomer-grade purity. Because these synthetic reactions generally require more than one step, and generally do not proceed in high yields, the synthesis of simpler, monofunctional biphenylene derivatives for use in crosslinking reactions was desirable. In this case the high purity of the biphenylene derivative is not necessary.

Polyquinolines can be crosslinked with bisbiphenylene agents such as 13.[28] In designing crosslinking agents

13

structures that mimicked the polymer to be crosslinked were synthesized in order to insure compatibility with the polymer. The crosslinking agents had the added advantage of being able to plasticize the polymer, thus facilitating processing. Unfortunately, because the biphenylene unit reacts only to a small extent by attack directly at an aromatic group, in this case by attack on a polymer main chain, this mode of crosslinking was not particularly effective.

A third approach to processing thermally stable polymers is the use of oligomeric prepolymers end-capped with a reactive group that undergoes a chain extension/ crosslinking reaction during the processing and subsequent post-curing. This approach has been used in certain resins, for example, Thermid-600[R] and the nadic end-capped PMR-15[R]. While these resins exhibit good performance, the former gives different types of linking structures that are potentially less stable than the parent polyimide,[29,30] and the latter has a crosslinking group that is a major contribution to the weight loss over time.[31]

Because of the good solubility of polyquinoline 3 in chloroform and its relatively low Tg, oligomers of this polymer capped with biphenylene, 14 were prepared according to the relationship $\overline{DP} = 1 + r / 1 + r - 2rp$, where r=moles of ketomethylene monomer/aminoketone monomer and p=extent of the reaction (Table 3). The oligomers before processing showed glass transition temperatures consistent with the relationship $Tg = T_\infty - k/Mn$, $(k \sim 2.5 \times 10^5)$.

The resins (14) containing a nickel catalyst could be melt pressed into thin films at 325°C under 500 psi for 15 minutes. After this relatively short processing time, all films were insoluble. Young's modulus above Tg had increased by an order of magnitude. The Tg (at 25°C) of the cured polymers were all relatively the same (234-243°C) as were the mechanical properties $(E' = 1-2 \times 10^8$ dynes/cm^2). This suggests that the chain extention reaction by conversion of biphenylene ends to

TABLE 3

Biphenylene End-Capped Oligomers 14

n^a	$\bar{M}n^b$	$[\eta]$	$\bar{M}n^c$	$\bar{M}n^d$	$\bar{M}w/\bar{M}n^d$	$T_g, ^\circ C$ [DSC]
3	2100	0.09	1100	600	5.7	153
11	6800	0.25	5000	6500	2.9	212
22	13,000	0.47	13,000	12,000	2.9	232
68^e	—	1.00	40,000	51,000	3.8	266
270^e	—	2.47	160,000	—	—	266

a. $n = 1+r/1-r$
    $r = AA/BB$

b. $n|590| + 2|151|$

c. $[\eta] = K\bar{M}n^a$

d. GPC

e. No End-Cap

tetrabenzocyclooctatetraene links proceeds until the different oligomers all reach about the same molecular weight, at which point further chain extension reactions by this mode stop due to the inaccessibility and reduced collision frequency of the biphenylene ends. Thereafter some crosslinking occurs by reaction of biphenylene units with the aromatic units in the polymer main chains.

Thus, graphite-reinforced composites were prepared from biphenylene end-capped polyquinoline 14, $\overline{DP}=22$. Pre-pregs were prepared by brush-coating a chloroform solution of the polymer either onto unidirectional Celion^R-6000 graphite fiber or Celion^R-3000 graphite cloth. The pre-pregs were processed at 330°C, 1500 psi, 2h (Ni catalyzed) or 390°C, 2000 psi, 3h (uncatalyzed). In both cases, high quality void-free laminates were obtained, but due to the moderate degree of actual crosslinking, thermoplastic breaks were observed in interlaminar shear tests, even after post-curing for 50 h at 330°C.

ACKNOWLEDGEMENT

This research was supported by the National Science Foundation and the U.S. Army Research Office, Research Triangle, N.C. We wish to thank W.B. Alston for his interest and help in the preparation and testing of the composites. The electrical properties and the isothermal ageing studies were carried out with the help of C.E. Sroog and R. Angelo, E.I. du Pont de Nemours and Co. Some of the tensile strengths and moduli of fibers were obtained with the help of F. Logullo, E.I. du Pont de Nemours and Co. This work would not have been possible without the enthusiastic efforts of the many graduate students and post-doctoral research associates, whose names appear in the references.

References

1.   J. K. Stille, Macromolecules, **14**, 870 (1981).
2.   J. K. Stille, Macromol. Chem., **8**, 373 (1973).
3.   Y. Imai, E. F. Johnson, T. Katto, M. Kurihara, and J. K. Stille, J. Polymer Sci., Polym. Chem. Ed., **13**, 2233 (1975).
4.   J. K. Stille, E. F. Johnson, J. Wolfe, M. F. Winegardner, and S. Wratten, Org. Coat. Plast. Chem., **33**(1), 127 (1973).
5.   J. K. Stille, "Proceedings of the International Symposium on Macromolecules", Rio de Janeiro, July 26-31, 1974; Elsevier:Amsterdam, 1975; p. 98.
6.   J. K. Stille, J. F. Wolfe, S. Norris, Y. Imai, E. F. Johnson, T. Katto, and M. Kurihara, "Advances in the Chemistry of Thermally Stable Polymers", Warszawa: 1977; p. 9.
7.   J. F. Wolfe and J. K. Stille, Macromolecules, **9**, 489 (1976).
8.   S. O. Norris and J. K. Stille, Macromolecules, **9**, 496 (1976).
9.   J. K. Stille, Pure and Appl. Chem., **50**, 273 (1978).
10.   W. H. Beever and J. K. Stille, J. Polym. Sci., Polym. Symp., No. 65, 41 (1978).
11.   F. A. Bottino, A. Mamo, A. Recca, J. P. Droske, and J. K. Stille, Macromolecules, **15**, 227 (1982).
12.   W. H. Beever and J. K. Stille, Macromolecules, **14**, 493 (1981).
13.   W. Wrasidlo and J. K. Stille, Macromolecules, **9**, 505 (1976).
14.   W. Wrasidlo, S. O. Norris, J. F. Wolpe, T. Katto, and J. K. Stille, Macromolecules, **9**, 512 (1976).
15.   R. M. Harris, S. Padaki, P. Sybert, and J. K. Stille, Polym. Prepr., Am. Chem. Soc., Div. Polym. Chem., **19**(2), 7 (1978).
16.   J. K. Stille, R. M. Harris, and S. M. Padaki, Macromolecules, **14**, 486 (1981).
17.   S. M. Padaki and J. K. Stille, Macromolecules, **14**, 888 (1981).
18.   R. M. Harris and J. K. Stille, Macromolecules, **14**, 1584 (1981).
19.   P. D. Sybert, W. H. Beever, and J. K. Stille, Macromolecules, **14**, 493 (1981).
20.   J. Lando and T. Hauschen, manuscript in preparation. An idealized structure is shown. Although the

recurring units are shown directly stacked over one another, chain slippage to give non-identically stacked chains is possible.

21. E. P. Goodings, Chem. Soc. Rev., 5, 95 (1976).

22. F. L. Vogel, Synthetic Metals, 1, 279 (1980).

23. T. C. Clark, K. K. Kanazawa, V. T. Lee, J. F. Rabolt, J. R. Reynolds, and G. B. Street, J. Polym. Sci., Polym. Phys. Ed., 20, 117 (1982).

24. J. Garapon and J. K. Stille, Macromolecules, 10, 627 (1977).

25. A. Recca, J. Garapon, and J. K. Stille, Macromolecules, 10, 1344 (1977).

26. A. Recca and J. K. Stille, Macromolecules, 11, 479 (1978).

27. J. K. Stille, Vysokomol. Soedin., 11, 2545 (1979).

28. W. Vancraeynest and J. K. Stille, Macromolecules, 13, 1361 (1980).

29. R. F. Kovar, G. F. L. Ehlers, and F. E. Arnold, J. Polym. Sci., Polym. Chem. Ed., 15, 1081 (1977).

30. J. M. Pickard, E. G. Jones, and I. Goldfarb, Macromolecules, 12, 895 (1979).

31. W. B. Alston, "Characterization of PMR-15 Polyimide Resin Composition in Thermooxidatively Exposed Graphite Fiser Composites", NASA TM-81565, 1980 and Proc. of 12th SAMPE National Technical Conference, October, 1980.

ADVANCES IN INORGANIC FIBER DEVELOPMENTS

Ashok K. Dhingra

I.E. Du Pont De Nemours & Co. Inc.
Pioneering Research Laboratory
Textile Fibers Department
Experimental Station
Wilmington, Delaware   19898

SYNOPSIS

Inorganic fiber based materials offer potential for the develpoment of next generation advanced materials beyond fiber reinforced plastics. Inorganic fiber needs and technology are briefly reviewed. The difference in structure and properties of alumina and alumina/silica fibers are discussed. The addition of silica to alumina increases the tensile strength but significantly lowers the modulus of the 100% alumina fibers. Next generation advanced aluminum alloys based on reactive Al-Li compositions are shown to be compatible with dense microstructure and relatively stable 100% alumina fibers to yield wellbonded alumina fiber reinforced aluminum matrix composites having predicted mechanical properties. Examples of potential applications for inorganic fibers materials in aerospace and automotive are shown. An important milestone in inorganic fibers technology was recently reached with the introduction of the world's first inorganic fiber reinforcedmetal matrix composite as a piston insert for an automotive diesel engine.

INTRODUCTION

I would like to give an overview of inorganic fibers technology, then discuss the developments and keys structural features of alumina and alumina/silica fibers and new products based on inorganic fibers. The development of inorganic fiber reinforced materials requires a systems approach to tailor the fiber structure to be compatible with the matrix and assure component fabricability and performance.

(1) **NEED SUITABLE HIGH PERFORMANCE REINFORCING INORGANIC FIBERS FOR**
- METALS
- CERAMICS
- THERMAL AND MECHANICAL COMPATIBILITY WITH THE SYSTEM

(2) **CATALYST SUPPORTS FOR EMISSION CONTROL SYSTEMS**
- HONEYCOMB/EXTRUDED STRUCTURES
- FIBROUS SUPPORT STRUCTURES OFFER VERY LARGE SURFACE AREA

(3) **DESALINATION AND FILTRATION DEVICES**
- HOLLOW INORGANIC FIBERS
- CORROSION RESISTANCE IN SEA WATER
- CONTROLLED POROSITY/MICROSTRUCTURE

(4) **BATTERY AND FUEL CELLS**

(5) **ELECTRONIC APPLICATIONS**
- LOW CONDUCTIVITY, HIGH PERFORMANCE CIRCUIT BOARDS

WHY INORGANIC FIBERS?

Figure 1.

NEED FOR INORGANIC FIBERS

Before discussing "Advances in Inorganic Fiber Developments" let us examine "why the world needs inorganic fibers" (Figure 1).

First and foremost, high performance inorganic fibers are
needed for the reinforcement of metals and ceramics. The
U.S. Department of Defense announced a major thrust for the
development of fiber reinforced Metal Matrix Composites (MMC)
in 1979 (1). NASA has initiated long term R&D programs for
the development of inorganic/ceramic composite materials (2).
Lightweight, high strength, high modulus fiber-reinforced
plastic composites based on graphite and Kevlar* aramid fibers
are creating a "Materials Revolution" and are increasingly
replacing metals, first in high value in use aircraft and
aerospace structural and semi-sturctural applications. I
believe inorganic fiber reinforced metal matrix composites
will be the next generation of advanced materials (Figures 2; 3)

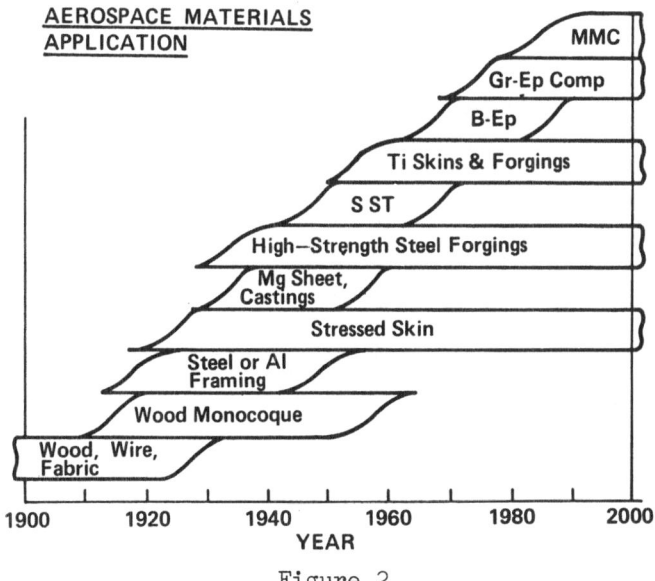

Figure 2.

---

* Du Pont Registered Trademark

showing evolution of new materials in aerospace applications
from 1900 to 2000. One of the major driving forces for the
development of inorganic fiber reinforced metal, carbon and
ceramic matrix composites is for high temperature applications to replace super alloys and minimize our dependence on
critical imported materials such as cobalt, chromium,
tantalum, etc., needed for production of super alloys used
in many critical applications, including jet engines and
defense weapon systems (4). MMC and inorganic composites
offer significant weight savings and increased temperature
capability vs. super alloys (Figure 3). There is also

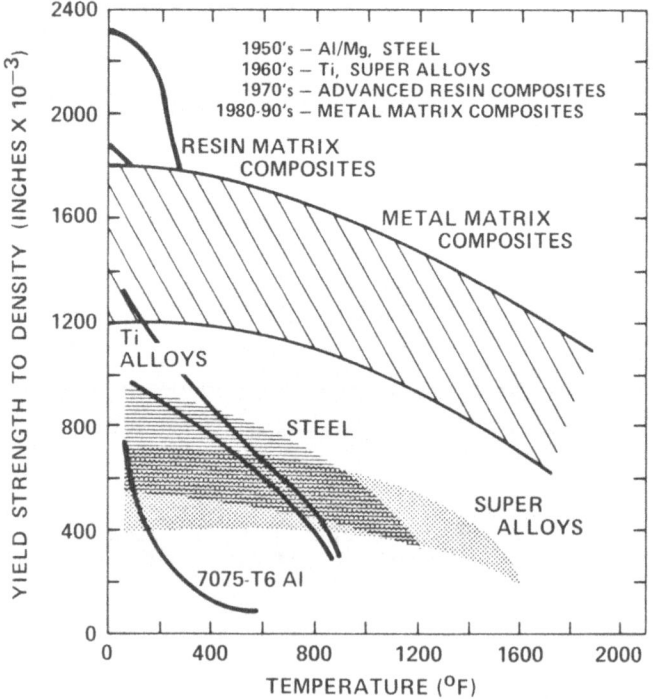

Figure 3.   Specific Strength of Fiber Reinforced Composites
and Metals at Elevated Temperatures.

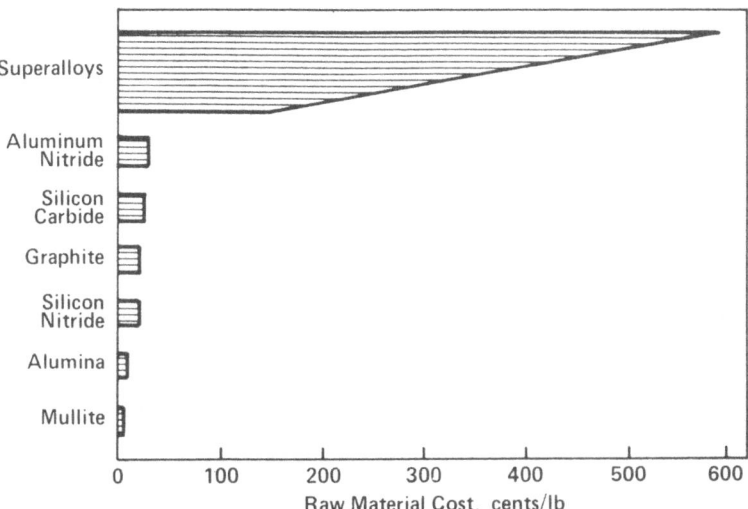

Figure 4. Raw Material Cost of Inorganic Fibers.

a strong economic incentive for the development of high
performance inorganic and ceramic fibers based on raw
material cost. For example, raw material cost of advanced
ceramics such as silicon carbide and silicon nitride is
only a fraction of that of super alloys (Figures 4; 5).

Beyond composites, several products such as catalyst supports for emission control systems, desalination, and filtration devices, battery and fuel cells, and circuit boards based on inorganic fibers technology have high potential. For example, catalyst supports used for emission control devices are made from ceramic honeycomb or extruded glass/inorganic structures. Catalyst supports from inorganic fibers can provide significantly higher surface area at reduced size for optimum emission control. Hollow inorganic fibers for use in desalination and filtration systems should not only provide superior corrosion resistance in salt water but also controlled porosity and structure for superior performance. Inorganic fibers, for example aluminum nitride, are needed for battery and fuel cells applications. Other potential materials for inorganic fibers in electronic application include high performance circuit board, radar transparent structures, and antenna windows.

INORGANIC FIBERS TECHNOLOGY

Various methods for producing inorganic fibers are reviewed in Figure 5. First high strength, high modulus fibers such as boron monofilaments were produced using chemical vapor deposition on a tungsten substrate. Other fibers in this category include silicon carbide coated boron (Borsic®) boron carbide and silicon carbide on tungsten or carbon substrate. Although the chemical vapor deposition yields high strength inorganic fibers exceeding one million psi, the high fiber production cost combined with relatively large diameter limit the market potential. High strength (H.S.), high modulus (H.M.) graphite fibers from pitch have excellent potential for metals reinforcement because of their outstanding mechanical properties and low density. However, the critical problem of galvanic corrosion coupling with metals due to conductivity of carbon needs to be resolved. especially for commercial application requiring long-term environmental stability. High modulus single crystal inorganic filaments have been grown from melt using crystal growing techniques. For example, Tyco Laboratories developed single crystal continuous filaments of sapphire (Figure 6) directly from alumina melt, but the filaments showed loss in tensile strength at elevated temperature (Figure 7) due to twinning and slippage along weak crystallographic planes (6). Short single crystal inorganic whiskers having nearly perfect structure and approaching theoretical strength have been

(1) <u>CHEMICAL VAPOR DEPOSITION (MONOFILAMENTS)</u>
- BORON ON W OR C SUBSTRATE
  ($BX_3 + 3H_2 \rightarrow 2B$ WHERE X IS A HALIDE)
- BORSIC® (SiC COSTED BORON)
- SiC ON W OR C SUBSTRATE
- B_4C

(2) <u>PYROLYSIS</u>
- H. S. GRAPHITE
- H. M. GRAPHITE

(3) <u>MELT</u>
- SUPER GLASS
- TYCO Al_2O_3 FIBER

(4) <u>WHISKERS</u>
- SHORT SINGLE CRYSTAL WITH HIGH ASPECT RATIO
  – HIGHEST STRENGTH AND MODULUS
    ✓ SiC WHISKERS FROM RICE HULLS

(5) <u>FIBER SPINNING</u>
- MARRIAGE OF FIBER SPINNING AND CERAMIC
  SINTERING TECHNOLOGIES
- SiC
- Al_2O_3; Al_2O_3/SiO_2

Figure 5.   Inorganic Fibers Technology.

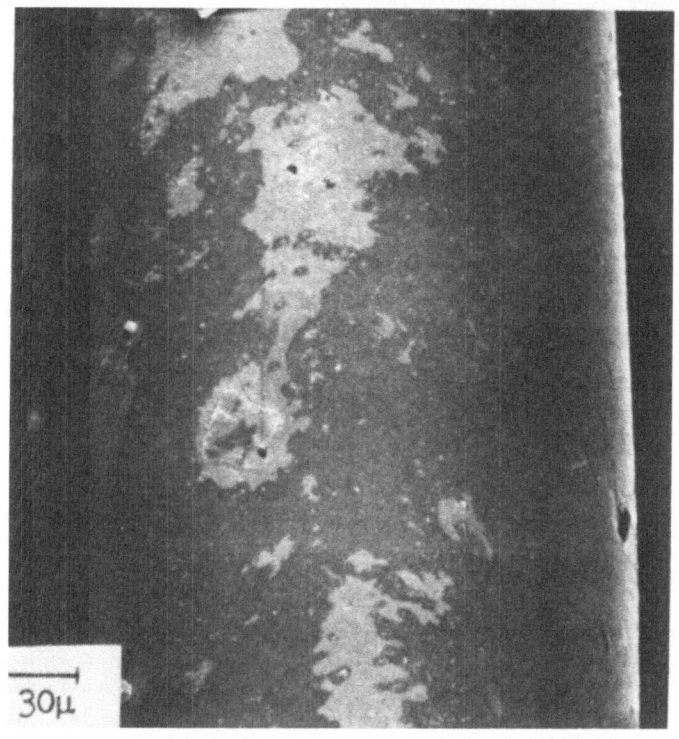

Figure 6.   Tyco Single Crystal Sapphire (Al$_2$O$_3$) Fiber.

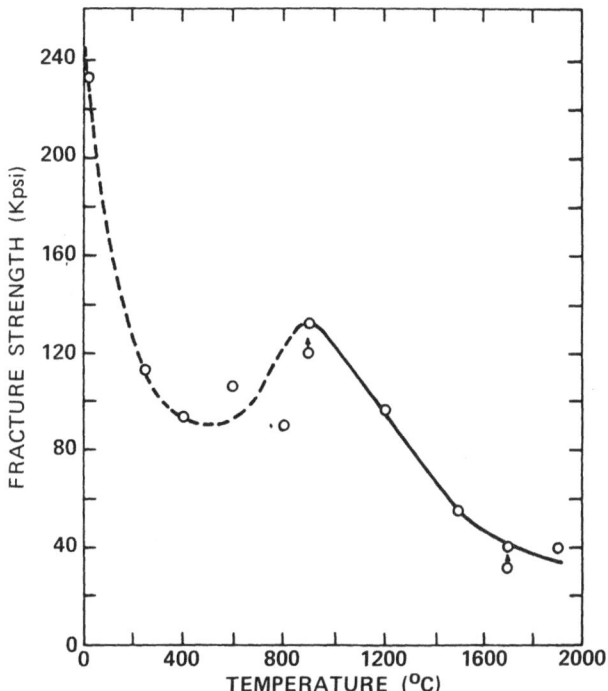

Figure 7. Tensile Fracture Strength of Sapphire Filament as
a Function of Temperature.

demonstrated (7). Arco Metals Company is producing silicon
carbide whiskers starting from rice hulls and using low cost
conventional metal-working technology to produce reinforced
aluminum products (8).

The continuous inorganic fibers such as silicon carbide,
alumina/silica, alumina, and silicon nitride produced from
direct spinning technology appear to have the best potential
based on properties, cost, composite fabricability, and environmental resistance. This inorganic fiber exploits the low
cost spinning technology of organic fibers combined with
advanced ceramic sintering techniques. For example, Nippon
Carbon Company of Japan produces silicon carbide fibers by
melt spinning polycarbosilane (9). The fibers develop their
strength via heat treatment and carbonization at 1200 - 1400
C. to produce β·SiC fibers.

ALUMINA FIBERS

Sumitomo Chemicals Company in Japan produces alumina/
silica fibers starting with organo-aluminum compounds (10)
dry spun and calcined at 1000 - 1200° C. to develop alumina/
silica fibers. Du Pont has developed essentially a 100%
alumina high modulus fiber called Fiber FP based on potentially low cost textile fiber spinning and advanced inorganic
sintering technology (11). A comparison between alumina/
silica and alumina fiber is shown in Figure 8. The effect
of silica addition on the tensile strength and the modulus
of alumina fibers is shown in Figures 9 and 10. It is seen
that silica addition to alumina increases the tensile strength
of alumina fibers but significantly reduces the fiber modulus.
The microstructures of Sumitomo alumina/silica fiber and
Du Pont alumina Fiber FP are shown in Figures 11 and 12.
Alumina/silica fibers have a smooth, glass like amorphous
structure having several ultra-fine crystalline regions.
Fiber FP essentially has a metal-like polycrystalline microstructure having an average grain size of about 0.5 micron.

The major advantages of 100% alumina fibers are: higher
modulus; higher temperature capability; long term stability
at high temperatures in metals (for example, Ni in Figure 13
and molten magnesium in Figure 14); and dense microstructure.
The advantages of alumina/silica fibers include higher
strength, less brittleness and better handleability.

CHARACTERISTIC	Al_2O_3/SiO_2	Al_2O_3
Composition (wt.)	~15% SiO_2 Balance Al_2O_3	>99% Al_2O_3
Density (gms/cc)	3.4	3.9
Tensile Strength (ksi)	230–370	225
Modulus (X 10^6 psi)	29–36	55
Elongation (%)	0.8–1%	0.4%
Diameter	10–15μ	20μ
Fiber Structure	Amorphous	Polycrystalline
Grain Size	Several 10Å crystallites 50Å crystallite regions	5000Å
Fiber Surface	Smooth Glassy	Rough Cobblestone

Figure 8. Comparison of Al_2O_3/SiO_2 and Al_2O_3 Fibers.

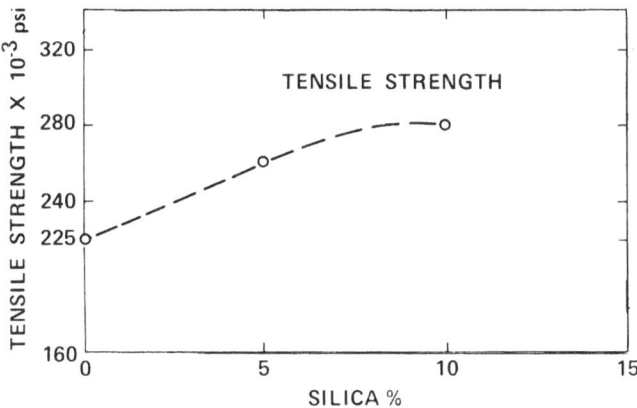

Figure 9. Effect of Silica on Tensile Strength of Alumina Fibers.

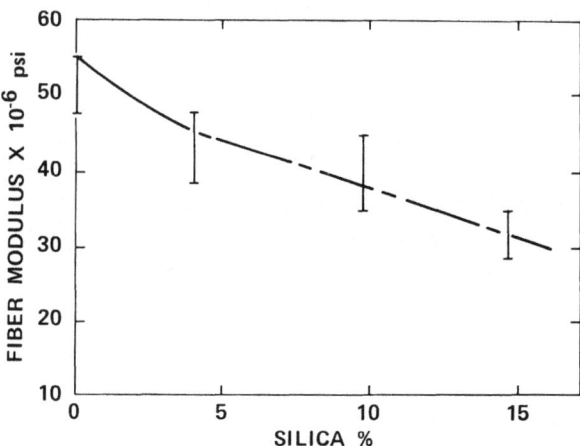

Figure 10.   Effect of Silica on Modulus of Alumina Fibers.

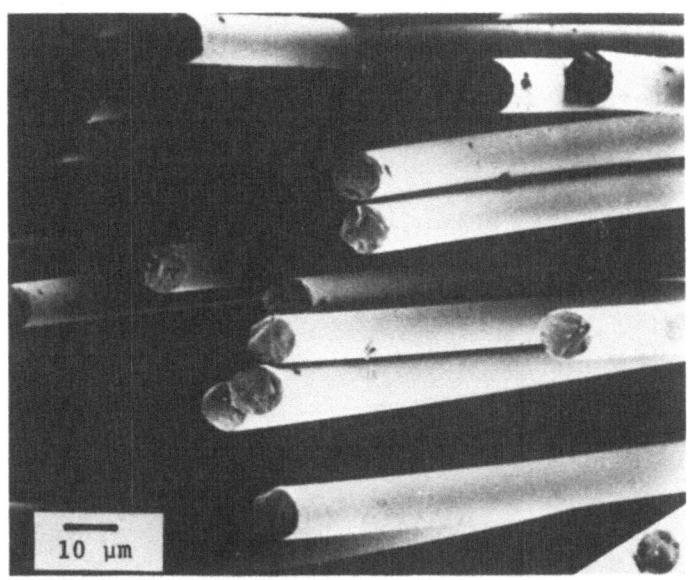

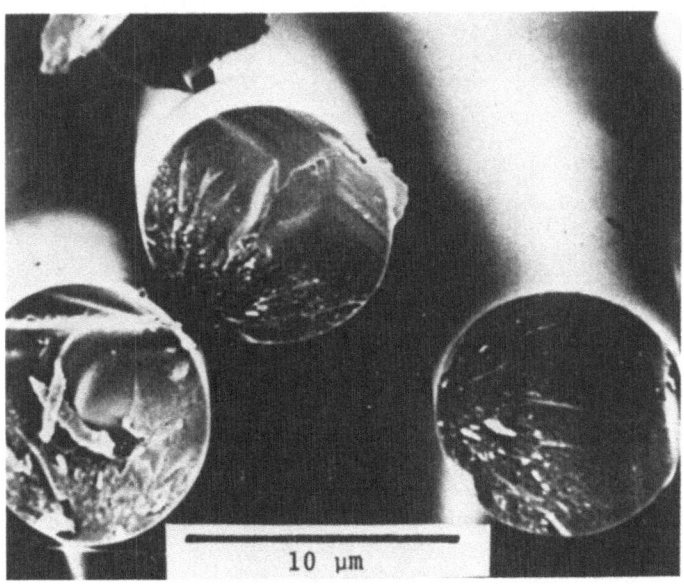

Fig. 11.   Scanning Electron Micrographs of Sumitomo's Alumina/
            Silca Fiber

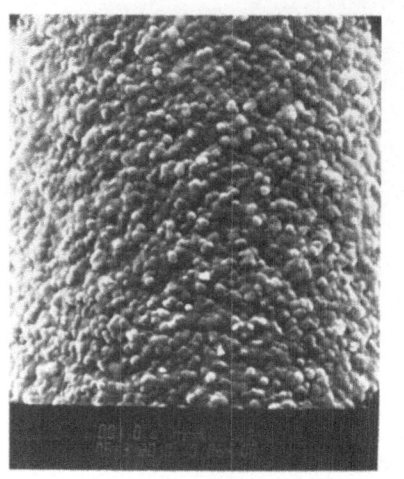

FIBER SURFACE

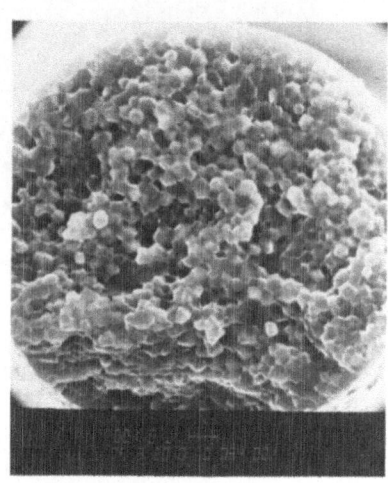

FRACTURE SURFACE

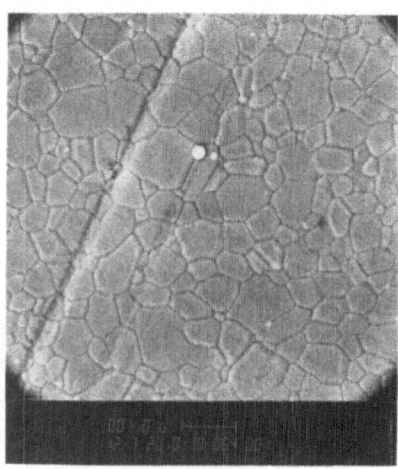

POLISHED & ETCHED SECTION

Fig. 12. Microstructure of Dupont Alumina Fiber FP

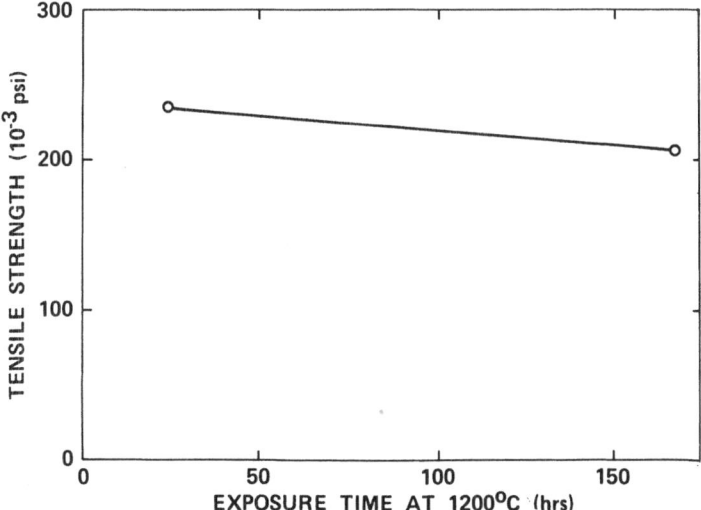

Fig. 13.   Tensile Strenght vs. Exposure Time of Fiber FP in
           Ni at 1200°C

FIBER CONDITION	TENSILE STRENGTH
Initial	1500 MPa (217 ksi)
Fibers Extracted From Casting (Casting Temp. 750°C)	1420 MPa (204 ksi)

Fig. 14. Retention of Fiber FP Strength in Magnesium Castings

COMPATABILITY WITH ADVANCED ALUMINUM-LITHIUM ALLOYS

    Advanced aluminum alloys based on aluminum lithium
(AL-Li) can be made 20% stronger and 10 to 15% lighter than
the best available commercial aluminum alloys (12). Alcoa
recently announced the development of three new aluminum-
lithium alloys called "Alithalite Alloys" which are expected
to reduce aluminum air frame weight from 10 to 15% when they
become commercially available in 1985 (13). Lockheed estimates an annual operational cost saving in excess of one-
million dollars per aircraft for long range aircraft using a
fuel factor of $1.00/gal. Advanced Al-Li alloys are there-
fore promising candidates for fiber reinforcement to further
improve their stiffness and performance. 100% alumina fiber
FP appears to be best suited for chemical and metallurgical
compatability with suitable Al-Li alloy matrices (14). At
proper Li concentrations and fabrication temperature reactive
Al-Li alloys wet dense microstructure Fiber FP without fiber
degradation (Figure 15). The addition of Li in advanced
aluminum alloys not only forms chemical bonding with relatively
inert alumina FP fiber but provides solid solution strength-
ening of potentially heat treatable Al-Li matrix with significant improvement in the transverse and shear strength of
composites. Commercially available, less stable, alumina/
silica fibers on the other hand show considerable fiber
degradation in aluminum-lithium alloys even at low lithium
concentrations (Figure 16). Fiber FP shows good bonding and
translation of fiber properties in FP/Al-Li composites as
shown by agreement between experimental data and theoretical
strength predictions at various fiber orientations in Figure
17.

 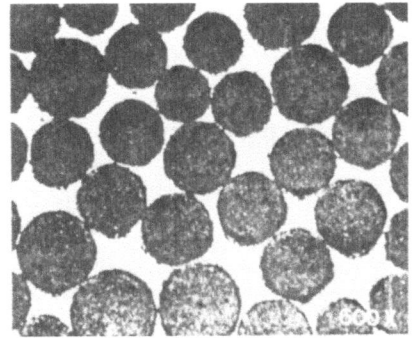

Poor Wetting Excessive Fiber-Matrix Inter-
Action Causing Loss in Fiber
Strength

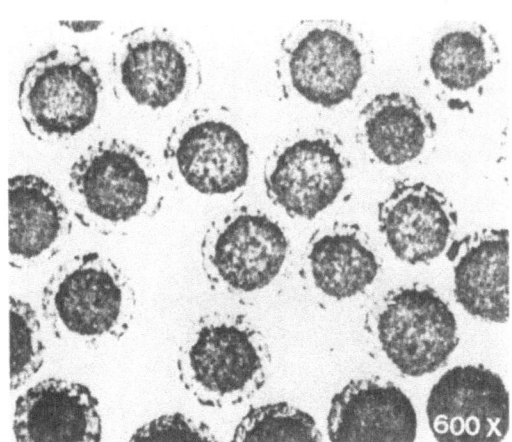

600X
Excellent Fiber-Matric Bonding
withMinimal Reaction with the
Fiber

Fig. 15. Wetting of Alumina Fiber FP
(Alumina-Lithium Alloys)

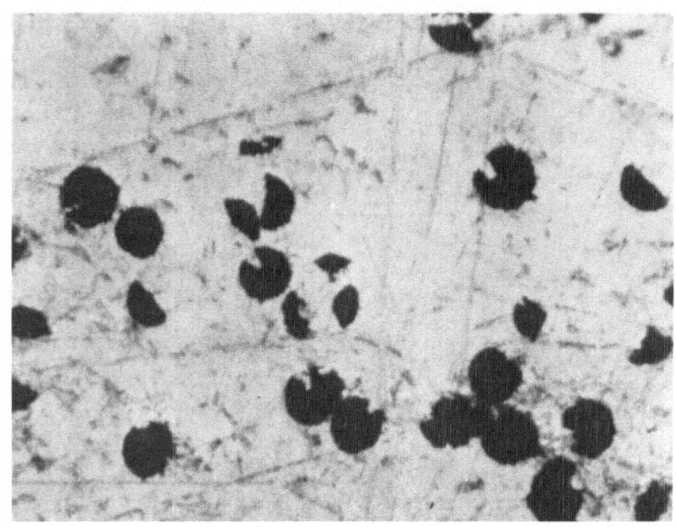

600X

1200X

Fig. 16.   Degradation of Alumina/Silica Fibers in Al-Li Alloys

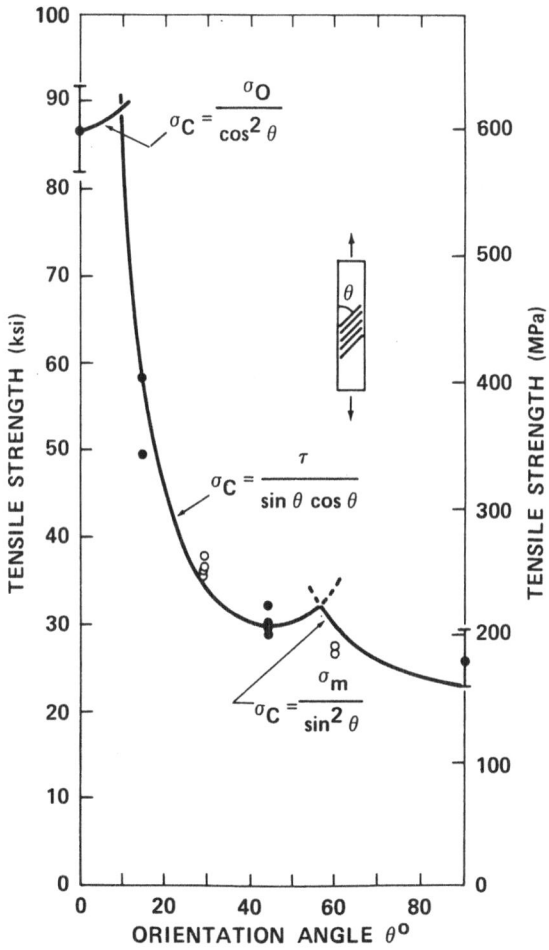

Fig. 17.   Effect of Fiber Orientation on Tensile Strength of
FP/Al Composites (vf=55%)

POTENTIAL APPLICATIONS

Inorganic fiber based composite materials offer potential for new applications in a variety of areas such as high temperature structures, lightweight, longer lasting automotive and jet engine materials, electronic structures, super conductors, fuel cells and battery plates. As an example, several new materials systems (FP/aluminum, FP/magnesium, FP/lead, FP Kevlar* 49 aramid hybrids, FP glass, and FP/ceramics) based on Fiber FP are being scouted for various applications (Figure 18). Some of the applications in the advanced development stage are helicopter transmission housings of FP/magnesium for Boeing-Vertol** (Figure 19); and battery plates of FP/pure lead (Figure 20). High purity lead is a desirable material for battery electrodes because it has higher corrosion resistance and service life than commercially available lead alloy. However, pure lead has low mechanical properties which can be significantly improved by even a small amount of Fiber FP reinforcement as shown in Figure 21 (15). Commercial potential for inorganic fiber based composite materials exists for automotive engine applications as shown by a few examples of automotive components in Figure 22. We have successfully demonstrated fabrication of complex FP/Mg and FP/aluminum connecting rods (Figures 23, 24, 25) which show 40 - 50% weight saving potential vs. steel (16).

TOYOTA DIESEL ENGINE PISTON APPLICATION

An important milestone in inorganic fiber reinforced metal matrix composite technology was reached recently when Toyota Motor Corporation announced the commercial production and use of the world's first metal matrix composite piston insert for the Toyota 2L Turbo Diesel Engine Car (17, 18) (Figure 26). The diesel piston insert made from commercial short alumina-silica fiber-reinforced aluminum matrix composite significantly improves wear, heat resistance and

---

*   Du Pont Registered Trademark

** Certain aspects of metal matrix composites technology fall under the purview of the United States Munition List and are subject to the export requirements of the Department of State as set forth in Munition Control Newsletter No. 74 of September, 1979.

cooling characteristics of diesel pistons compared to the incumbent cast iron insert (Figure 27). New material selection and substitution is based on several factors, including cost, and performance of competitive material systems. Figure 28 compares the performance of alumina-silica fiber-reinforced aluminum matrix composites with competitive ceramics and unreinforced aluminum alloys. Inorganic fiber reinforced metal matrix composites appear to have the best combination of properties for piston insert applications.

CONCLUDING REMARKS

In conclusion, high performance inorganic fiber based materials offer potential as the next generation of materials beyond fiber-reinforced plastics. However, successful development of inorganic fiber materials requires a systems approach to integrate design, material and processing from fibers to finished products. The development of an inorganic fiber system involves complex fiber-matrix interactions and fabrication technologies which will require strong inter-disciplinary R & D teams involving chemists, physicists, engineers, metallurgists and material scientists. Low cost inorganic fibers technology similar to glass and organic fibers will open new opportunities for large automotive and commercial materials applications beyond aerospace for the metals industry. The application of inorganic fiber based composite materials for diesel piston inserts should be considered the first of many in the large automotive market. I believe that such fibers will fill a need in the metals industry similar to that of fiberglass for the plastic industry.

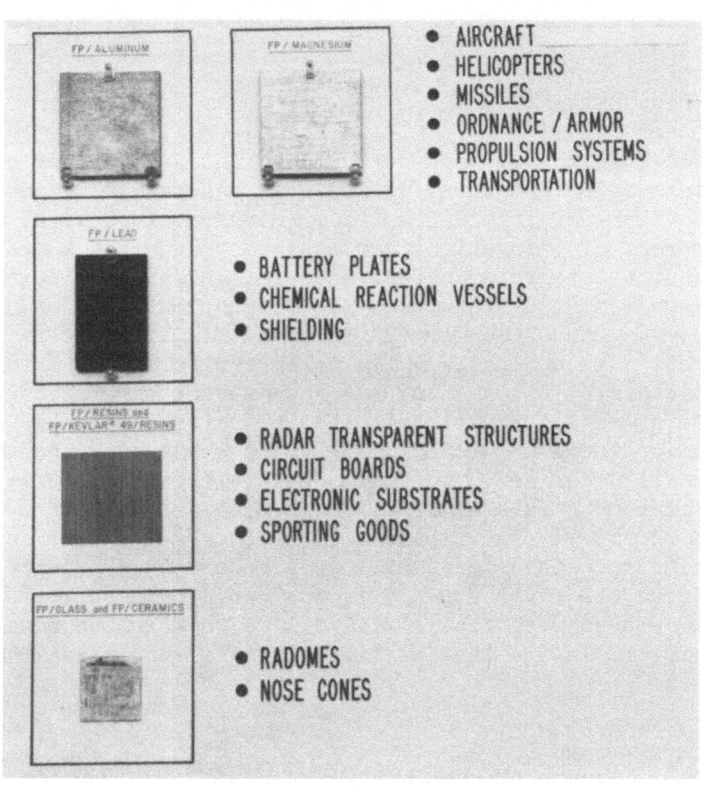

Fig. 18. Fiber FP Reinforced Materials
          Potential Applications

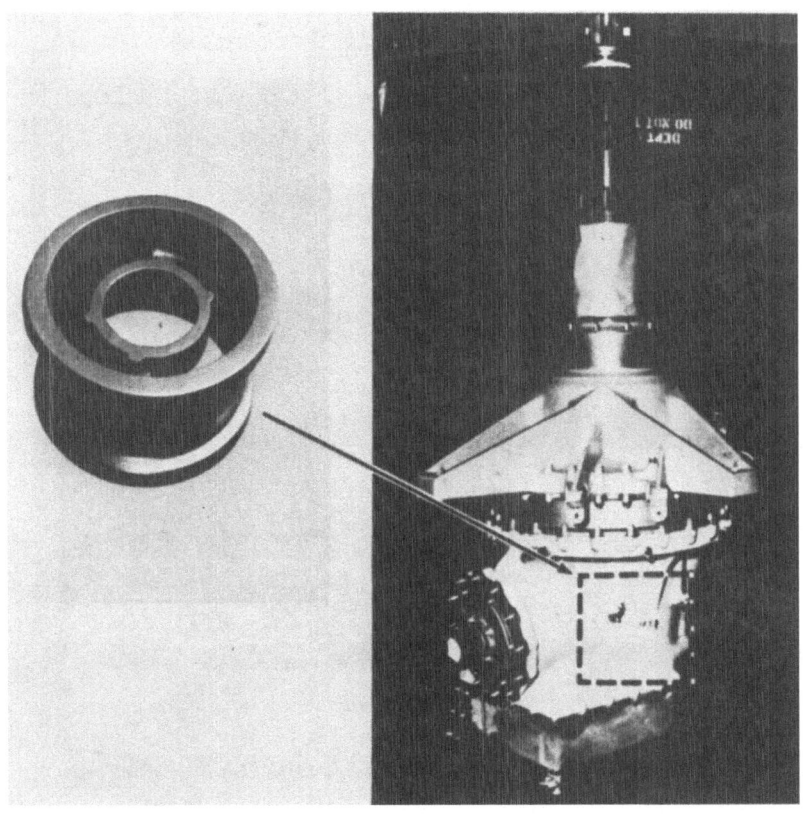

Fig. 19. Developmental Model and Boeing-Vertol Transmission
          Case

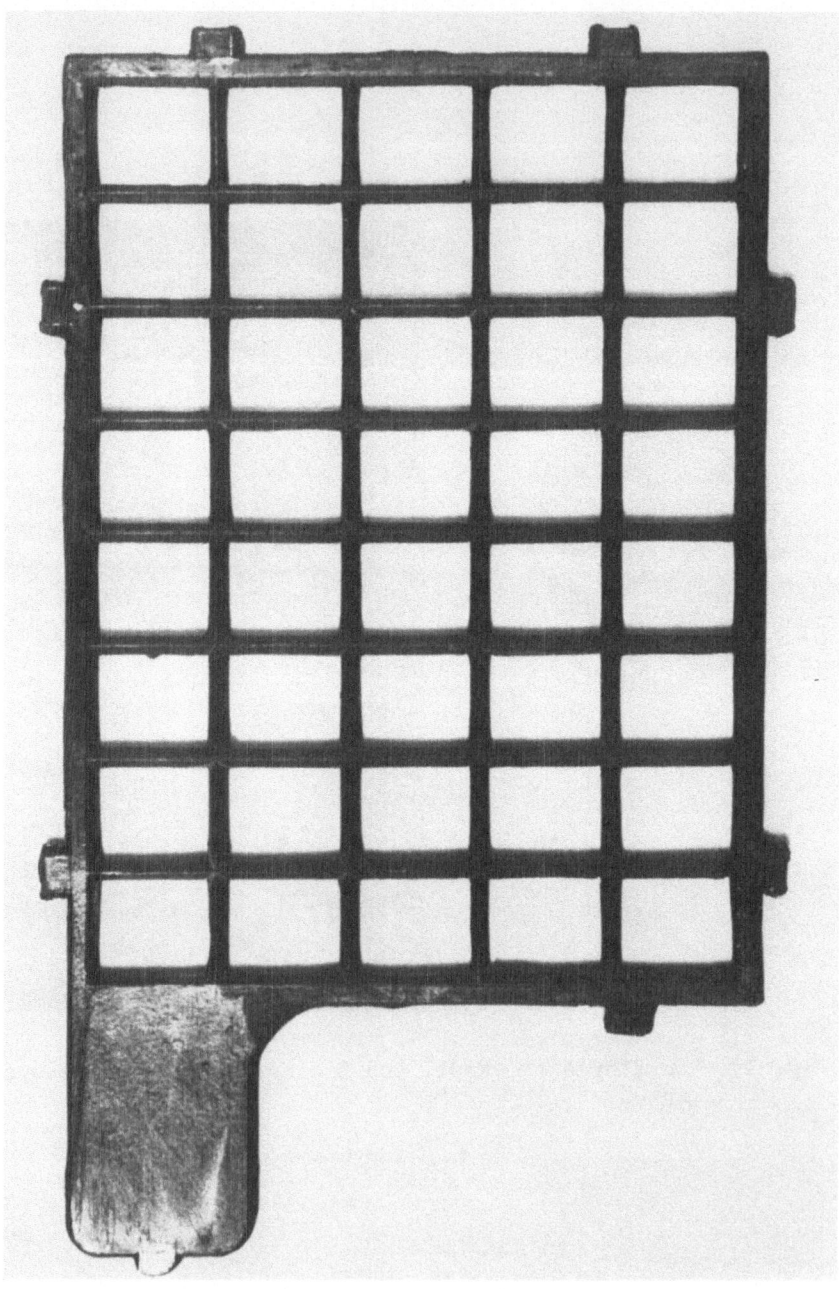

Fig. 20. FP/Pb Battery Plate

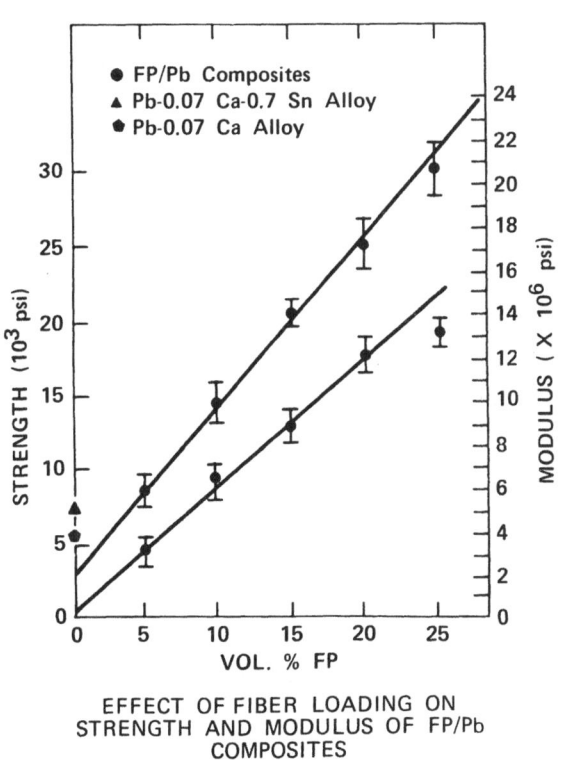

EFFECT OF FIBER LOADING ON
STRENGTH AND MODULUS OF FP/Pb
COMPOSITES

Fig. 21.   Tensile Strength of FP/Pb Composites

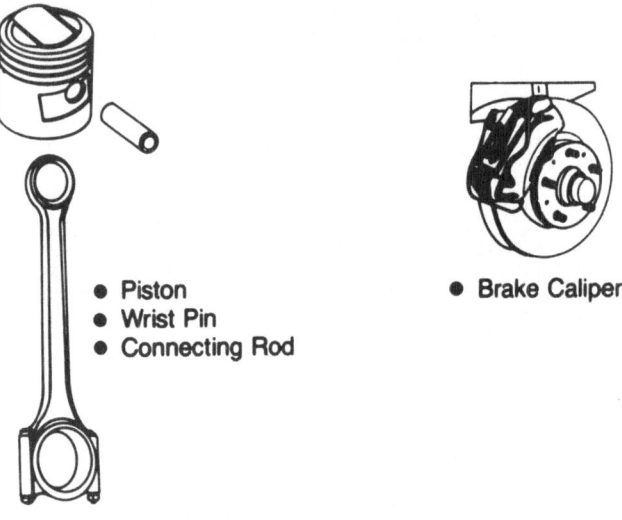

- Piston
- Wrist Pin
- Connecting Rod

- Brake Caliper

- Transmission Housing
- Gear Shift and Clutch Components
- Differential Housing

Fig. 22.   Potential Applications for FP/Al and FP/Mg in Automotive Components

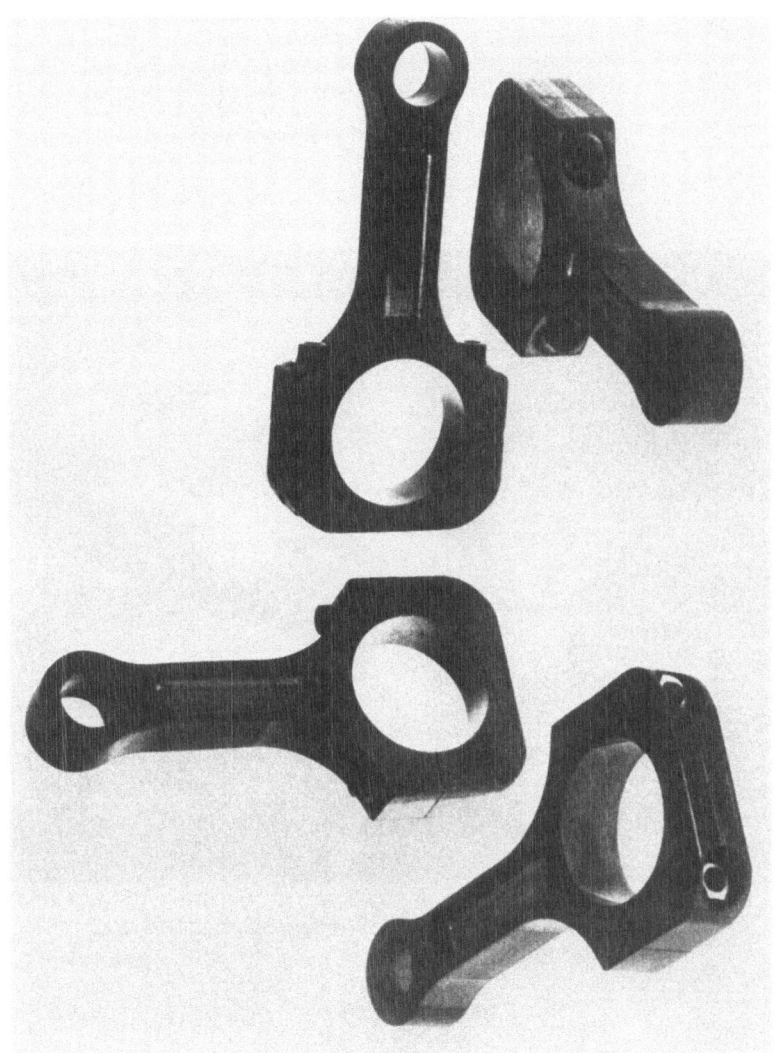

Fig. 23. Machined FP/Mg Connecting Rods

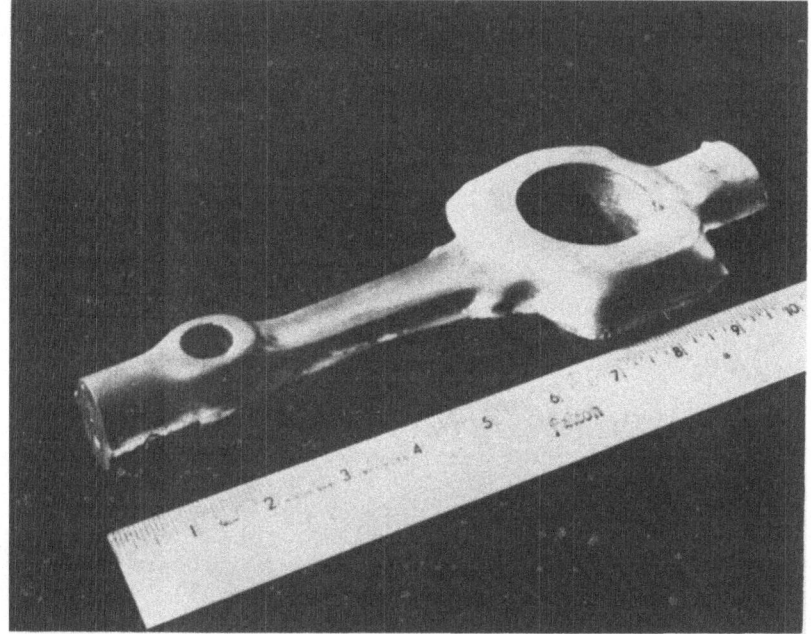

Fig. 24.   As-Cast Fiber FP Reinforced Magnesium Alloy Con-
           necting Rod

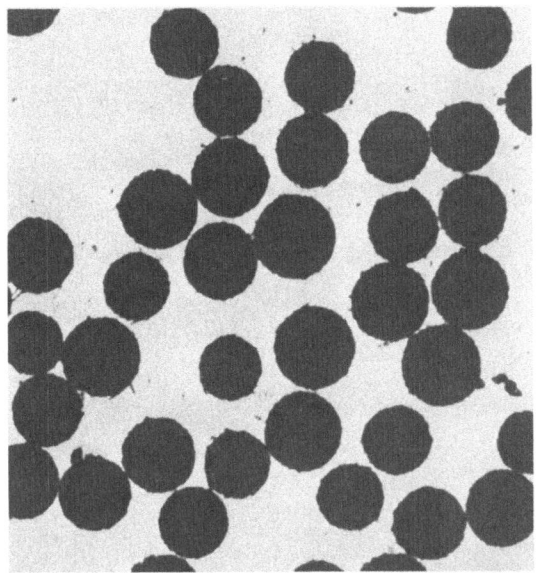

600X

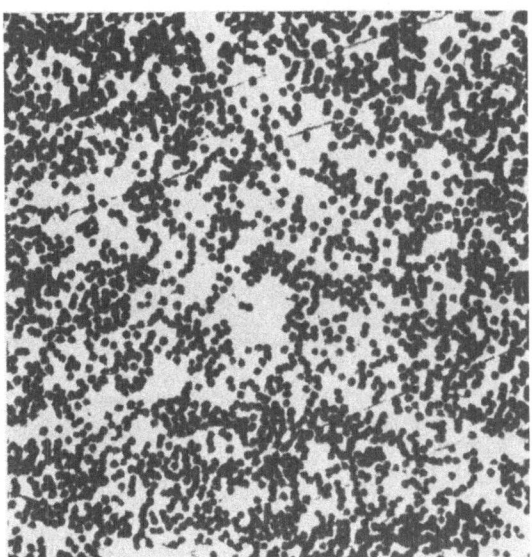

60X

Fig. 25. Typical Micrographs Showing Casting Soundness of Fiber
          FP Reinforced Magnesium Alloy Connecting Rod

- Engine type:

  OHC 4-cylinder
  diesel with
  türbocharger

- Total displacement:
  2,446 cc

- Compression ratio:
  20.0

- World's first metal
  matrix composite
  automobile parts,
  metal matrix
  composite piston.

Fig. 26.   Toyota 21 Turbo Diesel Engine

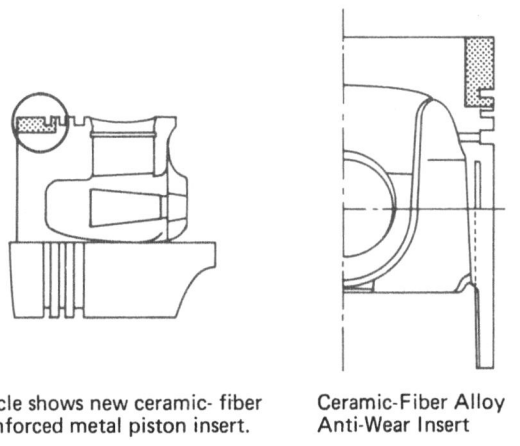

Circle shows new ceramic- fiber   Ceramic-Fiber Alloy
reinforced metal piston insert.   Anti-Wear Insert

**ALUMINA-SILICA INORGANIC FIBER REINFORCED
ALUMINUM PISTON INSERT IMPROVES ANTI-WEAR,
HEAT-RESISTANT AND COOLING CHARACTERISTICS
OF DIESEL PISTON COMPARED TO CONVENTIONAL
CAST IRON INSERT.**

(Automotive Engineering, October 1982)

Fig. 27.   World's First Fiber Reinforced metal (FRM) Motor
           Vehicle Part (TOYOTA)

Materials	Aluminum Alloys	Metal Matrix Composites	Ceramics
Toughness	□	□	X
Heat Radiation	O	O	X
Rigidity	Δ	O	□
Heat Resistance	Δ	O	□
Wear Resistance	Δ	□	□
Machinability	O	O	X

Key: □ = Excellent   0 = Good
        Δ = Fair   X = Poor

Fig. 28.   Comparison of Metal Matrix Composites (Piston
           Application) to Competitive Materials

REFERENCES

1. J. Persh, The Department of Defense Metal Matrix Composite Technology Thrust - Five Years Later, Proceeding of Fifth Metal Matrix Composites Technology Conference. P. xxiii, (May, 1983).

2. T. J. Miller and H. H. Grimes, Research on Ultra-High Temperature Materials - Monolithic Ceramics, Ceramic Matrix Composites and Carbon/Carbon Composite, Advanced Materials Technology, P. 275, NASA Conference Publication 2251, (Nov., 1982).

3. Composite and Rapid Solidification Technology Critical Materials Requirement of the U.S. Aerospace Industry, P. 186, U.S. Department of Commerce, (Oct., 1981).

4. Conservation and Substitution Technology for Critical Materials. Proceedings of Public Workshop sponsored by U.S. Department of Commerce/National Bureau of Standards and U.S. Department of the Interior/Bureau of Mines, June, 1981).

5. Arthur F. McLean, Ceramics in Small Vehicular Gas Turbine, Proceedings of the Second Army Materials Technology Conference, P. 12, (Nov., 1973).

6. G. F. Hurley, Harold E. LaBelle, Jr., A. I. Mlavsky, Mechanical Properties of High Strength, High Modulus to Density Continuous Alumina Filaments for Use as a Reinforcement in Structural Composites, AD 845 110/6, (Sept. 1968).

7. James Economy, Inorganic Refractory Fibers, Kirk Othmer Encyclopedia of Chemical Technology, Second Edition, Volume II, P. 654.

8. Charles B. Criner, H. J. Rack, Scale Up and Fabrication of Silicon Carbide Reinforced Aluminum Cylinders, Arco Metals Company, Silag Operation.

9. S. Yajima, K. Okamura, T. Matsuzawa, Continuous Silicon Carbide Fiber Reinforced Aluminum, P. 232, Japan - U.S. Conference on Composite Materials, Tokyo, (1981).

10. Y. Abe, S. Horikiri, K. Fujimura, E. Ichiki, High Performance Alumina Fiber and Alumina/Aluminum Composite, Proceedings of the Fourth International Conference on Composite Materials, P. 1427, Tokyo, (Oct. 1981).

11. A. K. Dhingra, Philosophical Transactions of the Royal Society, London, A294, 411, (1980).

12. John C. Bittence, New Aluminum Alloys Sighted on Horizon, Materials Engineering, P. 32, (June 1980).

13. New Aluminum Lithium Alloys to Reduce Aluminum Air Frame Weight by Up to 15 Percent, Alcoa News Release, (May, 1983).

14. A. K. Dhingra, Philosophical Transaction of the Royal Society, London 294, 559, (1980).

15. Hans S. Hartmann and Raymond S. Sutula, Journal of Electrochemical Society, Vol. 129. No. 8, P. 1749, (1982).

16. W. H. Krueger and A. K. Dhingra, New Composite Material and Technology, P. 13, American Institute of Chemical Engineers, Vol. 78, (1982).

17. Todashe Donomoto, Society of Automotive Engineers, International Congress and Exposition, Detroit, SAE #830252, (March, 1983).

18. Toyota Motor Corporation Exhibits, Fourth International Conference on Composite Materials, Tokyo, (Oct. 1982).

SYNTHESIS AND PROPERTIES OF POLYETHERIMIDE POLYMERS

H.M. Relles

General Electric Corporate Research and Development
Schenectady, New York 12301

SYNOPSIS

Polyetherimides are a new class of high performance polymers whose chemical compositions can be tailored to provide materials having glass transition temperatures ranging from ~150°C to ~300°C. Their outstanding thermo-oxidative stability and flame and solvent resistance are characteristic of aromatic polyimides but, unlike many other polyimides, they can also be processed by conventional techniques (e.g., injection molding). The key reactions in their various syntheses were nucleophilic aromatic nitrodisplacements by aryloxide ions. This and other chemical steps are discussed in detail along with a description of some of the physical properties of the polymers.

INTRODUCTION

The high performance characteristics of polyetherimide polymers and our desire to prepare them by an economically attractive scheme led us to explore several synthetic approaches. The key reaction in all of these was the formation of the diaryl ether linkage by a nucleophilic aromatic displacement reaction on suitably substituted (and activated) phthalic derivatives. Chloride displacement[1] proceeded relatively slowly and gave polymers having only modestly high molecular weights. Fluoride displacement[2] was quite facile, but processes based on the fluoro-deriviatives were economically unattractive. The synthetic and economic accessibility of appropriate nitro-substituted phthalic derivatives[3] and the high rates[4] of displacement of nitro groups by aryloxide ions led to a substantial effort to produce the polyetherimides using this reaction. Two approaches evolved as the best and were explored in depth. Each incorporated a nitro-displacement reaction but at different stages in the synthethic sequences. These approaches, "Nitro-Displacement Polymerization" and "Imidization Polymerization", are discussed in detail below.

NITRO-DISPLACEMENT POLYMERIZATION

In this method, bis-nitro-bis-imide monomers, 1, were produced from 3- or 4-nitrophthalic anhydride and a diamine in refluxing acetic acid. A number of these monomers are listed in Table I[5,6]. They were all produced in very high yield.

Bis-phenoxide salts, 2, used in the actual polymerizations with 1, were prepared in anhydrous form generally through the use of an azeotropic drying technique starting with the bis-phenol and an alkali hydroxide. Several of the starting phenols are shown in the first column of Table II.

The polymerization reactions themselves are carried out in dipolar aprotic solvents or mixtures of these with other non-interfering solvents such as toluene and chlorobenzene. The nitro-displacement reactions are very rapid and lead to high molecular weight polymers, 3.[5,6,7]

Table I.   Bis-Nitro-Bis-Imides $\underline{1}$

R	Isomer	Yield(%)	MP (°C)
	3	93.4	338–342
	4	94.7	358 Dec
	3	97.8	306–308
	4	96.5	322–323
	3	91.8	270–272
	4	96.4	317–320
	3	95.4	274–275
	4	98.2	278–279
—(CH$_2$)$_6$—	3	94.9	196–197
	4	95.9	212–213

Table II. Nitro-Displacement Polymerization

HO-AR-OH (2)	-R-	NO$_2$ Position	Polym. Solvent	Yield %	[n] (dl/g)	Tg (°C)
HO—⬡—OH (resorcinol)	4,4'Diphenyl-Ether	3,3'	DMF	100	016[a]	226
"	"	4,4'	"	79.1	<0.1[a]	–
HO—⬡—O—⬡—OH	"	3,3'	"	100	0.80[a]	239
"	"	4,4'	"	95.5	0.44[b]	215
"	(-CH$_2$-)$_6$	3,3'	DMSO	97.7	0.29[d]	128
HO—⬡—S—⬡—OH	4,4'-Diphenyl-Ether	3,3'	DMAc	100	0.37[c]	234
"	1,3-Benzene	3,3'	"	100	016[a]	209
HO—⬡—⬡—OH	4,4'-Diphenyl-Ether	3,3'	DMSO	95	0.60[e]	277

Column groups: "Bisimides 1" spans -R- and NO$_2$ Position; "Polymer" spans [n] (dl/g) and Tg (°C).

Table II (Continued)

HO-Ar-OH (2)	-R-	NO$_2$ Position	Polym. Solvent	Yield %	[η] (dl/g)	Tg (°C)
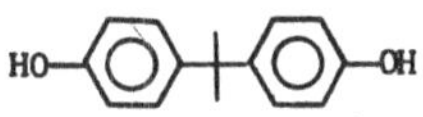 HO—⟨benzene⟩—C(CH$_3$)$_2$—⟨benzene⟩—OH	4,4'-Diphenyl-methane	3,3'	DMSO	100	0.28^d	230
"	4,4'-Diphenyl-Ether	3,3'	"	99.5	0.26^a	226
"	"	4,4"	"	98	0.23^d	196
"	4,4'-Diphenyl-sulfone	3,3'	"	79.5	0.13^a	-
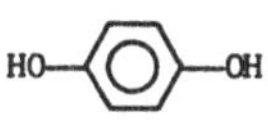 HO—⟨benzene⟩—OH	4,4'-Diphenyl-Ether	3,3'	"	76	0.45^e	237

^{a} Measured in DMF

^{b} Measured in H$_2$SO$_4$

^{c} Measured in CHCl$_3$

^{d} Measured in CH$_2$Cl$_2$

^{e} Measured in NMP

A number of specific examples are listed in Table II. These illustrate the versatility of the polymerization in terms of the bis-phenols, diamines (in the bis-nitro-bis-imides), and the possible positional isomers that were employed.

One problem associated with polymers made in this way is that they contain phenolic and aromatic-nitro end groups, **4** and **5**, which adversely affect their thermo-oxidative stability. In addition, one constraint on this synthetic approach is the critical requirement for extremely low levels of water in the polymerization reaction mixtures. Any water present under these conditions rapidly converts a nitro-imide to the corresponding nitro-amic-acid salt, **6**, from which the nitro-group cannot be displaced.[8] This limits chain growth and presents definite reproducibility problems.

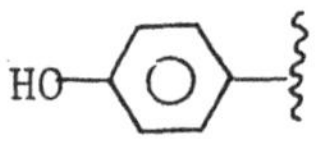

4

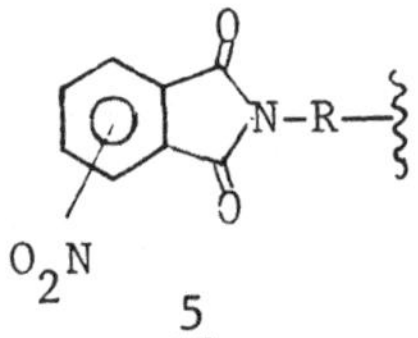

5

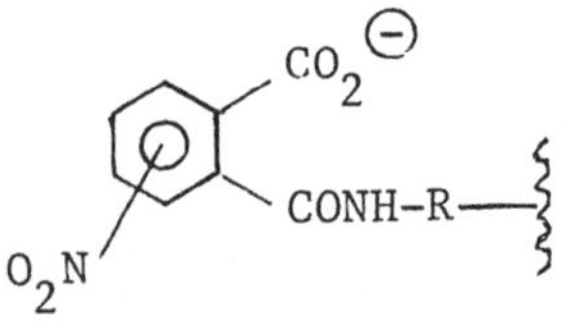

6

IMIDIZATION POLYMERIZATION

In this method, bis-ether-dianhydride monomers are prepared by a series of steps in which the key is again an aromatic nucleophilic nitro-displacement. Nitro-phthalic anhydrides themselves could not be used directly for the displacement reactions since aryloxides preferentially reacted rapidly at the carbonyl centers to give ring-opened ester-acid salts. This not only consumed aryloxide ions but it simultaneously generated materials from which the nitro-group could not be displaced and which underwent numerous undesirable side reactions[9].

However,   nitro-N-alkylphthalimides,   7   ("protected"
nitro-phthalic anhydrides), when treated with bis-phenoxide
salts, 2, underwent smoothe displacement to produce bis-
ether-bis-imides, 8. These can be readily purified to remove
all of the types of impurities which, in the prior method,
were built into the polymer as endgroups (see 4, 5, and 6
above). The bis-imides can then be converted to bis-ether-
dianhydrides, 9, by hydrolysis with base and neutralization,
to generate tetra-acid intermediates, and ring closure with
acetic anhydride.[7,10-12] A variety of these dianhydrides
are listed in Table III. In general, they were readily
obtained in pure form and in quite high yield with the
approach discussed.

Treatment of the dianhydrides 9 with the various diamines gave polyetherimides 3 in very high yield. These reactions, were carried out in solution between 140-180° or in
the melt at 260-280°.[5-7,10-12] A large number of examples of
polyetherimides produced in this way are displayed in Table
IV and are arranged in order of increasing Tg. One general
rule that emerges from this data is that for a given
bisphenol/diamine pair, the Tg of the polymer containing the
phthalimide-ring 3-isomer is always greater than that containing the 4-isomer (polymer numbers, ΔTg($^{\circ}$C) given): 13
and 8, 12°; 10 and 4, 17°; 15 and 9, 17°; 18 and 14, 26°; 20
and 16, 28°; 21 and 11, 48°. Overall, this tabulation indicates that, by the judicious choice of bisphenol, diamine,
and phthalimide-ring isomer, polyetherimides can be prepared
having any desired Tg.

With this polymerization scheme, end-groups can be
readily converted to thermo-oxidatively stable imide
moieties; anhydrides, by treatment with aniline, and amines,
by treatment with phthalic anhydride, give N-phenyl ether-
imide and N-aryl phthalimide end-groups, respectively. The
presence of small amounts of moisture is of no consequence
in these polymerizations since it is simply evolved along
with the rest that is formed during the imidization.

ULTEM^{R} POLYETHERIMIDE RESIN

Among all of the polyetherimide materials examined, one
turned out to be the best from the standpoint of physical
properties, environmental resistance, processability (especially injection moldabililty), and the commercial attrac-
tiveness of a process by which it could be made. This

Chart 1

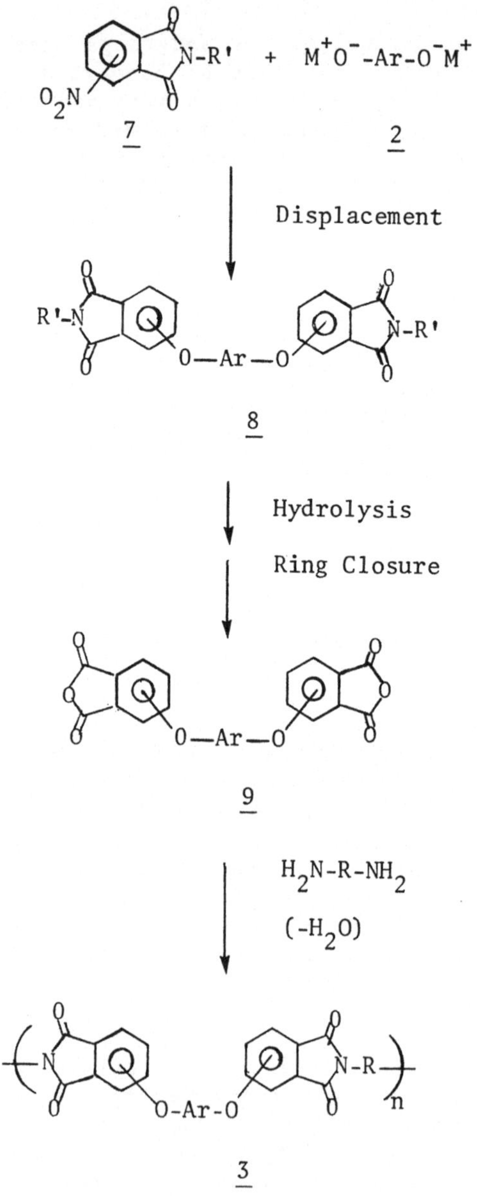

Table III. Bis-Ether-Dianhydrides 9

-Ar-	Isomer	Yield (%)	mp(°C)
1,3-Benzene-	3	100	228-229.5
	4	89.9	284.4-286
1,4-Benzene-	3	98.0	306-307
	4	91.2	265-266
4,4'-Biphenyl-	3	88.9	280-281
	4	100	285-286.5
4,4'-Diphenyl Ether-	3	98.9	254-255.5
	4	100	238-239
4,4'-Diphenylsulfide-	3	46.6	257-257.5
	4	97.0	189-190
4,4'-Diphenylsulfone-	3	57.9	230.5-231.5
	4	99.6	251.5-252
4,4'-Benzophenone-	3	59.0	278-279
	4	70.5	215-216
4,4'-(2,2-Diphenylpropane)-	3	95	186.5-187.5
	4	95	189-190

Table IV. Polyetherimides

POLYMER NUMBER	—Ar—	—R—	ISOMER	POLYMERIZATION	η(dl/g)	Tg(°C)
1	HO–⟨⟩–S–⟨⟩–OH	H₂N–⟨⟩–O–⟨⟩–NH₂ type	4	m-CRESOL / TOLUENE	0.60	178°
2	HO–⟨⟩–O–⟨⟩–OH		4	"	1.10	184°
3	HO–⟨⟩(OH) (resorcinol-type)		3	"	0.65	193°
4	HO–⟨⟩(OH) (catechol/resorcinol-type)		4	"	0.67	209°
5	HO–⟨⟩–S–⟨⟩–OH		4	"	0.45	209°
6	HO–⟨⟩–S–⟨⟩–OH		4	"	1.02	212°
7	HO–⟨⟩–O–⟨⟩–OH		4	"	0.97	215°
8	HO–⟨⟩–C(CH₃)₂–⟨⟩–OH	H₂N–⟨⟩–NH₂ type	4	MELT[a]	0.44[a]	218°

POLYMER NUMBER	—Ar—	—R—	ISOMER	POLYMERIZATION	η(dl/g)	Tg(°C)
9	*(structure)*	*(structure)*	4	m-CRESOL / TOLUENE	0.70	224°
10	*(structure)*	*(structure)*	3	"	0.44	226°
11	*(structure)*	*(structure)*	4	"	0.83	229°
12	*(structure)*	*(structure)*	3	PHENOL / TOLUENE	0.34	234°
13	*(structure)*	*(structure)*	3	MELT [a]	0.47 [a]	230°
14	*(structure)*	*(structure)*	4	m-CRESOL / TOLUENE	1.71	237°
15	*(structure)*	*(structure)*	3	"	0.57	241°
16	*(structure)*	*(structure)*	4	"	0.51	247°

(continued)

Table IV. Polyetherimides (Continued)

POLYMER NUMBER	—Ar—	—R—	ISOMER	POLYMERIZATION	η(dl/g)	Tg(°C)
17	HO—⬡—OH	H₂N—⬡—NH₂ (diamino)	4	m-CRESOL / TOLUENE	0.96	255°
18	HO—⬡—OH	H₂N—⬡—O—⬡—NH₂	3	"	2.25	263°
19	HO—⬡—SO₂—⬡—OH	H₂N—⬡—NH₂	4	"	0.70	265°
20	HO—⬡—⬡—OH	H₂N—⬡—NH₂	3	"	0.56	275°
21	HO—⬡—⬡—OH	H₂N—⬡—O—⬡—NH₂	3	"	1.70	277°
22	phenolphthalein-type bis-phenol	H₂N—⬡—NH₂	4	"	0.35	281°

[a]Phthalic anhydride added to control the molecular weight; see reference 4c.

material was introduced commercially in 1982 by the General Electric's Plastics Business Group and is being marketed under the name of ULTEM[R]. Its structure is shown below. In brief, some physical properties of this new material are: $Tg:217^\circ C$; H.D.T. at 264 p.s.i.: $200^\circ C$; tensile strength:15,200 p.s.i.; flexural modulus: 480,000 p.s.i.; Gardner impact: 320 in-lb.; oxygen index:47; specific gravity:1.27; water absorption:0.25%. A more complete tabulation of the physical properties of this new polymeric material is given in Table V.

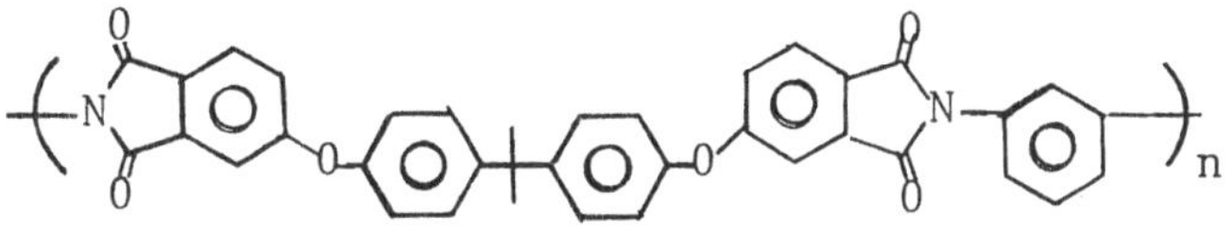

ULTEM[R] Polymer

CONCLUSION

Polyetherimides are a new family of condensation polymers. The key reaction step in each of their synthetic sequences is an aromatic nitro-displacement reaction which produces the diaryl ether linkages in high yield. Physical properties can be varied over a wide range depending on the choices of bis-phenol, diamine, and positional isomer incorporated into the backbone of the polymer. Our study of these materials has led to the commercial introduction of ULTEM[R] Resin as the first in a series of new high performance engineering thermoplastics.

EXPERIMENTAL

The experimental conditions for monomer and polymer syntheses by either of the approaches discussed above are given in detail in numerous published sources.[5,10-14] All structures were rigorously proven by elemental analysis and various spectroscopic techniques. The following polymerizations are included herein by way of example.

NITRO-DISPLACEMENT POLYMERIZATION

The BPA-dianion 11 was prepared from 77.84g (0.3409 m) BPA 10 and 54.56 g 50% aqueous NaOH solution (0.6818 m) in 680 ml DMSO and 200 ml toluene and then dried by azeotropic removal of water. The BPA dianion mixture was cooled to 55°C, 12 was added (187.0 g, 0.341 m) through a nitrogen bypass, and the system was stirred at 40°C for 2 h. Toluene, which had been dried by stirring over CaH_2, was added to facilitate stirring; a total of 350 ml of toluene was added during the reaction. The mixture was then cooled to ca. 25°C and 20 ml of glacial acid was added. The polymer was precipitated by addition to methanol in a 1-gal Henschel blender.

The polymer was washed with methanol, dried _in vacuo_ at 60°C, and reprecipitated by redissolving in $CHCl_3$ (containing some HOAc) and then adding to methanol. It was then dried as above. Yield of polymer 13 as a white powder was 222.0 g (94.5%); [η](25°C, DMF), 0.48 dl/g.

IMIDIZATION POLYMERIZATION

An intimate mixture of 3.000 g (0.00577 m) of the dianhydride 14 and 1.1418 g (0.00577 m) of the diamine 15 was heated at 290°C for one-half hour under nitrogen and for 1-1/4 hours in vacuo. The yield of amber, glassy polymeric product 16 was 4.0 g and it had an intrinsic viscosity of 0.65 dl/g (25°C, DMF).

14   15

16

Table V. Typical Properties of ULTEM^R Resin

MECHANICAL	ASTM Test	Metric Units	ULTEM 1000	English Units	ULTEM 1000
Tensile Strength, Yield	D638	N/mm^2	105	psi	15,200
Tensile Modulus, 1% Secant	D638	N/mm^2	3,000	psi	430,000
Tensile Elongation, Yield	D638	%	7-8	%	7-8
Tensile Elongation, Ultimate	D638	%	60	%	60
Flexural Strength	D790	N/mm^2	145	psi	21,000
Flexural Modulus, Tangent	D790	N/mm^2	3,300	psi	480,000
Compressive Strength	D695	N/mm^2	140	psi	20,300
Compressive Modulus	D695	N/mm^2	2,900	psi	420,000
Gardner Impact	-	N-m	36	in-lb	320
Izod Impact	D256				
notched (1/8")		J/m	50	ft-lb/in	1.0
unnotched (1/8")		J/m	1,300	ft-lb/in	25
Shear Strength, Ultimate	-	N/mm^2	100	psi	15,000
Rockwell Hardness	D785	-	M109	-	M109
Taber Abrasion (CS 17,1 kg)	D1044	mg wt.loss/ 1000 cycles	10	mg wt.loss/ 1000 cycles	10

THERMAL	ASTM Test	Metric Units	ULTEM 1000	English Units	ULTEM 1000
Deflection Temperature, Unannealed	D648				
@ 264 psi(1/4")		^{o}C	200	^{o}F	392
@ 66 psi (1/4")		^{o}C	210	^{o}F	410
Vicat Softening Point, Method B	D1525	^{o}C	219	^{o}F	426
Continuous Service Temperature					
Index (UL Buletin 746B)	–	^{o}C	170	^{o}F	338
Coefficient of Thermal Expansion					
(0 to 300°F), Mold Direction	D696	m/m–^{o}C	5.6×10^{-5}	in/in–^{o}F	3.1×10^{-5}
Thermal Conductivity	C177	W/m–^{o}C	0.22	Btu-in/h–ft^{2}–^{o}F	1.5
FLAMMABILITY					
Oxygen Index (0.060")	D2863	%	47	%	47
Vertical Burn (UL Bulletin 94)	–	–	V-0 @ 0.64mm	–	V-0 @ 0.025"
			5V @ 1.9mm		5V @ 0.075"
NBS Smoke, Flaming Mode (0.060")	E662				
D_s @ 4 min		–	0.7	–	0.7
D_{Max} @ 20 min		–	30	–	30

(continued)

Table V. (Continued)

ELECTRICAL	ASTM Test	Metric Units	ULTEM 1000	English Units	ULTEM 1000
Dielectric Strength (1/16")	D149				
in oil		kV/mm	28.0	V/mil	710
in air		kV/mm	33.0	V/mil	831
Dielectric Constant @ 1 kHz, 50% RH	D150	–	3.15	–	3.15
Dissipation Factor	D150				
1 kHz, 50% RH, 73°F		–	0.0013	–	0.0013
2450 MHz, 50% RH, 73°F		–	0.0025	–	0.0025
Volume Resistivity (1/16")	D257	ohm-m	6.7×10^{15}	ohm-cm	6.7×10^{17}
Arc Resistance	D495	seconds	128	seconds	128
OTHER					
Specific gravity	D792	–	1.27	–	1.27
Mold shrinkage	–	%	0.5-0.7	%	0.5-0.7
Water absorption	D570				
24 hours, 73°		%	0.25	%	0.25
equilibrium, 73°F		%	1.25	%	1.25

REFERENCES

1. JG Wirth and DR Heath, U.S. Patent 3,787,364, January 22, 1974.
2. FJ Williams, U.S. Patent 3,847,869, November 12, 1974.
3. NC Cook and GC Davis, U.S. Patent 3,933,852, January 20, 1976.
4. FJ Williams and PE Donahue, J.Org.Chem., 42, 3413 (1977).
5. DM White, T. Takekoshi, HM Relles, FJ Williams, et al., J. Polymer Science, 19, 1635 (1981).
6. T. Takekoshi, WP Hilling, GA Mellinger, JE Kochanowski, JS Manello, MJ Webber, RW Bulson and JW Nehrich, NASA CR-145007, July 1975.
7. T. Takekoshi, GA Mellinger, RW Bulson, JR Ladd and MJ Webber, NASA CR 145237, September 1977.
8. HM Relles, DS Johnson, and BA Dellacoletta, J. Org. Chem., 45, 1374 (1980).
9. FJ Williams, HM Relles, PE Donahue, and JS Manello, J. Org. Chem., 42, 3425 (1977).
10. DR Heath and JG Wirth U.S. Patent, 3,847,867, Nov. 12, 1974.
11. T. Takekoshi and JE Kochanowski, U.S. Patent 3,905,942, Sept. 16, 1975.
12. T. Takekoshi and JE Kochanowski, U.S. Patent 3,989,670, Nov. 2, 1976.
13. T. Takekoshi and JE Kochanowski, U.S. Patent 3,803,085, April 9, 1974.
14. JG Wirth and DR Heath, U.S. Patent 3,838,097, September 24, 1974.

REFLECTIONS ON THE DESIGN, THE STRUCTURE AND THE

PROPERTIES OF HIGHLY CONDUCTING POLYMERS AS

RELATED TO OTHER ORGANIC METALS

G. Wegner

Institut für Makromolekulare Chemie
Universität Freiburg
Stefan-Meier-Straße 31
D-7800 Freiburg, West Germany

SYNOPSIS

Structural requirements for the design and
synthesis of an organic metal are discussed using
the cation-radical salts of arenas, naphthalene
being an example. These cation radical salts serve
as models for conducting polymers. The chemistry
of the "doping" of polymers to the conductive state
is discussed in terms of redox reactions and redox
equilibria. The structure of oxidized poly(acetylene) is discussed based on available X-ray data
and in analogy to the cation radical salts of
arenes. A chain-chain-interaction is postulated to
play an important role in the conductivity mechanisms. The oxidation ("doping") is linked to a
phase transition by which regions of high conductivity in a less conductive matrix are formed.
Model compounds for poly(acetylene) and oxidized
poly(acetylene) are discussed with regard to the
explanation of the spectral features observed in
the conducting polymer.

INTRODUCTION

Polymers with rather high, metal-like conductivity can now be readily prepared. This is certainly one of the most interesting developments in modern polymer science. Although the search for conducting or even superconducting polymers has

attracted quite many scientists since quite some
time ago it was not until Heeger, Mac Diarmid, and
coworkers [1-4] inspired by the work of Ikeda, Shirakawa, and associates [5,6] first realized that po-
ly(acetylene) when treated with oxidizing or reducing agents becomes highly conducting and was, thus,
considered as the prototype conducting organic polymer. The ideas on the mechanism and origin of
the conductivity announced by the Philadelphia
group [4,7] helped to attract the interest of a
considerable number of groups from the area of
solid-state and theoretical physics. Further attention arouse when Mac Diarmid and coworkers first
described that poly(acetylene) may be oxidized or
reduced by electrochemical methods as well and that
this may be exploitet to construct polymeric storage batteries [4,7].

Unfortunately, poly(acetylene) is not as simple a material as some of the workers in the field
have first thought, [2] and as one might think just
looking to the chemical formula describing the
ideal chain in its cis- or trans-form. Similarly,
the chemical reactions and processes necessary to
bring about conductivity are not so clear as it is
sometimes assumed by those who try to provide the
explanation of the physical properties of the pristine or conducting polymer. It is, therefore, fair
to say that the area of conductive polymers is not
only one of the most interesting but also one of
the most controversial of all recent developments
in polymer science. A number of other polymeric
structures have been detected to show a conductivity of similar magnitude and quality soon after
the work on poly(acetylene) became known; [8-10] this
has, however, not contributed very much to solve
the existing controversies but rather has added
to the complexitiy of the problems. At the other
hand, considerable progress has been achieved in
the recent years in the design and understanding
of low molecular weight - so-called - organic metals. It will be necessary to consider these developments and to use the information available
from this area to better understand the properties of the conducting polymers.

In the following we want, therefore, to describe first the structure of the simplest organic
metals derived from as simple molecules as naphthalene or other arenes. These structures help to understand the type of intermolecular interactions
necessary to produce a quasi-metallic state in
organic systems. The structure and the structural
changes upon oxidation ("doping") of poly(acetylene) will then be described. A description of
these chemical reactions and their implications
for the electronic and vibronic spectra will
follow. Finally, some other conducting polymers
or oligomers will be described and the use of such
materials in electrochemical cells will be discussed as well.

HOW TO DESIGN AN ORGANIC METAL?

The structure of conducting cation-radical salts

The experimental and structural requirements
necessary to produce a so-called organic metal from
low molecular weight species are well known.[11] It
is necessary to start with a compound which can be
oxidized or reduced to an ion-radical state of a
lifetime long enough to form a complex with the
neutral precursor. This complex must then crystallize into an organic salt structure characterized
by an ordering of the ion radical complex into
stacks with a face to face arrangement of adjacent
molecules.[12] The complex formed between tetrathio-
fulvalene (TTF) and tetracyanoquinonedimethane
(TCNQ) was, for a long time, considered to be the
prototype organic metal until it was realized that
this complex is one of the most complicated structures to be investigated.

The simplest organic metals can in fact be
produced starting from naphthalene[13] and similar
arenes such as pyrene, perylene, fluoranthene or
triphenylene.[14-16] If these compounds are electrochemically oxidized in an otherwise inert solvent
containing a supporting electrolyte with anions of
low nucleophilicity, coloured crystals of high conductivity will grow at the anode. These crystals

have generally the stoichiometry $(arene)_2^+ X^-$, where
X^- designs the counterion which balances the positive charge located on the aromatic cation-radical
stack. Suitable anions are BF_4^-, ClO_4^-, PF_6^-, AsF_6^-,
SbF_6^- etc. In conjunction with tetraalkylammonium
cations they form the supporting electrolyte during the electrolysis. An electrochemical investigation of all the processes which finally lead to
the growth of the conductive crystals has revealed
that the total reaction may be split into at least
3 independent steps as follows:[15]

$$Ar \rightleftharpoons Ar^{+\cdot} + e^- \quad \text{electrode reaction} \tag{1}$$
$$\text{(oxidation)}$$

$$Ar + Ar^{+\cdot} \xrightarrow{\text{fast}} (Ar_2)^{\cdot +} \text{complexation} \tag{2}$$

$$(Ar_2)^{+\cdot} + X^- \rightleftharpoons \text{crystallization} \tag{3}$$

Reaction (1) describes the electrochemical oxidation of the arene (Ar) at the anode to give a
short lived intermediate cation radical which will
be stabilized by complexation to a dimer cation
radical as described by equ. 2. The conducting
crystals finally grow from the supersaturated solution of the dimer-complex on the electrode surface (equ. 3). Following this principle a large
number of different structures has been synthesized. A complicating factor arises when the structure allows for the inclusion of solvent. In these
cases solvent will replace some of the counterions X^- and - in order to keep electro-neutrality-
a deviation from the stoichiometray $(Ar)_2 X^-$ is observed;[16] in other words, the stack of the arenes
contains more neutral than positive centers.

One of the first materials of this type investigated and appreciated as an organic metal was
derived from naphthalene (Fritz et al. 1978).[13]
Fig. 1 shows a comparison of the crystal structures of neutral naphthalene (A) and the corresponding organic metal of the composition (Naphthalene)$_2^{+\cdot} PF_6^-$ (B). The typical features of the

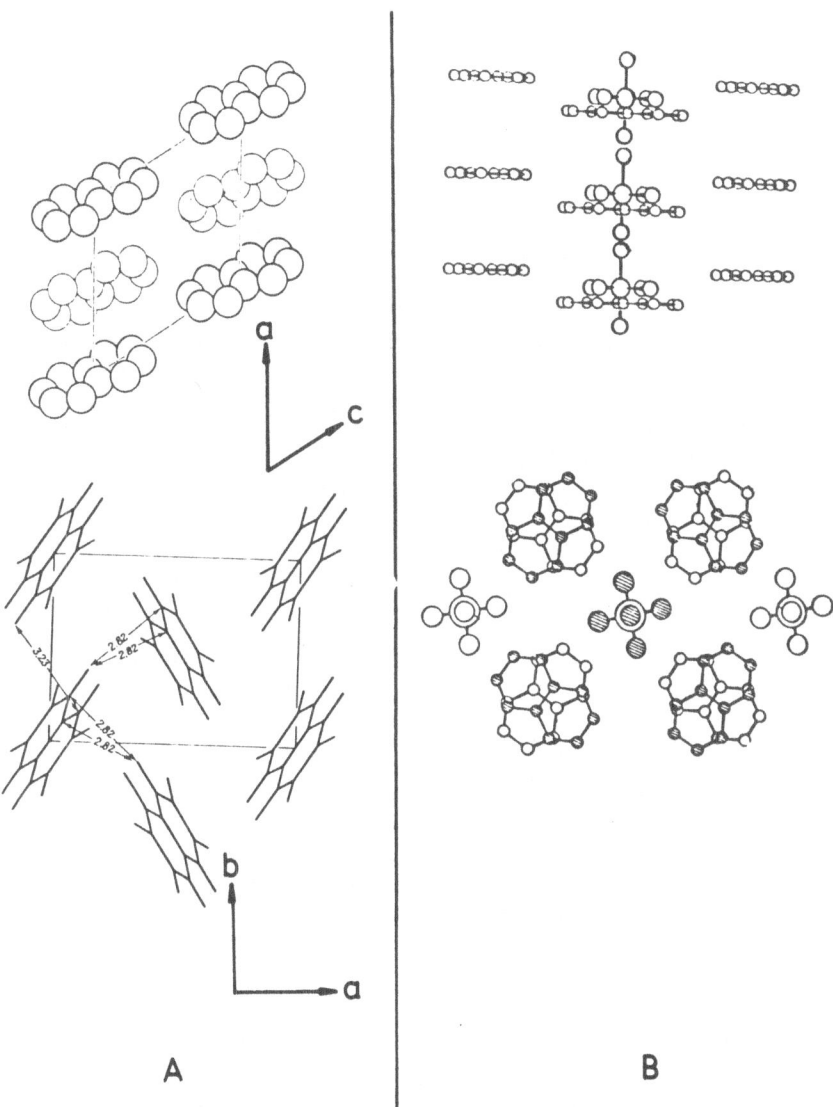

Fig. 1. The crystal structure of neutral naphthalene (A) and of the cation-radical salt
(naphthalene)$_2^+ \cdot PF_6^-$ (B).[13]

conductive structure are the columns of naphthalene molecules which are close packed as to leave
channels in which the counterions X^- are located.
Within the columns the individual naphthalene
units are spaced at a distance of 0.320 nm and adjacent molecules are rotated for $90°$ with regard
to each other. In a formal way every second molecule carries a positive charge and an unpaired
spin. In reality the charge and the spin state is
smeared throughout the whole stack and all molecules are crystallographically identical. The
counterions do not contribute to the electronic
properties of the crystal as far as the carrier
transport is concerned. They serve as "spectators"
and their only function is to keep the electroneutrality. As far as phase transitions and the relaxation behaviour of the crystals are concerned,
the counterions play a certain role,[17] which is of
little concern for our present discussion.

The room temperature conductivity of polycrystalline samples of the naphthalene cation radical
salts lies between 0.1 and 1 $(\Omega \text{ cm})^{-1}$. Single crystals measured along the stack direction show a
conductivity of several 100 $(\Omega \text{ cm})^{-1}$ depending on
the quality of the crystals and to some extent on
the nature of the counterions, solvent inclusion
etc. The naphthalin cation radical salts can be
stored for sometime at room temperature, if moisture is excluded. The corresponding radical ion
complexes of perylene, pyrene, fluoranthene, and
other arenes of higher number of fused rings exhibit a much greater stability and can be handled
under normal laboratory conditions.

Table 1 gives examples of some of the better
investigated compounds. The fluoranthene cation
radical salts have attracted the largest interest
because of the ease by which they can be prepared
and because of their stability and their interesting magnetic and electronic properties.[18]

Coming back to Fig. 1, it is revealing to
compare the crystal structure and the mode of
packing in neutral naphthalene and in its radical
ion complex.

Table 1. Radical cation salts of simple arenes

Arene	Anion	Composition	$\sigma_{300}/(\Omega cm)^{-1}$	Refer
Naphthalene	PF_6^-	$Naphth_2PF_6$	0.12 [a]	13
Perylene	AsF_6^-	$Pe_2(AsF_6)_{1.1}$ $(CH_2Cl_2)_{0.7}$	1000 [d]	16,18
Fluoranthene	AsF_6^-	FA_2AsF_6	700 [d]	15,17,18
Pyrene [b]	AsF_6^-	$\left\{ \begin{array}{l} Py_7(AsF_6)_4 \\ Py_4(CH_2Cl_2)_4 \end{array} \right\}$	500 [d]	18,19
Pyrene [c]	AsF_6	Py_2AsF_6	not de-termined	19

[a] isotropic powder samples

[b] structure contains 4 pyrene molecules not within the stack

[c] a second modification of a pyrene cation radical salt

[d] measured along the stack direction; $\sigma"/_\sigma$ >100

First of all, it becomes clear that the metal-like conductivity does not arise because of a particular molecular property of naphthalene but rather due to the mode of packing of the cation radicals in a columnar structure. It is further worthwhile to note that the stacks are held together by a CT-interaction involving the π-electron cloud of adjacent molecules. The obvious sign of this interaction is the shortening of the distance between the planes of neighboring molecules by more than 10 percent from the van der Waals distance encountered in neutral naphthalene.

Models for conducting polymers

In the language common to the field of conductive polymers going from neutral naphthalene (Fig. 1A) to the cation radical complex (Fig. 1B) may be called "doping" of naphthalene. In fact, the same type of complexes are obtained, if solid

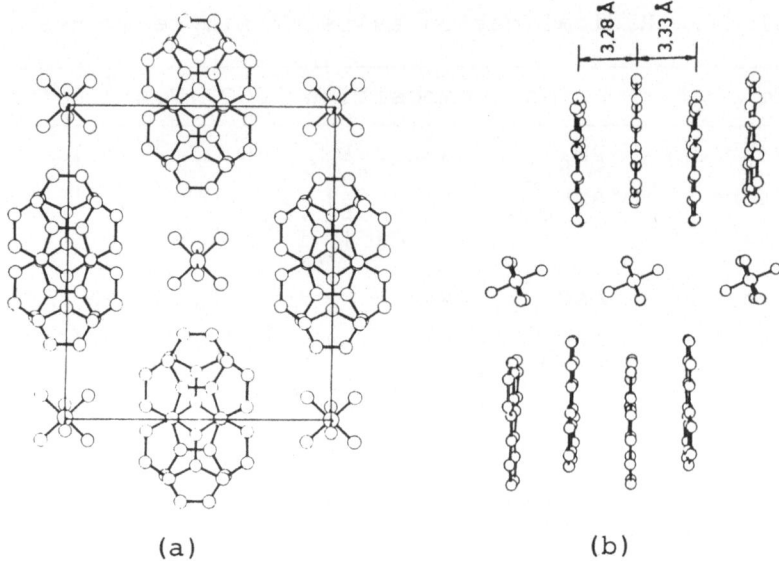

(a)   (b)

Fig. 2. Projection of the crystal structure of the fluoranthene cation radical salt FA$_2^+$·PF$_6^-$ at 298 K (a) onto the b,c-plane; (b) normal to the stack axes. Data cf. V. Enkelmann et al.[15]

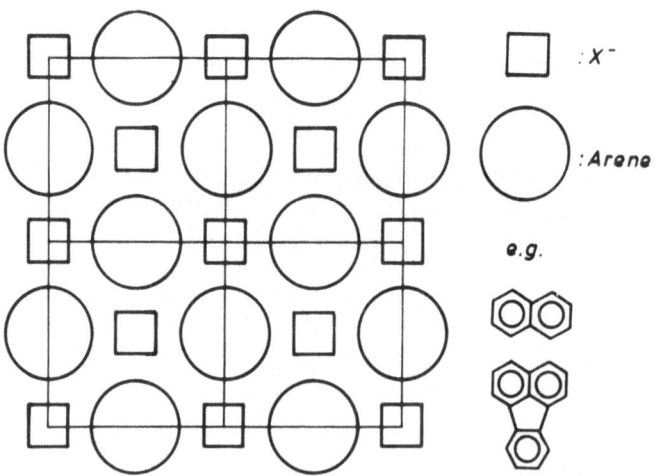

Fig. 3a. Packing scheme of cation-radical salts Ar$_2^+$·X$^-$

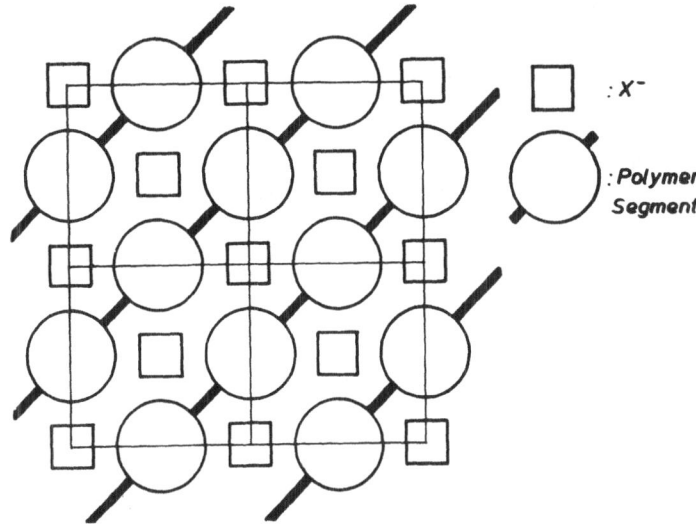

Fig. 3b. Proposed packing of polymer cation-radical salts $\left[(\text{segment})_2^{+}\cdot X^-\right]_n$

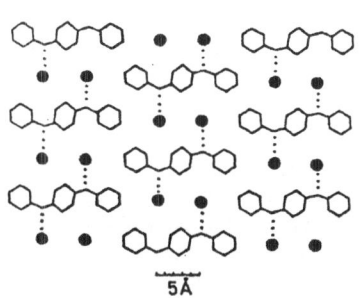

Fig. 4. Structure of the
cation-radical
salt of N,N-di-
phenyl-p-pheny-
lene diamine with
I_3^--counterions
(black dots);hy-
drogen bonds are
indicated by dot-
ted lines cf.
ref. 22.

5Å

naphthalene or other arenes are "doped" by gaseous AsF_5 or SbF_5. Shiny polycrystalline materials are obtained which do not have the purity and quality of the electrochemically produced crystals but exhibit similar properties and character.

In reality we deal with an oxidation of naphthalene as expressed by equ. 1-3 and, if we start from solid naphthalene, with a phase transformation as well. The structure of the conducting material and that of the neutral precursor have nothing in common except that both contain the naphthalene ring as a building block. It is important to remember these facts if discussing the "doping" of polymers such as poly(acetylene) and it will be worthwile to look for the occurence of phase changes in the course of the "doping" of polymers in order to draw the proper analogy to the case of the much simpler cation radical salts of arenes.

As a further example two projections of the crystal structure of the organic metal $FA_2^+ \cdot PF_6^-$ are shown in Fig. 2 to demonstrate the generality of the principle. Note that the distances between adjacent molecules along the stack exhibit a slight alternation in the room temperature (metallic) phase shown here. We will not discuss this feature further but merely like to point out, that it contradicts earlier theories on the necessity of a stacking at equal distance in organic metals.

The structure of the $FA_2^+ \cdot X^-$ complexes, especially the projection shown in Fig. 2a, may be generalized with regard to the possible desgin of polymeric organic metals. This is attempted by Fig. 3 showing in the upper part once more the general principle of the packing in arene cation radical salts. Several unit cells are depicted in order to better show how the structure can be understood as coming about by alternating layers of organic and inorganic ions. The organic layers in turn are composed of the columnar stacks of the cation radicals. If we now - in a Gedankenexperiment - connect the organic ions within a lattice plane and along a lattice direction by chemical

bonds, we create the model of a polymeric organic
metal shown in the lower part of Fig. 3. The circles which in the upper part of Fig. 3 symbolize
the individual arene molecules serve now to depict
a segment of the polymer chain connected to the
adjacent other segments by some bonds symbolized
by the black bars. Note, that the stacking of the
layers and the columnar structure is still the
same as in the upper part. We have thus created
the prototype packing of a polymeric organic metal.

It is important to realize that conductivity
in this model arises along the same principles as
in low molecular weight organic metals by electronic interactions between the segments of adjacent
chains and not because of interactions other than
steric interactions within the same chain. An immediate prediction from this model is that the
value of conductivity should be rather independent
of the degree of polymerization and if some effects
are seen, they can only arise because of crystal
defects created by the chain ends.

It should further be feasible to prepare a
polymerhomologous series of organic models thereby
going from the low molecular weight models to the
polymeric material.

At this point the excellent work of Hadek and
coworkers (1965-1969) must be mentioned.[20-24]
In a series of papers these authors described the
preparation, structure and electronic behaviour of
a series of cation-radical salts of oligomeric
poly(p-phenylene amine)s (PPA) of the general formula.

PPA	n
a	2
b	3
c	4
d	6

PPA

The insulating oligomers were reacted with iodine
to create a new, cation-radical salt structure of
high conductivity. The crystal structure of the
iodine complex of N,N'-diphenyl-p-phenylene diamine $(PPAa)_5I_{12}$ was partially resolved by Huml and
Hadek, and Fig. 4 shows a projection onto (010).
The structure contains iodine chains running normal to the layers in which the oligomers are extended. The distance between these layers amounts
to 3.77 Å and the distance between the ring planes
of oligomers in adjacent layers to ca. 3.3 Å.

From the diffuse layer lines observed in the
single crystal diffraction pattern the presence of
I_3^- -ions was concluded which form chains however
disordered with regard to the cation sublattice
normal to (010).

Table 2 summarizes some of the data reported
by Hadek [21,24] on the isotropic (powder) conductivity of various of these oligomeric cation-radical
iodides. Note the similarity in behaviour as far
as the non-linear dependence of the conductivity
on the iodine content is concerned with the features commonly observed in the "doping" of polymers. Moreover, a chain length dependence of the
conductivity cannot be deduced from the available
data.

It may be worthwhile mentioning that recent
work in our laboratory has shown, that the same
oligomers can also be electrochemically oxidized
to give the cation radical salts with various
counterions as described for the arene salts.[25]
A second example demonstrating the heuristic value of the considerations contained in Fig. 3 is
provided by the electrochemical oxidation of oli-
gophenylenes. J. Eiffler, G. Wieners and V. Enkelmann[26] have demonstrated in our laboratory that
these oligomers, e.g. terphenyl and quaterphenyl,
give rise to cation-radical salt structures under
the same experimental conditions as described for
fluoranthene which exhibit the conductivity behaviour of an organic metal. A full crystal structure analysis of these materials is in progress
(by V. Enkelmann). They may serve as models for
"doped" poly-p-phenylene.

Table 2. Electric properties of the iodine-complexes of some oligo(p-phenylene amines)[a] according to Hadek et al. (1969)[24]

Oligomer[a] n	$[I]$[b]	$\sigma_{296}/(\Omega cm)^{-1}$	E_A/eV
2	2.4	$3.4 \cdot 10^{-1}$	0.06
3	1.7	$3.9 \cdot 10^{-4}$	0.18
3	3.4	$3.8 \cdot 10^{-1}$	0.05
4	1.0	$9.3 \cdot 10^{-2}$	0.09
4	3.0	$9.6 \cdot 10^{-1}$	0.05
6	1.1	$5.8 \cdot 10^{-2}$	0.08
6	6.0	$1.3 \cdot 10^{0}$	0.021

[a] structure:

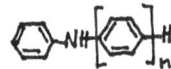

[b] number of iodine atoms per oligomer

Table 3 reports the unit cell parameters of two of these salts. Even in the absence of a full structure analysis these data allow to say that the packing is very much along the lines depicted by Fig. 3.

Finally - as a word of caution - it is to emphasize that the concept summarized by Fig. 3 is not meant to exclude other possible packing modes which may lead to the electronic interactions necessary to create an organic metal. It is nothing but a helpful concept to point out a packing mode which has often been encountered and to emphasize the possibilities and consequences of electronic interchain interactions in "doped" polymers. The model does not say, that conductivity arises solely from interactions between segments of adjacent chains but it stresses the importance of contributions of such interactions to the total electric and electronic behaviour.

Table 3. Unit cell parameters of terphenyl(TP) and quaterphenyl (QP) and of their cation-radical salts $(TP)_2(SbF_6)(CH_2Cl_2)$ and $(QP)_2(SbF_6)_2(CH_2Cl_2)$ according to V. Enkelmann 19,26)

Compound	TP	QP	$(TP)_2(SbF_6)(CH_2Cl_2)$	$(QP)_2SbF_6)(CH_2Cl_2)$
a/Å	8.08	8.05	20.42	20.43
b/Å	5.60	5.55	11.12	11.01
c/Å	13.59	17.81	13.07	17.63
α/°			95.8	98.7
β/°	92	95.8	99.9	98.7
γ/°			93.4	86.7
	$P2_1/a$	$P2_1/a$	$C\bar{1}$	$C\bar{1}$

THE STRUCTURAL BACKGROUND OF CONDUCTIVITY IN POLY(ACETYLENE)

Poly(acetylene) as a redox-polymer

Pristine polyacetylene in either cis- or transform is an insulator.[27,28] Treatment of the solid polymer in form of films or sheets with oxidizing or reducing reagents leads to an exponential increase in the specific conductivity $\sigma(\Omega^{-1}cm^{-1})$ with increasing extent of the reaction, i.e. with the number of charges per molar unit of the polymer transfered in the redox process.[1-4] A polymeric salt is formed where positive or negative charges, depending on whether the polymer has been oxidized or reduced, reside on the polymer and are neutralized by the appropriate number of counterions in order to satisfy electroneutrality.[29,30] The redox process may also be carried out electrochemically such that the pristine polymer in contact with an electrode is either anodically

oxidized or reduced at the cathode.[4,7] The conductivity of the oxidized or reduced poly(acetylene)
is very reproducibly found at ca. 500 $(\Omega^{-1}cm^{-1})$.
In other words, poly(acetylene) may be considered
as a special case of a redox-polymer which can be
reversibly either oxidized or reduced to the res-
pective ion-radical state as follows:

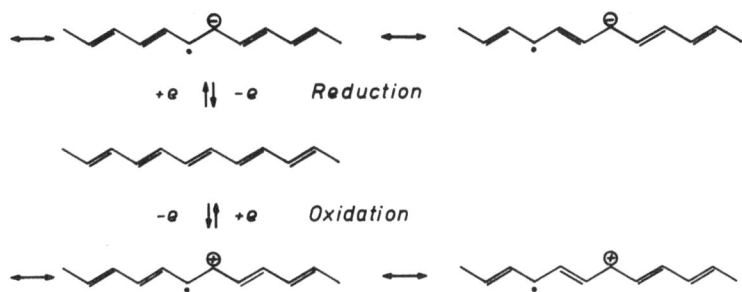

A cis-trans isomerization takes rapidly place if
the initially cis-polymer is submitted to the redox-reaction.
Further oxidation or reduction to the dication or
dianion may be possible and would explain some of
the magnetic anomalies seen in the epr of the partially and fully oxidized polymer.[31,32] Alternatively, the di-ions may always be formed by a disproportionation reactions as it is well documented
for the case of the equilibrium between the radi-
calanions and dianions of diphenyloligoenes.[33]
In the special case of oxidation of polyacetylene
by AsF_5 or I_2 the reaction may be formulated:[30]

$$2-(CH=CH)_n + 3AsF_5 \longrightarrow 2-(CH=CH)_n^{+\cdot} + 2AsF_6^- + AsF_3 \quad (5)$$

and

$$2-(CH=CH)_n + 3I_2 \longrightarrow 2-(CH=CH)_n^{+\cdot} + 2I_3^- \quad (6)$$

The index n in equ. 5 and 6 indicates the number
of double bonds or, even better, designates the
segment of the polymer chain which 1) serves as
the redox partner in the reaction and 2) acts as
the building element of the salt structure which

will be formed as a consequence of the reaction.

This raises the question how much charge per unit length of the polymer backbone may be transfered in the redox process. The answer to this problem is very relevant discussing the possible use of poly(acetylene) in storage batteries where the capacity of the cell is determined by the exact stoichiometry of the redox reaction. The analysis of the X-ray structural data, despite their limitations due to the limited accuracy of the X-ray patterns, seems to indicate that the charge density on the chain is essentially limited by the packing density of the counterions X^- in the case of the oxidized polymer. The disproportionation reaction may then be formulated by

$$2-(CH=CH)_n^{+\cdot} \rightleftharpoons -(CH=CH)_n + -(CH=CH)_n^{2+} \qquad (7)$$

a process which may occur either inter- or intramolecularly. The reduction is described using the reaction of the polymer with sodium naphthalide as an example:

$$-(CH=CH)_n + \left[\bigodot\bigodot\right]^{\overline{\cdot}} Na^+ \longrightarrow -(CH=CH)_n^{\overline{\cdot}} + Na^+ + \bigodot\bigodot$$

$$(8)$$

and everything said so far on the oxidation is valid for the reduction as well.

Finally, the electrochemical reactivity of the polymer is described by the appropriate half-cell reactions 9a,b which add up to the electrochemical equilibrium described by equ. 10.

Here a situation is described first realized by Mac Diarmid et al.[4,7] in which the polymer is submitted to the electrochemical redox-reaction in the presence of a supporting electrolyte. This electrolyte must consist of a salt whose components will later act as the counterions in the polymer radical ion salts. Consequently, the ions must be selected as to be inactive with regard to substitution, addition and abstraction reactions toward the unsaturated polymeric ion radicals. Typical examples of salts which have been successfully used in chlorinated hydrocarbons, cyclic ethers or esters as the solvents are $LiClO_4$, NaI, tetraalkylammonium salts of the anions BF_4^-, PF_6^-, SbF_6^-, AsF_6^-, $SbCl_6^-$, etc.

$$+CH=CH+_n \; + \; R_4N^+X^- \longrightarrow [-(CH=CH+_n^{+\cdot} \; X^-] + R_4N^+ + e^- \tag{9a}$$

$$e^{\mp} \; +CH=CH+_n + R_4N^+X^- \;-\; [-(CH=CH+^{-\cdot}R_4N^+] + X^- \tag{9b}$$

$$2 \; +CH=CH+_n \; + \; R_4N^+X^- \longrightarrow [\; +CH=CH+_n^{+\cdot}X^-] + [+CH=CH+_n^{-\cdot} \; R_4N^+] \tag{10}$$

The emf of a cell described by equ. 10 i.e. oxidized vs. reduced polymer was found to be ca. 3.4 v.[34]

Another possibility to create the cation radicals of poly(acetylene) simply consists in the treatment of the polymer with strong acids especially with Lewis-acids in the presence of solvents like nitromethane or CH_2Cl_2. The so-called "acid-doping" has been observed and reported by numerous workers in the field without giving it the proper description as a redox process.[4,35,36]

It may, however, be remembered that $AlCl_3$ in nitromethane has been commonly used by many workers for oxidation of unsaturated and aromatic molecules to radical cations for the purpose of spectroscopic investigations.[37] Even $AlCl_3$ dissolved in CH_2Cl_2 is a suitable oxidizing reagent to create the cation radicals of organic molecules whose first ionization potential is below 8 eV,[38] i.e. of most of the polymers known to be converted to organic metals.

Redox reactions of organic molecules are rarely fully reversible due to the high reactivity of the ion radical state created in the first electron transfer. Although the ion radical state is

stabilized by the formation of a lattice with a
defined salt-structure in the case of poly
(acetylene) certain side reactions may be expected
to occur from a simple consideration of the chemical nature of the products.

The most prominent side reaction expected may
be the dimerization of radical sites on adjacent
chains to give permanent crosslinks.
A final remark concerns the application of poly
(acetylene) in electrical storage batteries. It is
now quite clear, that poly(acetylene) is an example of yet another redox-polymer - a term first
used by H. G. Cassidy[39] many years ago - to describe a class of polymers which can be reversibly
oxidized and reduced and may thus be used as exchange resins etc. The prototype polymer of this
class is poly(vinylhydroquinone) which can be reversibly oxidized to poly(vinylquinone) the semi-
quinone radical being an intermediate.

The differences between these redox-polymers
and the now much discussed "polymeric metals" is
the additional property being conductive in the
ion-radical state. This is a feature most welcome
in the construction of batteries but as such not
a necessary requirement. The possibilities offered
by the redox polymers were of course realized
many years ago and have been described by
Cassidy.[39] Batteries in which one or both electrodes were a redox polymer in nature have been
patented.[40,41] Due to the insulating nature of
these early redox polymers the electrodes were
prepared from a mixture of the finely divided polymer and a conductor such as graphite, metal
dust or activated carbon. Additionally or alternatively a redox electrolyte was added to act as the
mediator between the insoluble and insulating redox polymer and the metal electrode. Since the
oxidized or reduced form of the redox polymer can
be regenerated and the process recycled, the resulting unit is a secondary cell.

The structure of oxidized poly(acetylene)

If polyacetylene is oxidized by either of the methods described in the proceeding section, the reaction product has a crystal structure different from the pristine polymer. This was first observed for the oxidation by treatment with iodine, AsF_5 etc. by R. H. Baughman et al.[42] The diffraction patterns of electrochemically oxidized poly(acetylene) containing AsF_6, ClO_4, and SbF_6 as the counterions were investigated by Monkenbusch et al.[43] Further work on the diffraction pattern of iodinated polyacetylene is available from a number of other sources, especially from Shimamura et al. on the diffraction from oriented samples.[44]

The patterns of the different samples show largely similar features. A new bragg-peak arises toward small angles the d-spacing of which scales with the size of the counterion X^- as first realized by Baughman. This observation was taken as evidence for an intercalation of the "dopant" into the lattice of the polymer similar to the interpretation of the diffraction patterns of intercalated graphite.[42]

All authors agree on the fact, that other structures then those of the initial and of the oxidized polymer are not observed. In other words, intermediates as expected, if the intercalation was to proceed homogeneously to form a solid solution of oxidized segments in the lattice of the initial polymer, have never been observed. The only exception are recent reports by R. H. Baughman who reports the occurence of new diffraction peaks in iodinated samples annealed for a very long time at room temperature.[45]

The prove that the oxidation of the polymer (and the reduction resp.) is linked to a phase change is of greatest relevance for the understanding of the change of the electrical properties observed as depending on the extent of the reaction.

A proper experiment in that direction was recently carried out by M. Monkenbusch in our laboratory.[46] He investigated the evolution of the new structure by measuring the changes in the X-ray pattern simultaneously with the change in conductivity, when a sheet of polyacetylene was exposed to gaseous iodine.

Fig. 5 shows the dependence of the integrated net intensity of the "intercalation" peak from the polyacetylene iodine complex compound at $2\vartheta \approx 10.4°$ on the iodine content. A linear relationship is obtained. Note that there is a threshold-value of iodine content at y=2.3 percent I/CH. Below this level an intercalation peak is not formed but the background intensity increases from the initial product up to this level due to iodine absorbed and/or reacted at the surface of the sample. The specific surface area of such free standing sheets of poly(acetylene) is of the order of $100 m^2 g^{-1}$. The results of this study lead to the conclusion that poly(acetylene) oxidized by iodine essentially consists of three different components: unreacted bulk polymer, iodine covered surfaces and/or amorphous regions containing iodine and a metallic conducting polymer-iodine complex salt structure with a new lattice composed of rearranged chains together with I_n^--chains. The conductivity data indicated a percolation process involving the metallic conductive parts of the sample as component of a system with a complex texture.

It thus becomes clear, that two regimes of the reaction corresponding to two different regimes of the electrical behaviour have to be discussed: a) the regime below the threshold value of the new structure and b) the regime where the properties are related to the appearance of the new polymer salt structure. In the first regime an exponential increase in conductivity to a level of about $10^{-2} (\Omega^{-1} cm^{-1})$ is observed, obviously due to surface reactions and/or reactions in the amorphous domains of the sample. The transition to the "metallic" regime of conductivity seems, however, to be linked to the formation of the new structure. The increase in conductivity with iodine-concentration follows the law expected for a 3-dimensional percolation in this regime.[30,46]

By and large the same features are observed if samples oxidized electrochemically to different extents of reaction are investigated by X-ray diffraction.[47]

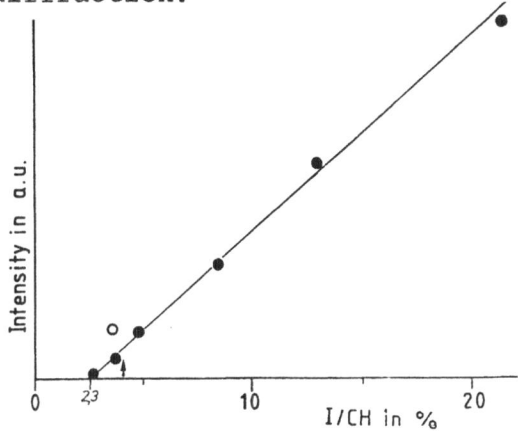

Fig. 5. Integrated intensitiy of the "intercalation" peak at $2\theta=10.4°$ vs. the iodine content of iodine treated poly(acetylene). 46

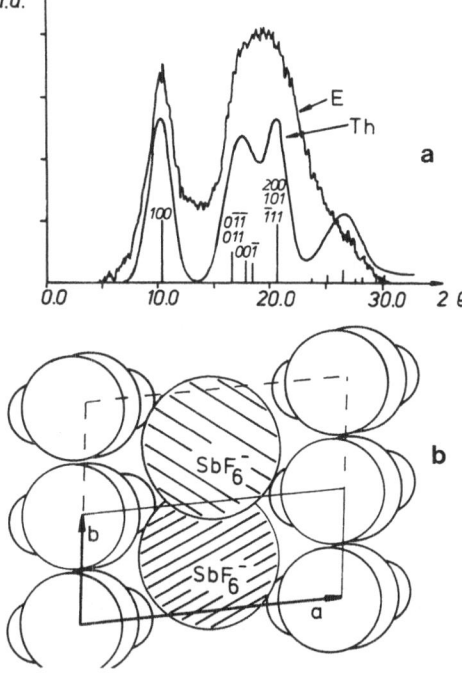

Fig. 6.
(a) Diffraction pattern(E), expected position of Bragg-peaks and calculated scattering intensity(Th) of the poly(acetylene)cation radical salt structure shown in Fig. 6b.
(b) Structural model of a polyacetylene cation radical salt.

Fig. 6 shows the actual diffraction pattern
of a polymer sample oxidized electrochemically as
to form the SbF_6^- -salt. The sample had a conductivity $400\Omega^{-1}cm^{-1}$.

The "intercalaction" peak, here indexed as
(100) in terms of the structure depicted in the
lower part of Fig. 6 is very well resolved. In
addition there is a very broad peak with maximum
at about $2\vartheta=20°$. All features of the initial polymer structure have disappeared. The second peak
may be decomposed in contributions from various
reflections as indicated by the curve marked
"Th". It is actually the expected intensity distribution for the peaks the position of which is
indicated as well and which are assumed to be
broadened by a gaussian profile due to limited
crystal size and defects.

All peak positions and intensities relate to
the structure shown in the lower part of Fig. 6
which has been optimized by packing calculations.
The details of the packing calculations will be
published elsewhere.[47]

In the context of the present discussion we
want merely to point out that a structure like
the one in Fig. 6 fits the observed scattering
pattern reasonably well and agrees with regard
to the packing interactions completely to the picture developed for the low molecular weight organic metals derived from arenes (comp.Figs.1-4).

As a model, it was assumed that the poly(acetylene) chains pack together in sheets with the
zig-zag plane normal or nearly normal to the plane
of the sheet as to create the interaction between
the π-electron clouds of adjacent chain elements
discussed earlier. The counterions are assumed
to pack densely in between the sheets. The interactions between the inorganic ions and the organic sublattice are taken to be the same as observed in the cation radical salts discussed in
section 2.

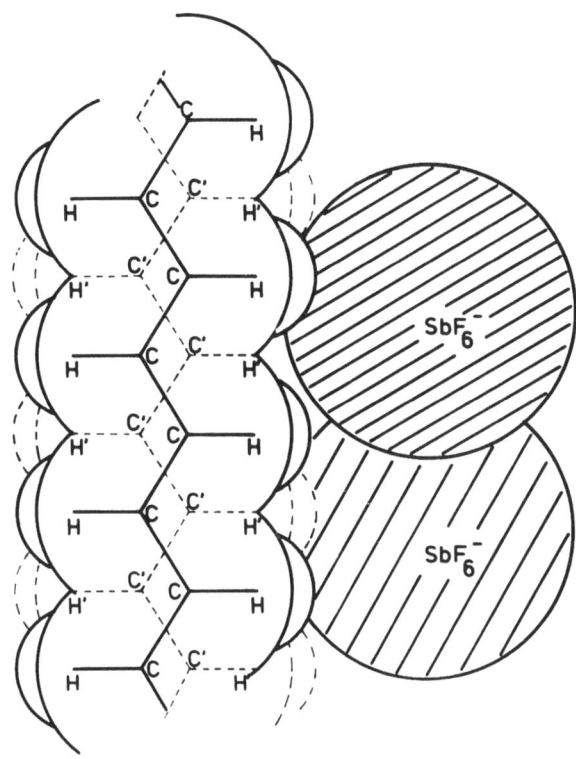

Fig. 7. Same structure as in Fig. 6 projected onto
        the plane of the zig-zag-chains.

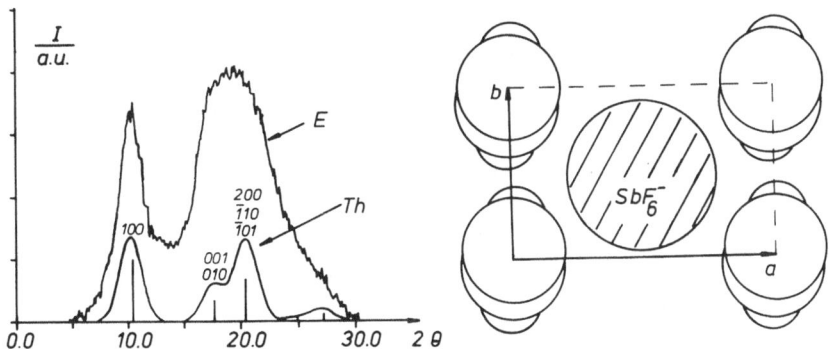

Fig. 8. Alternative model of the poly(acetylene)
        cation radical structure (see text).

A projection of the structure optimized with
regard to packing calculations in a direction
normal to the polymer chain direction reveals
further similarities to the structures shown in
Fig. 1-4. A dense packing of the counterions
SbF_6^- results in a stoichiometry of 6-HC=CH-
groups per one counterion. Coming back to the
question raised earlier as to what is the charge
density on the polymer backbone this can now be
answered. The cation radical formed by equ. 4
(comp. also equ. 5 and 6) comprises 3 double bonds.
It will be stabilized by formation of a charge-
transfer complex with neutral chain segments of
adjacent chains as indicated by

and by Fig. 3. Thus the charge counterbalanced by
one SbF_6^--ion is distributed in a formal way over
two chains, in reality, of course, the charge will
be spread out over the whole sheet of interacting
polymer chains.
The composition calculated for a maximum packing
density in terms of Fig. 6 and 7 coincides very
well the analytical data of the polymer oxidized
to its optimum conductivity. Some workers in the
field favour an alternative structural model which
consists again of sheets of polymer chains but
with the zig-zag-plane in the plane of the sheet.
This model is based on the assumption that there
will be an appreciable interaction between the
π-electrons of the chain and the counterions,
notably with I_n^- -ions in the case of iodine-oxi-
dized poly(acetylene).[44,48,49] At the same time
this model negates appreciable electronic interactions between adjacent chains.

Fig. 8 shows a packing arrangement in terms
of this model. Again, the structure was optimized by packing calculations and the expected
position of the Bragg-reflections and intensities
was calculated.[47] The results are seen in the

upper part of Fig. 8 and may be compared with the
experimentally observed diffraction pattern. The
(100) peak is similar as in the previous model,
but the fit between calculated and observed data
is rather poor in the angular region between
$15 < 2 > 20°$. In addition, a dense packing in terms
of this structure requires 4-HC=CH-groups per
SbF_6 -ion, a value which is much higher than the
experimentally observed one.

Although we tend to favour the model depicted by Fig. 6 and 7 we must admit that the data
presented are not yet convincing enough to settle
the problem. Further work is necessary and especially fibre diagrams of the oxidized polymer
would be of great help to refine our picture of
the structure of the conducting phase of oxidized polyacetylene.

Spectroscopic investigations relevant to the conduction mechanism

Pristine poly(acetylene) in either cis- or
trans-form exhibits optical spectra typical for a
semiconductor. The absorption peak of the cis-
polymer is structured with maxima at 2.1, 2.3 and
2.4 eV, the absorption coefficient is about
$4 \cdot 10^5$ cm^{-1}. The structureless and very broad absorption of the trans-polymer peaks at 1.9 eV with
an absorption coefficient of $3 \cdot 10^5$ cm^{-1}. The large
halfwidth of the peak is generally explained as
being due to the rather broad chain length distribution; in addition, the small crystallite size
and crystal defects may contribute to the broadening. A discussion is still ungoing whether the
peak is linked to a direct interband transition
or whether it is of an excitonic character.[2-4,
7,32]

The absorption spectra in the visible and near
IR show strong changes when the samples are oxidized or reduced. The absorption peak of the pristine polymer decreases and concurrently a new
band builds up centered between 0.5 and 0.8 eV; an
isosbestic point is observed, if one starts from
the trans-polymer indicating a transition between

two, and only two, well defined states of the system.
The appearance of the new peak is practically independent of the oxidizing reagents used and must,
therefore, be related to a change of the electronic structure of the polymer backbone. The new
peak in the near IR is generally called a "midgap"
absorption since its energy is approximately situated at half of the interband (or excitonic) transition of the initial polymer. Its intensity is
proportional to the extent of the reaction ("the
dopant level") over the whole range of conversion.

In addition to the appearance of the new peak
in the near IR, 5 new bands are formed in the IR
region at 1400, 1290, 1177, 900 and 814 cm^{-1}.[50]
The exact position of these very intense bands depends somewhat on the extent of the reaction (a
shift to lower wave numbers is observed with increasing conversion) but the general features remain
the same to full conversion, i.e. the highest levels of conductivity. If deuterated poly(acetylene)
is oxidized, the IR-bands show large isotope-shifts
(e.g. the 1400 cm^{-1} band shifts to 1088 cm^{-1}.[50,51]
This helps to assign the IR-bands to molecular vibrations of segments of the chains which carry a
charge counterbalanced by a polarizable counterion
in its neighbourhood. In principle, without going
into detail, the bands are assigned to an allylic
type carbocation as it was already formulated in
equ. 4-7. Recent work of Harada et al. on the visible and IR-spectra of iodine-treated trans-β-
carotene revealed that in essence the same features
are seen in this molecule as in trans poly(acetylene), thus supporting the notion that the spectral
data show the presence of allylic type carbocations
in oxidized poly(acetylene).[51] An alternative explanation specifically involving a soliton-like -
that is a highly mobile - defect mode has also
been proposed and attempts to explain the observed
IR-data on this basis have appeared in the literature.[52,53]

In the light of the work of Rabolt et al. and
of Harada et al. this explanation is not very likely, especially since the IR-features persist into

a level of conversion where solitons are not believed to exist anyway.

The soliton concept of conductivity[2-4,52,53]
is a very interesting one and needs some further
consideration.

The soliton model

It is well documented that trans-poly(acetylene) as obtained by thermal isomerization from
the cis-form contains a fairly large number of
free radicals. The spin concentration amounts to
10^{-4} to 10^{-3} spins per carbon atom (100-1000 ppm).
These spins are assumed to be highly mobile because they correspond to a free radical situated
on the chain extending over several bonds. The
phase of the bond length alternation reverses at
the site of the topological defect created by the
presence of the radical on the chain. The chain
segments to the left and to the right of the defect are energetically degenerate; consequently no
energy is involved in moving the spin along the
chain and its movement may be described in the
manner of a solitary wave. 7,52,53).

The determination of the spin mobility from
experimental data rests on a number of assumptions
and is rather difficult especially in the light
of the ongoing discussion on the morphology and
complex texture of poly(acetylene). It suffices
to say that the soliton concept relates to a
strictly one-dimensional situation where the in-
finetely long chains are stretched out in vacuo.
Chain ends, chemical and conformational defects
and chain-chain-interactions (lattice interactions)
are not considered which all would lead to a trapping or binding of the soliton. The free radical
on the chain is called a "neutral" soliton because
it may be oxidized or reduced to the corresponding carbanion or carbocation resulting in a
"charged" soliton.

Contrary to the neutral soliton, the charged
soliton although degenerate with regard to the bond
alternation cannot move freely along the chain.
It is pinned by the Coulomb attraction to the
counterions introduced into the sample by the redox reaction necessary to "charge" the soliton.
Consequently the presence of free radicals in
trans-poly(acetylene) creates a second set of redox states in addition to the ones discussed in
section 3.1 as follows:

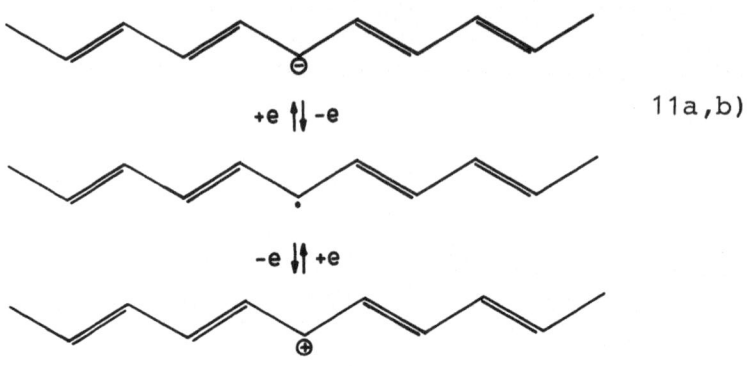

11a,b)

The situation described by equ. 11a, b was used by
S. Kivelson[54] as the basis of a theory which describes the conductivity and the electronic behaviour of "lightly doped" poly(acetylene), i.e. the
material which was treated with little oxidant
(or reductand) as to bring the conductivity from
the insulating state to the level of a good semiconductor.

The theory explains among others the exponential increase of the conductivity at small levels
of counterion content and the frequency dependence
of σ.[55] The principal assumption of the theory is
that whenever a highly mobile "neutral soliton"
encounters a "charged soliton" which is assumed to
be localized a charge transfer reaction between
adjacent chains becomes possible as indicated in
equ. 11a,b. The mobility of the charge carrier is
thus linked to the mobility of the neutral soliton

although the important step involves a kind of
a hoping process between chains. Once again, the
discussion of the experimental data in terms of
Kivelson's theory needs certain assumptions. e.g.
that the counterions are homogeneously distributed
throughout the sample and that the transport properties are not affected by internal phase boundaries just to mention to factors which need consideration in the light of the complex morphology of
poly(acetylene).

In the meantime it has been shown that the
frequency dependence of σ measured by Epstein et
al.[55] and thought to be a prove for Kivelson's
theory (and thus for the presence of solitons) may
be explained by a variable range hoping between
localized ion radical sites as well.[56]

The presence of charged solitons, i.e. carbocations and carbanions of the type shown in
equ. 11a,b in "lightly doped" samples was deduced
from the appearance of a strong optical transition
between 0.5 and 0.8 eV, the so-called mid-gap absorption.[7,57] A look to the literature on the optical spectra of available model compounds reveals, however, that this explanation is not
unique.

The spectra of a series of α,ω-diphenyloligo-
enes, of their anionradicals and dianions (1a-c)
are available from the work of Hoijtink et al.[33]

$$\text{Ph-}(\text{CH=CH})_{\overline{n}}\text{Ph} \quad [\text{Ph}(\text{CH=CH})_{\overline{n}}\text{Ph}]^{\cdot-} \quad [\text{Ph}(\text{CH=CH})_{\overline{n}}\text{Ph}]^{2-}$$

$$\underline{1a} \qquad\qquad\qquad \underline{1b} \qquad\qquad\qquad \underline{1c}$$

They may serve as models for the possible oxidation states of poly(acetylene) as described by
equ. 5-8 in Sect. 3.1. Models for the "charged
soliton" described by equ. 11a,b are available
from the work of Hafner et al.[58] and of Kuhn
et al.[59] who studied the spectra of diphenylcarbo-
cyanine cations 2a and corresponding anions 2b.

$$[\text{Ph}-(\text{CH}=\text{CH})\underset{n}{\cdots}\text{CH}-\text{Ph}]^{+} \quad [\text{Ph}(\text{CH}=\text{CH})\underset{n}{\cdots}\text{CH}-\text{Ph}]^{-}$$

$$\underline{2a} \qquad\qquad\qquad\qquad \underline{2b}$$

The very interesting compounds 2a may be easily
prepared from the corresponding unsaturated secondary alcohols by acid treatment, the anions 2b by
deprotonation of the corresponding hydrocarbon as
indicated by equ. 12a,b.

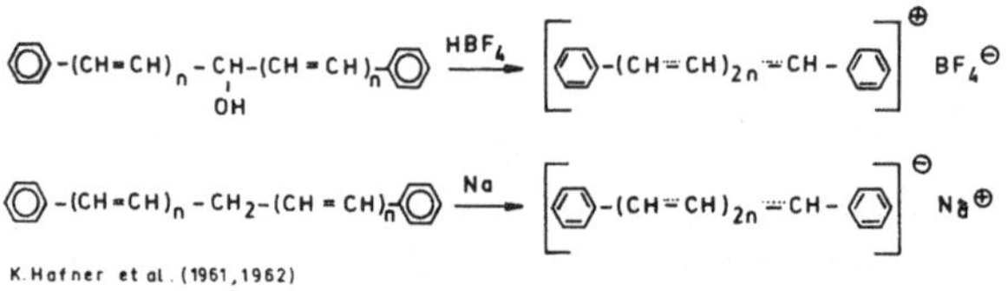

12a,b)

K. Hafner et al. (1961,1962)

A plot of the energy of the peak maximum of the
main optical transition of these model compounds
vs. the reciprocal effective conjugation length
n_{eff} is shown in Fig. 9.

It is obvious that the spectra of radical
ions, carbocyanine ions and di-ions at given n_{eff}
peak approximately at the same energy. Consequently, the extrapolation to infinite chain length
does not allow to differentiate between these
states in poly(acetylene). The carbocyanine anions
2b are the only exception for which the data extrapolation to infinite chain length gives a value at about 1.4 eV.

Unfortunately, the extrapolation of the data
shown in Fig. 9 is not of high accuracy due to the
limited number of model compounds in each case.
Nevertheless, the extrapolated value for polyacetylene between 0.8 and 1.1 eV is not too bad in comparison with the experimental data.

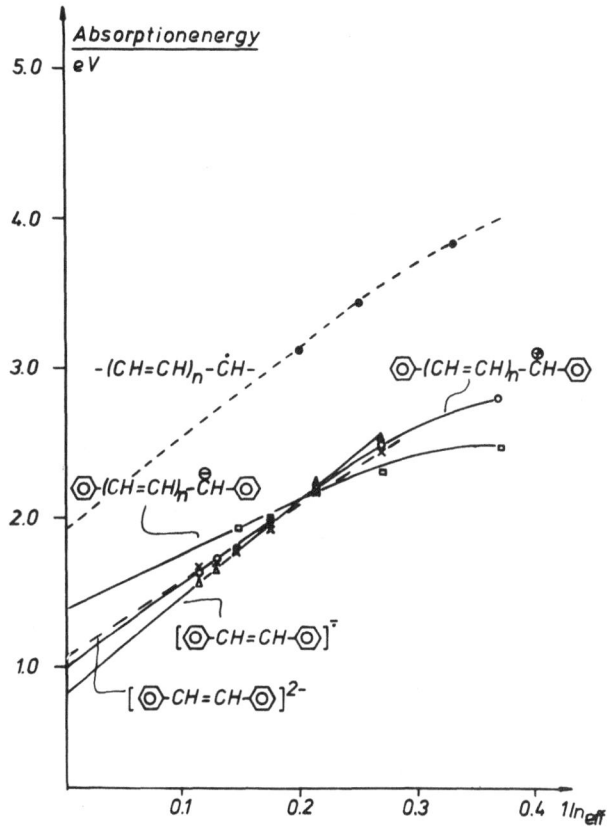

Fig. 9. Energy of the longest wavelength absorption of some model compounds vs. the reciprocal effective conjugation length.

Fig. 9 shows data on the optical transition of allylic radicals as well, which serve as models for
the "neutral" soliton. The spectra of these structures are available from the study of the radioly-
sis of paraffines.[60] Again, the extrapolation to
the case of polyacetylene is not very accurate but
helpful in the search for the expected peak in the
optical spectrum. As it looks, it should be hidden
or contribute to the very broad absorption of
trans(polyacetylene) around 1.9 eV.

The work on carbocyanines available from the
literature[58,59] may also be helpful in the search
for the reactivity and determination of crosslinks
in polyacetylene. Two possible (and expected)
structures of crosslinks are shown to undergo reactions with "dopants" which would lead to carbocyanine   structures in the "doped" polymer in equ.
13a,b. The possible relevance of these reactions
in the context of the optical and electrical properties of "doped" poly(acetylene) has been discussed previously.[61]

Morphology

The morphology of poly(acetylene) as synthesized by a Ziegler catalyst ("Shirakawa method")
or by Luttinger's method[62] is complex. The material
is fibrous in appearance and the so-called free
standing films used in most of the studies of the
electrical and optical studies have a tissue-like
texture. Depending on the reaction conditions the
morphology may change from purely fibre-like to
lamellar, in most cases fibres with extensive lamellar overgrowth are obtained. The actual crystal
size of transpoly(acetylene) as derived from the
width of the diffraction lines is rather small and
in the same range as observed for most semicrys-
talline polymers. The coherence length in chain
direction is certainly less than 200 Å. It is interesting to note that so far a relation between
morphology and electrical data has not been obtained. Similarly, the parameters which control the

formation of the so-called free-standing films and the interparticle contacts necessary for mechanical and electrical perfomance do not seem to have attracted much interest.

In this context it is worthwile to report experiments carried out by W. Müller in our laboratory[63] by which it was shown, that free standing films may be prepared from a suspension of individual particles and that the mechanical strength as well as the electrical properties are linked to a interparticle crosslinking reaction which seems to occur during the drying of the particle suspension.

A suspension of poly(acetylene) was prepared carrying out the polymerization as described by either Luttinger's or Shirakawa's method under rapid stirring of the catalyst solution into which acetylene was introduced by a gas inlet.
The suspension was left to sedimentation, excess solvent was decanted and the concentrated suspension was spread onto a glass plate. Residual solvent was evaporated under reduced pressure, but in contact with air. A film similar in appearance to the films obtained by in situ film formation was obtained which could be removed from the substrate. It had similar mechanical strength and electrical properties after oxidation as the usual films.
The essential steps of the procedure are sketched in Fig. 10 and a comparison of electrical data is given in Table 4. It contains data on the behaviour of pellets as well which were obtained from the suspension as well after centrifugation and compression in a KBr-press.
The data compiled in Table 4 clearly show that the surface of the sample accessible to the chemical reaction in the "doping" is one of the most important parameters. They also show that the catalyst system used (Shirakawa vs. Luttinger) and the method of preparation of the free standing film (in situ preparation vs. successive polymerization and film formation) is of little consequence as far as the level of conductivity is concerned.

Sedimentation   *Drying*
                                   (O_2)

Suspension

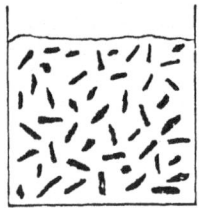   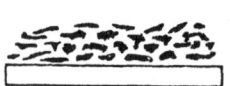   *Film*

              Substrate (e.g.glas)

Fig. 10. Scheme describing the preparation of free
         standing films of poly(acetylene) from
         suspensions of the pristine polymer by
         sedimentation and drying. Comp. Tab. 4 for
         a comparison of such films with the ones
         prepared in situ.

Table 4. Comparison of samples of poly(acetylene) prepared by Shirakawa's ("S") and Luttinger's ("L") method with regard to their conductivity after doping under exactly the same conditions

Sample type	Method	Conductivity $\sigma(\Omega cm)^{-1}$ after oxidation by	
		Iodine	Electrochemical/SbF$_6$ counterions
Free standing film synthesized in situ	L	120	450
	S	110	490
Film obtained from a suspension (c.f.Fig.10)	L	80	no exp.
	S	70	
Powder from suspension pressed to pellet	L	1.3	9
	S	4.8	15

ACKNOWLEDGEMENT

The author is greatful to Drs. V. Enkelmann, M. Monkenbusch, W. Müller, and J. Eiffler for their permission to report on parts of their yet unpublished work or from their theses. The work reported from the authors laboratory was supported by a grant from the Stiftung Volkswagenwerk (cation radical salts) and by a grant from the BMFT (conducting polymers).

REFERENCES

1. H. Shirakawa, E. J. Louis, A. G. Mac Diarmid,
   C. K. Chiang, A. H. Heeger, J. Chem. Soc. Chem.
   Comm., 578 (1977).
2. A. G. Mac Diarmid, A. J. Heeger, in: W. E. Hat-
   field, Ed. Molecular Metals, Plenum Press,
   New York 1979, p. 161 f.
3. A. G. Mac Diarmid, A. J. Heeger, Synth. Met.
   1, 101 (1980).
4. A. G. Mac Diarmid, A. J. Heeger, in: L. Alca-
   cer, Ed., The Physics and Chemistry of Low-
   Dimensional Solids , D. Reidel Publ., Dordrecht, NL, (1980) p. 393 f.
5. H. Shirakawa, S. Ikeda, Polymer J. (Japan), 2,
   231 (1971).
6. T. Ito, H. Shirakawa, S. Ikeda, J. Polymer
   Sci., Polymer Chem. Ed. 12, 11 (1974).
7. A. J. Heeger, A. G. Mac Diarmid, Mol. Cryst.
   Liq. Cryst. 77, 1 (1981).
8. L. W. Shacklette, R. R. Chance, D. M. Ivory,
   G. G. Miller, R. H. Baughman, Synth. Met.
   1, 307 (1979).
9. R. R. Chance, L. W. Shacklette, G. G. Miller,
   D. M. Ivory, J. M. Sowa, R. L. Elsenbaumer,
   R. H. Baughman, J. Chem. Soc. Chem. Comm.,
   348 (1980).
10. K. K. Kanazawa, A. F. Diaz, R. H. Geiss, W. D.
    Gill, J. F. Kwak, J. A. Logan, J. F. Rabolt,
    G. B. Street, J. Chem. Soc. Chem. Comm.,
    854 (1979).
11. M. Pope, Ch. E. Swenberg, Electronic Proces-
    ses in Organic Crystals, Oxford Science Publi-
    cations, Clarendon Press, Oxford 1982.
12. D. J. Sandman, Mol. Cryst. Liq. Cryst. 50,
    235 (1979).
13. H. P. Fritz, H. Gebauer, P. Friedrich, P.
    Ecker, R. Artes, U. Schubert, Z. Naturforschg.
    B33, 498 (1978).
14. Ch. Kröhnke, V. Enkelmann, G. Wegner, Angew.
    Chem., Int. Ed. Engl. 19, 912 (1980).
15. V. Enkelmann, B. S. Morra, Ch. Kröhnke, G.
    Wegner, J. Heinze, Chem. Phys. 66, 303 (1982).
16. H. J. Keller, D. Nöthe, H. Pritzkow, D. Wehe,
    M. Werner, P. Koch, D. Schweitzer, Mol. Cryst.
    Liq. Cryst. 62, 181 (1980).

17. W. Höptner, M. Mehring,J. U. von Schütz, H.C. Wolf, B. S. Morra, V. Enkelmann, G. Wegner, Chem. Phys. in press
18. P. Koch, D. Schweitzer et al. Mol. Cryst. Liq. Cryst. 86, 87 (1982).
19. V. Enkelmann, Habilitationsschrift, Freiburg 1983.
20. J. Honzl, K. Ulbert, V. Hadek, M. Tlustakova, Chem. Comm. 19, 440 (1965).
21. V. Hadek, J. Chem. Phys. 49. 5202 (1968).
22. K. Huml, Acta Cryst. 22, 29 (1967).
23. V. Hadek, phys. stat. sol. 30, 275 (1968).
24. V. Hadek, P. Zach, K. Ulbert, J. Honzl, Coll. Czech. Chem. Commun. 34, 3139 (1969).
25. Ch. Kröhnke, unpublished.
26. J. Eiffler, G. Wieners, V. Enkelmann, unpublished. J. Eiffler, Thesis, Univ. of Freiburg 1982.
27. H. Kiess, W. Meyer, D. Baeriswyl, G. Harbeke, J. Electron. Mater. 9, 763 (1980).
28. J. M. Pochan, D. F. Pochan, H. Rommelmann, H. W. Gibson, Macromolecules 14, 110 (1981).
29. T. C. Clarke, G. B. Street, Synth. Met. 1, 119 (1980).
30. G. Wegner, Angew. Chem. 93, 352 (1981); Angew. Chem. Int. Ed. Engl.
31. S. Ikehata et al. Phys. Rev. Lett. 45, 1123 (1980).
32. D. Baeriswyl, G. Harbeke, H. Kiess, W. Meyer, in: J. Mort, G. Pfister, Eds. Electronic Properties of Polymers , J. Wiley, New York, N. Y. (1982), p. 268 f.
33. G. J. Hoijtink, P. H. van der Meij, Z. Phys. Chem. N. F. 20, 1 (1959).
34. A. Feldblum, J. H. Kaufman, S. Etemad, A. J. Heeger, T.-C. Chung, A. G. Mac Diarmid, Phys. Rev. B26, 815 (1982).
35. A. G. Mac Diarmid, A. J. Heeger, Synth. Met. 1, 101 (1979/80).
36. E. K. Sichel, M. F. Rubner, S. K. Tripathy, Phys. Rev. B26, 6719 (1982).
37. W. F. Forbes, P. D. Sullivan, H. M. Wang, J. Am. Chem. Soc. 89, 2705 (1967).
38. H. Bock, W. Kaim, Acc. Chem. Res. 15, 9 (1982).

39. H. G. Cassidy, K. A. Kun Oxidation-Reduction
     Polymers , Interscience Publ., New York,
     N. Y. 1965.
40. P. Farber, German Patent 1,035,227 (1958).
41. S. W. Mayer, D. E. McKenzie, US-Patent 3,
     185,590 (1965).
42. S. L. Hsu, A. J. Signorelli, G. P. Pez, R.H.
     Baughman, J. Chem. Phys. 69, 106 (1978);
     R. H. Baughman et al, in W. E.Hatfield, Ed.
     Molecular Metals, Plenum Press, 1979, p. 187.
43. M.Monkenbusch, B. S. Morra, G. Wegner, Makro-
     mol. Chem. Rapid Comm. 3, 69 (1982).
44. K. Shiramura, F. E .Karasz, J. C. W. Chien,
     J. A.Hirsch, Makromol. Chem. Rapid Comm. 3,
     269 (1982); J. C. W. Chien, F. E. Karasz,
     K. Shimamura, Macromolecules 15, 1012 (1982).
45. R. H. Baughman, lecture presented at the 28th
     IUPAC-Symposium on Macromolecules, Amherst
     1982.
46. M. Monkenbusch, Makromol. Chem. Rapid Comm.
     3, 602 (1982).
47. M. Monkenbusch, G. Wieners, unpublished.
48. C. Riekel, H. W. Hässlin, K. Menke, S. Roth,
     J. Chem. Phys. in press.
49. M. Stamm, J. Hocker, A. Axmann, Mol. Cryst.
     Liq. Cryst. 77, 125 (1981).
50. J. F. Rabolt, T. C. Clarke, G. B. Street,
     J. Chem. Phys. 76, 5781 (1982).
51. I. Harada, J. Chem. Phys. 73, 4746 (1980).
52. W. P. Su, J. R. Schrieffer, A. J. Heeger,
     Phys. Rev. Lett. 42, 1698 (1979).
53. S. Etemad et al. Phys. Rev. B23, 5137 (1981).
54. S. Kivelson, Phys. Rev. B25, 3798 (1982).
55. A. J. Epstein, H. Rommelmann, M. Abkowitz,
     H. W. Gibson, Mol. Cryst. Liq. Cryst. 77,
     81 (1981); Phys. Rev. Lett., 47, 1549 (1981).
56. S. Roth, Lecture on occasion of the Int.
     Conf. Conducting Polymers, Les Arcs, France
     1982.
57. C. R. Fincher, Jr., M. Ozaki, A. J. Heeger,
     A. G. Mac Diarmid, Phys. Rev. B19, 4140 (1979).
58. K. Hafner, H. Pelster, Angew. Chem. 73, 342
     (1961); K. Hafner, K. Goliasch, Angew. Chem.
     74, 118 (1962).

59. J. Sondermann, H. Kuhn, Ber. dtsch. Chem. Ges.
    99, 2491 (1966).
60. D. C. Waterman, M. Dole, J. Phys. Chem. 74,
    1906 (1970).
61. G. Wegner, Makromol. Chem. Suppl. 4, 155
    (1981).
62. G. Lieser, G. Wegner, W. Müller, V. Enkelmann,
    Makromol. Chem. Rapid Commun.1, 621 (1980).
63. W. Müller, Ph.D. Thesis, University of Frei-
    burg 1982.

POLYMERS AS ELECTRONIC MATERIALS - TODAY'S POSSIBILITIES

AND TOMORROW'S DREAMS*

R. H. Baughman

Allied Corporation
Corporate Research Center
Morristown, N.J. 07960

SYNOPSIS

Semiconducting and metallic organic polymers can be obtained by the addition of either electron donors or electron acceptors to insulating organic polymers. A status report is made for such presently available highly-conducting polymers, emphasizing both properties aspects which might find applications and problem areas. From this starting point, predictions are made of properties which might ultimately be obtainable - in part, by eliminating the gross structural and chemical irregularities of today's materials. The properties include conductivities comparable with copper, ultimate strengths and modulii higher than for strong steels, and the possibility of superconductivity. Finally, the applications horizon is critically examined, both by comparison with competing materials for today's technologies and by examining the materials needed for future technologies.

INTRODUCTION

In many ways the materials which we use today are not too different from those which early man used to improve the quality of his life. Consider for example the vast array of synthetic plastics and metals compared with naturally-occuring products such as silk, wood, flax, wool, leather, amber, naturally-occuring metals, or the steel of biblical

---

*An abbreviated version of this work was published in Japanese in the journal KOBUNSHI (March 1984).

times. Organic polymeric materials known to the ancients
were obtainable in forms having the general properties profiles of todays commercial plastics - processibility, high
strength, low density, high anisotropy in properties, and
very low electrical conductivity. In contrast, the metals
available in antiquity had high density, low anisotropy,
high electrical conductivity, and even the property of
superconductivity.

The organic polymeric conductors to be discussed provide an important opportunity - the application of materials
which have fundamentally different properties profiles than
either the organic linear polymers, the metals, or the composite materials which mankind has previously known. These
are materials which can have the high anisotropy, processibility, high strength, and low density of organic polymers -
while at the same time having electrical properties which
can be controllably varied from that of insulators to semiconductors and metals and, perhaps in the future, to super-
conductors.

The next section deals with the present state of the
art in conducting charge-transfer organic polymers. These
polymeric conductors are quite different from plastics made
conducting because of metal or carbon fillers, where the
plastic functions largely as a matrix to hold a conventional
conductor. We will be concerned with the variety of organic
backbone polymers which can be made highly conducting
(conductivity greater than 1 S/cm) by the addition of
electron donors or acceptors (doping), general properties
aspects, and problem areas. One will see that the total
properties profiles of conducting organic polymers are not
yet optimized and that major deficiencies exist. Such a
situation is to be expected because of the embryonic nature
of this research and development area.

The third section speculates on the properties which
might ultimately be obtained for highly conducting organic
polymers. A variety of key property aspects are dominated
by extraneous structural features resulting from the synthesis techniques now available. In the future it will be
possible to synthesize doped polymers without the inter-
crystallite and interfibril boundaries, limited conjugation

lengths, gross intracrystallite imperfections, and poor
degree of chain alignment typical for todays polymer
complexes. Dramatic increases in chain-direction properties
(such as conductivity, strength, and modulus) will result
from this improvement in structure.

The last section briefly examines the applications
potential of highly conducting organic polymers. Some of
the applications, such as conducting polymer batteries,
might be near at hand. Other applications depend upon
obtaining the materials improvements contemplated in the
third section.

HIGHLY CONDUCTING POLYMERS FROM TODAY'S TECHNOLOGY - A
STATUS REPORT

The conducting organic polymers shown in Table I are
formed by the addition of either electron donors or electron
acceptors to insulating polymers.[1-24] In the first case a
n-type material (electron conductor) results and in the
second case a p-type material (hole conductor results). By
varying the dopant level or by the "compensation" of p-type
and n-type dopants the electrical conductivity of polymers
such as poly(p-phenylene) can be controllably varied from
less than 10^{-16} S/cm to over 500 S/cm. As expected, dramatic increases in the absorption characteristics (infrared
and longer wavelength) are associated with this increase in
electrical conductivity. Although there is indirect evidence that a number of highly doped organic polymers are
metals, they typically evidence a nonmetallic temperature
dependence of electrical conductivity (positive temperature
coefficient). This is believed to result from either interparticle contact resistances or the contribution of high
resistance domains which are nonmetallic.

Polymers which become highly conducting upon doping
fall into two groups.[25] The first group are those polymers
which can be made highly conducting via simple charge
transfer between the polymer chain and a dopant species. In
this case no irreversible chemical changes, such as covalent
bond formation, are required for the polymer backbone.
Examples of polymers in this group are polyacetylene and
poly(p-phenylene) doped with selected agents such as the

Table I. Polymers which Form Highly Conducting Compositions

Table II. Highly Conducting Compositions from the Polymers in Table I

PARENT POLYMER	R	R'	TYPICAL DOPANTS OR DOPANT IONS (CONDUCTIVITY, S/cm)	REF.
1	-	-	Li^+ (200), ClO_4^- (970)	1 - 4
2	$p-C_6H_4$	-	AsF_5 (3)	5
	O	-	BF_4^- (80)	6
	NH	-	BF_4^- (100)	6, 7
	S	-	ClO_4^- (100)	0,8,9
3	$(CH_2)_3$	-	Iodine (0.1)	10
4	-	-	AsF_5 (500), K^+ (7)	11-13
	S	-	AsF_5 (10)	14-18
5	H	H	BF_4^- (80)	6, 7
	CH_3	H	Acceptor (4)	6, 7
	CH_3	CH_3	Acceptor (10)	6, 7
6	-	-	Iodine (0.45)	19
7	-	-	TCNQ (0.41)	20
8	-	-	AsF_5 (18.5)	21
9	-	-	SbF_5 (0.18)	22
10	-	-	Iodine (1)	23
11	-	-	H_2SO_4 (2), K^+ (1)	24
12	-	-	H_2SO_4	24

alkali metals (n-type doping) or AsF$_5$ (p-type doping).
Group I polymers are singularly important for applications
where the reversibility of the doping process is a key
feature, as in secondary batteries, visual displays, and
variable transmission shields. These polymers are generally
characterized by a broad valence band and low ionization
energy for formation of p-type conductors and a broad conduction band and high electron affinity for formation of n-
type conductors.

The second group are polymers which become highly conducting only as a consequence of covalent bond formation
produced by either the doping process or later treatments,
such as thermally annealing the doped polymer. Good
examples of group II polymers are poly(p-phenylene sulfide)
and poly(m-phenylene sulfide). In the absence of chemical
modification of the polymer backbone, conductivity appears
to be limited to about 10^{-2} S/cm for poly(p-phenylene
sulfide) and to much lower values for poly (m-phenylene
sulfide). Dopants such as AsF$_5$ induce formation of diben-
zothiophene linkages in the polymer backbone via covalent
bond formation. Corresponding to this dopant-induced chemical change, the polymer ionization energy is decreased, π
band widths are broadened, and observed conductivities for
the doped state are increased by about four orders of magnitude for poly(p-phenylene sulfide). The highly conducting
composition can then be characterized as a simple charge-
transfer complex between the polydibenzothiophene polymer
and a p-type dopant. While group II polymers are unsuitable
for applications which require a reversible doping-dedoping
process, we will see that they can have advantages from the
viewpoint of processibility.

Convenient processibility is a property-related aspect
which is clearly important for the application of conducting
polymers. In this regard, group I polymers such as polyacetylene and poly(p-phenylene) have limitations, since neither
high molecular weight polymer is soluble or fusible below
temperatures at which degradation occurs. Fortunately,
polyacetylene can be readily obtained in a gel-like form
which contains microscopic fibers with high aspect ratio.
As originally shown by Shirakawa and coworkers,[26-29] films
having useful properties can be obtained by removal of the

carrier liquid from the gel. High molecular weight poly(p-phenylene) is processible despite infusibility and insolubility because this polymer has sufficient thermal stability (ca. 500°C in vacuum) that high strength parts can be molded by high pressure sintering, analogous to powder metallurgy.[30]

Polymers of group II tend to be more readily processible than those of group I. This is because structures with high degrees of electronic delocalization (broad bands, low ionization potential, and high electron affinity) tend to be relatively rigid. This rigidity means that the enthalpy cost for obtaining entropy increases of melt or solvent states is too high until temperatures at which degradation occurs. However, there are examples of group I polymers which can be processed from fluid states to obtain high strength objects, such as films. Examples are the benzimidazobenzophenanthroline ladder polymers (processible from methane-sulfuric acid solution)[24] and the poly(N-methyl 3', 3-carbazolyl)[23] shown in Table I. The group II polymer poly(p-phenylene sulfide) is commercially fabricated from the melt (melting temperature of about 290°C). However, recent research by Frommer et al.[18] has shown that poly(p-phenylene sulfide) or a dopant-modified version of this polymer can be directly cast as a flexible film from a solution or gel state using a AsF_5/AsF_3 solvent mixture. Conductivities as high as 200 S/cm have been reported, which are more than an order of magnitude higher than obtained by solid-state doping processes. The doping process of Frommer et al.[18] coincides with the film fabrication process, which is important because it is difficult to uniformly dope thick polymer articles by diffusion in the solid state. The solubility necessary for the solution process appears to result from energetically favorable charge-transfer and associated solvation in the fluid phase.

Environmental instability and/or thermal instability is the major problem which has till now precluded a variety of important applications of organic conducting polymers. With the exception of p-doped polypyrrole and related polymers, which appear to have a relatively stable ambient state conductivity,[6,7] highly conducting polymer complexes generally decrease in conductivity upon exposure to the

atmosphere.[31-34] The stability of p-doped polypyrrole
appears to result from the very low ionization energy of
this polymer (ca. 4eV), but the oxidation of this polymer is
a complex, multistep process which can involve eventual
chemistry at the nitrogen atoms without dramatic changes in
conductivity.[35]

Quantum chemical calculations can provide considerable
guidance in the search for polymers which will form environ-
mentally stable highly conducting complexes. For example,
Bredas and Baughman[36] have predicted that the following
biphenylene polymer has an ionization energy (3.4eV) even
lower than for polypyrrole (ca. 4.0eV) or graphite
(4.39eV)[37].

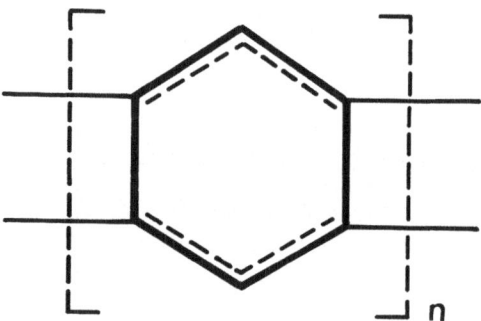

These calculations indicate other promising features: a very
small indirect gap (ca. 0.12eV), a high electron affinity of
order 3.3eV (compared with ca. leV for polypyrrole), and a
highest occupied band nearly as wide as in polyacetylene.

Acceptor-doped complexes of polyacetylene typically
have low thermal stability.[31-34] However, Delannoy et
al.[38] have found that alkali-metal complexes of polyacetylene have surprisingly high thermal stability. More specifically, potassium-doped polyacetylene maintains high
conductivity and does not undergo measureable decrease in
structural perfection after a 16 hour anneal in inert
atmosphere at 200°C. In fact, the potassium-doped cis-
polyacetylene increases electrical conductivity by about a
factor of five during high temperature thermal annealing and

the apparent activation energy for conduction substantially decreases. Furthermore, the anneal process causes pressure contacts to become ohmic and the contact resistance decreases by a factor of about 25.

HIGHLY CONDUCTING POLYMERS FROM FUTURE TECHNOLOGIES – SPECULATION ON ULTIMATELY OBTAINABLE PROPERTIES.

We are here concerned with developing upper limit estimates for the properties obtainable as a consequence of eliminating the gross structural imperfections of todays conducting polymers. We will further speculate on the possibility of obtaining properties such as superconductivity, as well as on synthetic methods which might be important in the future.

The first question relevant here is "What is the maximum electrical conductivity which might eventually be obtained for highly conducting organic polymers?". Since conductivity σ is $nq\mu$ (where n is the carrier density, q is the charge on the electron, and μ is the mobility), the above question reduces to the estimation of the maximum values for n and μ. We will specifically focus on complexes of polyacetylene.

The simplest part of this problem is evaluating the maximum concentration of carriers. The maximum value might result for a small ion such as Li^+. Since the radius of interchain voids in polyacetylene is comparable with the radius of Li^+, relatively little volume expansion will occur upon doping if solvent coinsertion is avoided.[39] Hence, to good approximation, the carrier density in $(CHLi_{0.30})_x$ will be 0.3 electrons in the volume of a CH group in undoped polyacetylene (18.6 $\overset{\circ}{A}{}^3$) or n = 1.6 x 10^{22} electrons/cm^3. For comparison, the carrier concentration in copper (8.5 x 10^{22} electrons/cm^3) is a factor of five higher. Lower maximum carrier densities would result for larger dopants such as iodine, which is largely present in heavily-doped polyacetylene as I_5^-.[40,41] Using structural results for this complex,[25,39] n is 2.3 x 10^{21} electrons/cm^3 or a factor of seven lower than for the lithium complex. Note that these estimates of carrier concentration might grossly underestimate the carrier concentration if the doping process eliminates bond alternation in the polyacetylene backbone. In this case, it is conceivable that each π electron in the polymer contributes to the electron carrier density, which would itself provide $n = 5.4 \times 10^{22}$ electrons/cm^3 (neglecting dopant-induced lattice expansion).

W. Röss et al.[42] have used measurements of collinear magnetoresistance and coaxial Corbino resistance at 4.2°K to derive the "effective carrier mobility" in unoriented iodine-doped polyacetylene. Values obtained slightly decreased with increasing dopant levels and depended upon doping method, but an observed value of 120 cm^{2}V^{-1}s^{-1} was obtained even for a I/C ratio as high as 0.23. Uncertainties exist in the meaning of this mobility because of the microfibular nature of the polyacetylene investigated. However, the derived decreasing mobility with increasing doping level at high dopant levels is consistent with results obtained for graphite complexes[43-46] and a plausible explanation for such high mobility (relative to the observed conductivity level of 270 S/cm) is that interfibril resistances limit the observed conductivity. The presence of gross intracrystallite imperfections and some reaction of iodine with the polymer chain[33,47] means that the intracrystallite mobility of the presently available material is far below that which will be eventually obtainable for improved materials.

In addition, a large anisotropy in carrier mobility is expected for polyacetylene complexes, especially those involving electron acceptors. This is consistent with the observation of electrical anisotropy (a conductivity greater than 2000 S/cm in the direction of chain orientation for $[\text{CH(AsF}_5)_{0.10}]_x$, which is a factor of about 15 higher than for orthogonal directions) and optical anisotropy for even poorly-oriented, doped polyacetylene.[48-50] Consequently, the above mobility results obtained for unoriented samples are expected to grossly underestimate the chain-direction mobility. Using a simple model for electron-phonon interaction, Guinea[51] has estimated that the chain-direction mobility for electrons is 10^3 cm^{2}V^{-1}s^{-1} in defect-free polyacetylene. This result is consistent with measurements of chain-direction mobility in high perfection single crystals of polydiacetylenes,[52-55] which have a backbone structure that has somewhat analogous properties with that of polyacetylene.[56] Chain-direction electron mobilities of 6×10^3 $cm^2V^{-1}s^{-1}$ have been experimentally derived from dark-field carrier injection and from the initial interpretation of pulse and steady-state photoconductivity measurements on polydiacetylene crystals.[52-54] More recent interpretation of the latter results, considering that drift velocity saturates at relatively low fields, provides an even higher mobility ($>2 \times 10^5$ $cm^2V^{-1}s^{-1}$).[53,54] The value 6×10^3 $cm^2V^{-1}s^{-1}$ is roughly consistent with electroreflection measurements[55] and theoretical analysis[57], so this value will be used for further discussion of the plausible electron mobility for polyacetylene complexes. Calculated hole mobilities for polydiacetylene are dramatically lower (26 $cm^2V^{-1}s^{-1}$).[57]

The most optimistic and perhaps most naive estimate of the maximum chain-direction electrical conductivity which might be obtainable for an essentially defect-free crystal of a polyacetylene complex results from the product of the above chain-direction electron mobility experimentally derived for polydiacetylene crystals and the calculated carrier concentration in $(CHLi_{0.3})_x$. The resulting calculated electrical conductivity is 1.7×10^7 S/cm or over an order of magnitude higher than the room temperature conductivity of copper. This estimate for a defect-free crystal is probably overly optimistic for several reasons. As discussed by Fischer for graphite complexes,[44-46] the degree of chemical ionization need not correspond with the degree to which free carriers are added to the conduction band or the valence band. This effect for graphite is not very significant for LiC_6 (10% deviation between the above quantities) or for weak acceptors such as bromine, but can be quite large for strong acceptors. A major decrease in carrier mobility for the polymer complex compared with that for the undoped polymer could dramatically decrease the obtainable conductivity for an essentially defect-free crystal. At low dopant levels, coulomb attraction will localize carriers on the polymer chain - thereby reducing the effective mobility. At high dopant levels, such coulomb attraction will be more effectively screened. The magnitude of reduction of carrier mobility in the highly-doped, fully-ordered polymer is unclear at the present time. However, note that even if this reduction in mobility were several orders of magnitude, chain-direction conductivities comparable to that of copper would be obtainable for polymer complexes having high structural perfection. For comparison, graphite complexes are known in which the high mobility characteristic of the basal plane is not degraded by more than an order of magnitude by complex formation (i.e., $\mu(300°K) > 1000$ $cm^2V^{-1}s^{-1}$ in the complexes).[46] The basal-plane conductivities of certain graphite complexes are comparable with that of copper.[46]

While the calculation of carrier concentration used here is for lithium, this doped polymer might not be the best choice for maximizing conductivity. First of all, theoretical calculations indicate that lithium will be statistically distributed at room temperature until high dopant concentrations. This is not the case for larger dimension donor or acceptor dopants which can aggregate in columns in place of displaced polymer chains. Also, the basal-plane Hall mobilities for graphite heavily doped with electron acceptors are much higher than for comparable dopant levels of the alkali metals. Presently observed maximum conductivities for acceptor-doped polyacetylene are also higher than the maximum observed for alkali metal complexes, but the difference is only a factor of 2 to 4 and might arise from differences in high resistance domains and interparticle contacts for the acceptor and donor dopants.

Applications of effective medium theory to optical data of Fincher et al.[50] on partially-oriented AsF_5-doped polyacetylene provided a dc conductivity above 2×10^4 S/cm for the highly disordered material. Park et al.[49] obtained even a higher value (4×10^4 S/cm) by analysis of the temperature dependence of dc conductivity. It is certainly conceivable that the order of magnitude increase required to obtain a conductivity comparable with that of copper (6×10^5 S/cm) could be obtained by increasing the perfection of the material.

The high anisotropy of carrier mobility in conjugated organic polymers might have useful applications. Siddiqui and Wilson[58] obtained a 10^3 higher mobility in the chain direction than for orthogonal directions in polydiacetylene crystals. As is the case for graphite complexes,[43-46] higher anisotropy is expected for acceptor complexes of polymers such as polyacetylene than for alkali metal complexes.[59-60] This anisotropy (ratio of in-plane to orthogonal conductivity) increases from 2500 in undoped graphite to above 10^6 for the C_8AsF_6 complex, but is much lower for the alkali metal graphite complexes (ca. 30 for the potassium complex)[43-46]. Likewise, Park et al. report[49] that the experimentally determined electrical anisotropy of both AsF_5-doped polyacetylene and iodine-doped polyacetylene increases with doping level.

Extremely high modulus and tensile strengths in the chain direction are expected for high perfection fibers of conjugated polymers. Specifically, the per-chain modulus and per-chain tensile strength obtainable for polyacetylene should equal those which have been obtained for polyethylene[61,62]. Since lithium doping causes little lattice expansion, nearly defect-free $(CHLi_{0.30})_x$ should have a modulus of about 2.9×10^{12} dyne/cm^2 and an ultimate strength in excess of 4.9×10^{10} dyne/cm^2. These values are derived from experimentally determined per-chain properties in polyethylene[61,62]. Use of corresponding measured chain-direction parameters for polydiacetylene crystals[63] provides the same modulus, but an even higher ultimate tensile strength for the conducting polymer (1.1×10^{11} dyne/cm^2). Such modulus and ultimate strength values are higher than those typical for conventional metals (2.1×10^{12} dyne/cm^2 and 2×10^{10} dyne/cm^2, respectively, for piano wire steel).[64]

The search for conducting polymers which undergo a transition to the superconducting state will be one of the more interesting adventures of the next decades. There is naturally special interest in the discovery of compositions which display new mechanisms for superconductivity, such as proposed by Little[65] in his theory of excitonic superconductivity, because of the possibility of high temperature superconductivity. Both organic charge transfer complexes (di-tetramethyltetraselanafulvalenium perchlorate, $(TMTSF)_2ClO_4)$[66-68], inorganic polymers $((SN)_x$ and the $(SN)_x$-bromine complex)[69], and graphite complexes $(C_8K, C_8Rb,$ and $C_8Cs)$[70] are known which become superconducting at

atmospheric pressure, but transition temperatures are low
and there is no clear evidence for an entirely new mechanism
of superconductivity. On the other hand, tunnel spectroscopy, infrared spectroscopy, and the sharp dependence of
electrical conductivity on magnetic field for the
(TMTSF)$_2$ClO$_4$ salt has been interpreted in terms of electron
pairing on one-dimensional stacks occuring in the 20-40°K
range.[67-68] If correct, this result would denote a mean-
field superconducting transition temperature in the range
10-20°K, which is far above the observed three-dimensional
transition (ca. 1.5K). Grant[67] concludes from semiempirical
molecular orbital calculations that this complex is quasi
two-dimensional and that this extra dimensionality stabi-
lizes the salt against a transition to the Peierls insulating state. The behavior of (TMTSF)$_2$ClO$_4$ was found remini-
scent of (SN)$_x$, where the normal state scattering process is
determined by mechanisms other than electron-phonon effects
and where the dependence of these effects upon unit cell
parameter (electron-hole scattering for (SN)$_x$) determines
the pressure and temperature dependence of the
conductivity.[67,69]

Alkali metal doped polymers, and in particular the
polyacetylene complexes with these donors, provide likely
candidates for the first superconducting organic polymer,
albiet with low expected transition temperature. The reason
for this choice is the observation of bond distances consistent with similiar hybridization of alkali metal s orbitals with carbon π orbitals in the polyacetylene and the
graphite complexes.[59] The decreased electrical anisotropy
expected from such interactions is believed important for
explaining the existence of superconductivity in the alkali
metal-graphite complexes.[46] Further encouragement for the
existence of superconductivity in metallic polymer complexes
is the notable absence of instabilities as a function of
temperature resulting in metal-insulator transitions. Since
the transition temperature is much higher for the potassium-
graphite complex than for the rubidium and the cesium
complexes, the potassium complex of polyacetylene is the
most reasonable choice for an initial search for superconductivity in the polymer complexes.[46,70]

THE APPLICATIONS HORIZON

In discussion of the applications potential of conducting polymers, we might anticipate the existence of two extreme grades of such materials. In the first extreme (economy grade), where cost is a main consideration, the materials synthesis and processing methods might be analogous to those used in todays polymer technologies. In the second extreme (device grade), where optimal electronic and/or mechanical properties are required, advanced synthetic methods must be developed which result in high structural perfection (low trap density, high mobility) materials. Here we could be in the high cost regime of semiconductor crystals.

An inexpensive method for forming highly conducting polymers might be the coaddition of monomer and doping agent during a step involving both polymerization and product fabrication. Polymerization methods are already claimed for insulating polymers, whereby fibers are spun directly from the monomer at speeds of up to 5000 yards per minute at ambient temperatures without use of any additional liquid such as a solvent.[71] Likewise, processes might evolve which are analogous to todays RIM technology (Reaction Injection Molding), wherein conducting polymers are directly fabricated into shaped articles by combining monomer and dopant in a mold. In this regard it is noteworthy that monomers like terphenyl,[12] acetylene,[72] and thiophene[73] can be directly polymerized to form highly conducting polymer compositions (complexes of poly(p-phenylene), polyacetylene, and polythiophene, respectively) merely by the addition of a dopant such as AsF_5. Unfortunately, the conductivity and/or mechanical properties of these materials is presently inferior to those obtained by doping the polymer synthesized by conventional techniques.

The formation of the best "device grade" conducting polymers would likely involve (1) matrix or substrate-controlled polymerization of monomer in the presence of dopant or (2) successive reaction, doping, and crystal formation at a growing interface.[25] Nakaniski and coworkers have already attempted the direct synthesis of conducting polymers by solid-state (matrix-controlled) polymerization

of doped diacetylene monomer crystals.[74] Unfortunately the
resulting polydiacetylenes did not provide highly conducting
charge-transfer complexes.

The application of conducting polymers as simple conductors largely falls in three categories: the replacement
of (1) elemental metals and metal alloys, (2) graphite and
related carbonaceous materials, and (3) polymer–metal and
polymer–carbon composites. Looking into the future, we must
consider the possible importance of conducting polymers in
each of these areas.

As discussed in the last section, it is conceivable
that chain-direction electrical conductivities comparable to
copper could be obtained for nearly defect-free polymer
complexes. The synthesis of such materials would not
require the high temperatures necessary for the preparation
of elementary metals and carbons. Furthermore, the low density, high modulus, and high ultimate strength obtainable
for the conducting polymer complexes could be of crucial
importance. While an obtainable chain-direction modulus and
strength higher than for strong steels is predicted for
nearly defect free $(CHLi_{0.30})_x$, the density of the polymeric
composition would be about one-fifth that of steel. The
high electrical anisotropy obtainable for polymer complexes
could also be of major importance. In the future it might
be possible to use certain conducting polymer fibers
(acceptor complexes) as essentially "self-insulating wires".
Conducting polymer films might be obtainable that, because
of high anisotropy of conduction, function as an array of
minute, self-insulated wires for addressing molecular-scale
circuit elements.

Assumming that conducting polymers with useful property
profiles can be obtained by processes analogous to those now
used for engineering plastics, the energy crisis will provide impetus for the displacement of conventional metals.
This benefit could result even for conducting polymers
having conductivities much lower than those of conventional
metals, since many applications require no higher conductivities than already obtained for conducting polymers.
Although the manufacture of organic polymers requires energy
resources both as raw materials and for synthesis processes,

the total energy cost here is typically much less than for
producing an ingot of a conventional metal.[75] Furthermore,
the lower energy requirement for product fabrication can
provide an additional advantage for the polymers.

The advantages of using conducting polymers to replace
conductor-loaded insulating plastics might be several-fold.
In order to obtain as high electrical conductivities for
conventional composites as are presently obtained for highly
conducting polymers, it is necessary to employ high concentrations of the conducting component, whether it is an
elementary metal or a carbon composition. Depending upon
the specific composite and the fabrication technique, this
can cause severe problems for processibility and mechanical
properties (embrittlement).[76-77] Some of the most
interesting applications are those when the inhomogeneous
nature of conducting composites limits application. For
example, carbon-loaded insulating polymers are currently
used as conductor shields and as insulator shields for
underground electrical transmission lines. The purpose of
these shields is to eliminate electric field inhomogeneities
resulting in premature dielectric breakdown. However, since
the conventional composites are inherently nonuniform in
electrical properties, aggregation of the conductor particles results in regions of grossly nonuniform electric
field. According to a report prepared for the Electric
Power Research Institute,[78] operating fields (2 to 3.5
kV/mm) must be several orders of magnitude lower than the
observed dielectric strength for the crosslinked polyethylene insulation (787 kV/mm). The most important factor for
this difference is irregularities in the electric smoothness
of the conductor shield, which can result from aggregation
of carbon particles.[78] Uniformly doped conducting polymers
might solve this problem, which requires relatively low
electrical conductivities (ca. 10^{-7} S/cm). Similarly low
electrical conductivities are quite sufficient for many
applications of conducting polymers which require static
charge elimination (carpeting, materials for explosion-proof
industrial and hospital environments, and for protection of
discharge-sensitive electronic devices).[77] Ambient stability is much less of a problem in this low conductivity
range. Also, impurity levels present in "undoped" trans
polyacetylene can result in higher conductivities. In fact,

small bandgap polymers (such as the polybiphenylene polymer
mentioned in the second section) should have a higher conductivity even in the pristine, dopant and impurity-free
state.

Conducting charge-transfer polymers might also replace
metal-polymer composites used in polymer thick-film technology for circuit boards. In todays technology, polymer
inks containing metal particles are printed on substrates
such as plastics, glass, or paper - followed by a heating
step to eliminate solvent or to set resins. A variety of
electronic circuit elements (such as resistors, capacitors,
fuses, and connecting wires) can be conveniently formed at
speeds approaching that of conventional printing processes.
The less costly inks used today (containing nickel or etched
copper) are expensive (8 to $13/oz.)[79] and do not provide
either as high a conductivity or the uniform conductivity
which would be obtainable using the conducting polymer
charge-transfer complexes. Hocher et al.[80-82] have already
demonstrated deposition of adherent, uniform coatings of
polyacetylene on paper, glass, and metals by deposition from
a liquid/polymer mixture.

Problems with electrically conducting adhesives provide
another possible opportunity for conducting polymer-dopant
compositions. Today's adhesives, such as used to replace
soldered connections, conduct because of metal or carbon
fillers. While high loadings are necessary to obtain high
conductivities, these loadings dilute the mechanical properties of the adhesive and the conductive particles serve as
possible nucleation sites for crack initiation.

Highly conducting organic polymers are of considerable
interest for shielding sensitive electronic equipment from
the disturbing effects of electromagnetic radiation, as well
as for shielding potential sources of electromagnetic
radiation. Recent regulations in a number of countries
against environmental pollution by electromagnetic interference and problems with existing technologies are strong
motivating factors. The application possibilities are enormous, as housings for everything from office equipment, computers, cash registers, automotive electronics, and motors
to electronic games. The application of conductive finishes

is most commonly used at present to obtain EMI (electromagnetic interference) shielding for plastics.[76] Associated
problems relate to the cost of such secondary steps in manu-
facturing housings and the susceptibility of the surface
coatings to flaking, peeling, or corrosion. Use of composites of conventional plastics conductors is of increasing
commercial interest, but is hampered by non-uniformity in
shielding due to inhomogeneous dispersion, degradation of
mechanical properties, decreased moldability, and cost.
Since the absorption coefficient goes as the square root of
conductivity, bulk conductivities of 1 S/cm are sufficient
for many applications. A bulk conductivity of 1 S/cm for a
typical thickness housing can provide a shielding effec-
tiveness of 30-40 db corresponding to blocking 99.9-99.99%
of the interference.[76] Hence, the conductivities of the
presently available, highly conducting organic polymers are
quite suitable. Highly conducting polymers having the
desired processibility, mechanicals, or ambient stability
for this application can be obtained from the present art,
but no presently reported highly conducting organic polymer
has all of these features.

The application of highly conducting polymers as battery electrodes provides exciting applications possibilities
which are being aggressively pursued by companies in the
United States, Japan, Germany, and France. Interest in specific conducting polymers for this application (such as
polyacetylene and poly(p-phenylene)) relate to their low
density, insolubility, high obtainable surface area, reversible dopability with both n-type and p-type dopants, and
high electrical conductivity. Extremely high power densities are obtainable for conducting polymer batteries,
while maintaining high energy densities, and the polymeric
nature of the active battery materials permits development
of novel battery designs.[83-89]

The reversibility of the electrochemical doping process
is of interest for a variety of applications other than batteries. For example, since the visual appearance of polymers such as polypyrrole dramatically change on doping,
electronic displays have been made using this polymer.[90-92]
Similarly, it is feasible to electrochemically switch a
conducting polymer film from states which are transparent to

those which are opaque for microwave or infrared radiation –
so as to provide a variable window for these radiations.
One example might be dynamic "solar-age" windows for the
control of room temperature via changing the relative
transparency and reflectivity of window panes. As is true
for the battery use, long cycle life and rapid electrochemical changes can be important for specific applications. The
use of special foam-like doped polyacetylene as an efficient
microwave absorber has been described by Feldbaum et al.[93]

Opportunities exist for applications which utilize
simultaneously the insulating nature of the undoped polymer
and the highly conducting nature of the doped polymer. For
example, many of the circuits used today employ plastic
substrate film or board upon which conventional metals as
deposited for connecting circuit elements. It is possible
to use an undoped polymer as such a substrate and "write"
the conducting circuit elements via the addition of dopant
to selected regions of the polymer. Clarke et al.[94] have
demonstrated a photoinduced reaction that renders UV exposed
regions of polyacetylene film highly conducting. Likewise,
Allen et al.[95] and Mazurek et al.[96] have shown that ion
implantation can be used to render conducting exposed region
of an insulating film.

Highly conducting organic polymers have been of considerable interest in semiconductor device applications.
Specific examples are as either the metal or the semiconductor part of Schottky diodes or as the n-type or the p-type
material in n-p juctions.[97-104] The performance of solid-
state devices in which the conducting polymer is used as a
semiconductor has been rather disappointing. For example,
Schottky diodes which use polyacetylene as the semiconductor part of the junction display relatively low ratio of
forward-to-back resistance and high values for the diode
quality factor (above about 3). Conversion efficiencies
were low for Schottky barrier solar cells using
poly(p-phenylene sulfide) or polyacetylene as the semiconductor.[99,102,103] In the case of polyacetylene, the
maximum energy conversion efficiency was about 1% (based on
energy absorbed in the polymer for 7 mW/cm^2 incident intensity in the barrier region) and saturation occurred at low

light levels.[102] These results are not surprising considering the gross structural inhomogenities and the high
impurity levels of presently available materials.

The question to be addressed in future work is whether
or not the quality materials can be made that would be competitive with the well-developed materials technologies of
conventional inorganic semiconductor materials. While
nearly defect-free polymer crystals might have mobilities in
the chain direction which are comparable with those of
inorganic semiconductors, presently available conducting
polymers have high trap concentrations, chain disorder, and
internal resistance domains that result in low effective
mobilities and severe trapping. The dream of conducting
polymers as inexpensive semiconducting plastic sheets (for
example, as large area photocells) must face the necessity
for dramatic improvement in methods used for polymer synthesis. When such methods are developed, it remains to be seen
whether or not there is a cost-performance advantage compared with alternate semiconductor technologies, such as for
amorphous silicon. In addition, dopant mobility can pose a
severe problem for many types of junction applications of
conducting polymers. For this reason it is desireable to
employ dopants whose diffusion coefficient in the solid
state is so low that the doping process must be done either
(1) during polymer synthesis, (2) during solution or melt
fabrication processes, (3) by chemical transformation of
mobile dopant agents to immobile dopant species, (4) by ion
implantation, or (5) by high temperature processes. On the
positive side, polyacetylene has an absorption spectrum
which closely matches the solar spectrum, a very large
absorption coefficient (10^5 -10^6 cm^{-1} between 1.5 and 3.0
eV), and a band gap (ca. 1.5 eV) at which optimum energy
conversion can be obtained.[97,102,103]

The prospects for using conducting polymers as metals
in junction devices might not require the material improvements of the semiconductor application. The major advantage
here is that p-doped conducting polymers can provide work
functions that are much higher than for the elemental
metals. Ohmic contacts to p-type wide-bandgap semiconductors require high work function metals and the barrier
height with n-type semiconductors increases with increasing

work function of the metal. For polypyrrole, which has an
unusually low work function for a organic polymer, observed
barrier heights for the doped polymer on n-type semiconductors are consistent with a work function equal to that of
gold.[105-106] At comparable acceptor levels per carbon the
work function of poly(p-phenylene) will be about 0.7 eV
higher than for polyacetylene[25] and the work function of
heavily-doped polyacetylene (5.7 eV by photoemission studies
for AsF_5 dopant)[107] already well exceeds that of gold.[98]
The behavior of $(SN)_x$ in Schottky barrier devices is consistent with a work function at least 0.5 eV higher than for
gold.[100,101] As a consequence, Schottky solar cells using
$(SN)_x$ as the metal on GaAs provide an open circuit voltage
which is over 40% higher than for the analogous junction
using a gold contact.[100]

The use of conducting polymers as metals to protect
semiconductor surfaces in photoelectrochemical solar cells
has been of special interest and might afford application
possibilities in the future. This application addresses the
most serious obstacle in the use of semiconductors in photochemical solar cells: the susceptibility of narrow-band-
gap n-type semiconductor anodes to oxidative degradation.
Dramatic increases in semiconductor stability has been
obtained using p-doped polypyrrole as a protective layer on
n-Si, n-GaAs, and hydrogenated amorphous silicon
electrodes.[105,106,108,109] The high work function obtained
for the doped polymer is important in the case when the
polymer is not penetrated by the electrolyte, since rec-
tification at the semiconductor-polymer interface facilitates charge transport away from the semiconductor.

Dopable conjugated polymers have been investigated as
both chemical and biochemical sensors - wherein changes in
electrical conductivity of the polymer provide the device
response.[110-113] Initial efforts in this area dealt with
smoke detectors,[110-112] where the motivation was to eliminate the radioactive sources of ionization gauges. An immediate problem in this and in related applications is both
reversibility and the selective response to desired agents.
Malmros has addressed the problem of selectivity in his use
of polyacetylene as the active element in a semiconductor
device for performing immunoassays.[113] The polymer is

exposed to an antibody which interacts with the polymer and thereby changes electrical conductivity. Subsequent exposure to the corresponding antigen provides a further change in electrical conductivity related to the concentration of this antigen.

In a new materials area it is easy to be too constrained in thinking of application possibilities. Various applications might result that do not directly depend upon electrical conductivity. For example, both donor and acceptor complexes of graphite have been of interest for some time as chemical reagents and as catalysts.[114-115] High catalytic activity for doped polyacetylene in causing electrolyte breakdown has been a major problem area for battery development, but might be of value for other applications. Wnek and coworkers[116] have shown that the catalytic activity of doped polyacetylene can be used to polymerize a number of monomers so as to produce polyacetylene composites with insulating polymers.

In these early days of research and development in highly conducting polymers area it is important to have many ideas of what might be possible in the future. As more knowledge accumulates, we expect to find that some of these ideas are not feasible and many will not be commercially exploitable. The growth of conducting polymers as engineering materials will depend upon the continual modification of goals and the generation of new dreams for the future based on the evolving knowledge of the synthesis, properties, and cost of these and competing materials. There is opportunity in the highly conducting polymers area, both in the realization of the dreams we dream today and those which mankind will envision tomorrow.

ACKNOWLEDGEMENTS

The author thanks P. Delannoy, L. W. Shacklette, and R. L. Elsenbaumer for valuable comments on this work.

REFERENCES

1. C.K. Chiang, M.A. Druy, S.C. Gau, A.J. Heeger, E.J.
   Louis, A.G. MacDiarmid, Y.W. Park, H. Shirakawa, J.
   Am. Chem. Soc. 100, 1013 (1978).
2. H. Shirakawa, E.J. Louis, A.G. MacDiarmid, C.K. Chiang,
   and A.J. Heeger, Chem. Commun. 1978, 578.
3. P.J. Nigrey, A.G. MacDiarmid, A.J. Heeger, J.C.S. Chem.
   Comm. 594 (1979).
4. A.G. MacDiarmid and A.J. Heeger, Synthetic Metals 1,
   101 (1980).
5. G.E. Wnek, J.C.W. Chien, F.E. Karasz, and C.P. Lillya,
   Polymer 20, 1441 (1979).
6. G. Tourillon and F. Garnier, J. Electroanal. Chem. 135,
   173 (1982).
7. K.K. Kanazawa, A.F. Diaz, R.H. Geiss, W.D. Gill, J.F.
   Kwak, J.A. Logan, J.F. Rabolt, and G.B. Street,
   J.C.S. Chem. Comm. 854, (1979); G.B. Street, T.C.
   Clarke, R.H. Geiss, V.Y. Lee, A. Nazzal, P. Pfuger,
   and J.C. Scott, Journal de Physique, Colloq., in
   press.
8. K. Kaneto, K. Yoshino, and Y. Inuishi, Solid State
   Comm. 46, 389 (1983).
9. T. Yamamoto, K. Sanechika, and A. Yamamoto, J. Polym.
   Sci.: Polym. Lett. Ed. 18, 9 (1980).
10. H.W. Gibson, F.C. Bailey, J.M. Pochan, A.J. Epstein,
    and H. Rommelmann, Organic Coatinga and Plastics
    Chemistry 42, 603 (1980).
11. D.M. Ivory, G.G. Miller, J.M. Sowa, L.W. Shacklette,
    R.R. Chance, and R.H. Baughman, J. Chem. Phys. 71,
    1506 (1979).
12. L.W. Shacklette, H. Eckhardt, R.R. Chance, G.G. Miller,
    D.M. Ivory, and R.H. Baughman, J. Chem. Phys. 73,
    4098 (1980).
13. L.W. Shacklette, R.R. Chance, D.M. Ivory, G.G. Miller,
    and R.H. Baughman, Synthetic Metals 1, 307 (1980).
14. R.R. Chance, L.W. Shacklette, G.G. Miller, D.M. Ivory,
    J.M. Sowa, R.L. Elsenbaumer, and R.H. Baughman,
    J.C.S. Chem. Comm. 347 (1980).
15. J.F. Rabolt, T.C. Clarke, K.K. Kanazawa, J.R. Reynolds,
    and G.B. Street, J.C.S. Chem. Comm. 348 (1980).
16. L.W. Shacklette, R.L. Elsenbaumer, R.R. Chance, H.
    Eckhardt, J.E. Frommer, and R.H. Baughman, J. Chem.
    Phys. 75, 1919 (1981).

17. T.C. Clarke, K.K. Kanazawa, V.Y. Lee, J.F. Rabolt, J.R. Reynolds, and G.B. Street, J. Poly. Sci.: Polym. Phys. Ed. 20, 117 (1982).
18. J.E. Frommer, R.L. Elsenbaumer, and R.R. Chance, Organic Coatings and Applied Polymer Chemistry 48, 552 (1983).
19. T. Yamamoto and K. Sanechika, preprint.
20. H. Anzai, M. Tokumoto, and T. Ishiguro, preprint.
21. R.L. Elsenbaumer and L.W. Shacklette, J. Polym. Sci.: Polym. Phys. Ed. 20, 1781 (1982).
22. M. Sato and K. Kaeriyama, preprint for Autumn Meeting of Chemical Society of Japan (Niigata, Oct. 2, 1982).
23. S.T. Wellinghoff, T. Kedrowski, S. Jenekhe, and H. Ishida, J. de Physique, Colloq., in press.
24. Oh-Kil Kim, J. Polym. Sci.: Polym. Lett. Ed. 20, 663 (1982).
25. R.H. Baughman, J.L. Bredas, R.R. Chance, R.L. Elsenbaumer, and L.W. Shacklette, Chem. Rev. 82, 209 (1982).
26. H. Shirakawa and S. Ikeda, Polym. J. 2, 231 (1971).
27. H. Shirakawa, T. Ito, and S. Ikeda, Polym. J. 4, 460 (1973).
28. T. Ito, H. Shirakawa, and S. Ikeda, J. Polym. Sci.: Polym. Chem. Ed. 12, 11 (1974).
29. T. Ito, H. Shirakawa, and S. Ikeda, J. Polym. Sci.: Polym. Chem. Ed. 13, 1943 (1975).
30. D.M. Gale, J. Appl. Polym. Sci. 22, 1955 (1978).
31. T. Inoue, J.E. Osterholm, H.K. Yasuda, and L.L. Levinson, Appl. Phys. Lett. 36, 101 (1980).
32. J.E. Osterholm, H.K. Yasuda, and L.L. Levenson, J. Appl. Polymer Sci. 27, 931 (1982).
33. A. Pron, P. Bernier, M. Rolland, S. Lefrant, M. Aldissi, F. Rachdi, and A.G. MacDiarmid, Materials Science Vol VII, 305, No. 2-3 (1981).
34. M. Rolland, S. Lefrant, M. Aldissi, P. Bernier, E. Rzepka, and F. Schue, J. Electronic Materials 10, 619 (1981).
35. P. Pfluger, M. Krounbi, G.B. Street, and G. Weiser, J. Chem. Phys. 78, 3212 (1983).
36. J.L. Brédas and R.H. Baughman, J. Polym. Sci.: Polym. Lett. Ed. 21, 475, (1983).
37. P. Delhaes, Materials Science and Engineering 31, 225 (1977).

38. P. Delannoy, G.G. Miller, N.S. Murthy, C.E. Forbes, H.
    Eckhardt, R.L. Elsenbaumer, and R.H. Baughman,
    Presentation at the March 1983 Meeting of the
    American Physical Society (Los Angeles, CA).
39. R.H. Baughman, N.S. Murthy, L.W. Shacklette, G.G.
    Miller, and H. Eckhardt, J. de Physique, in press.
40. T. Matsuyama, H. Saki, H. Yamaoka, and Y. Maeda, Solid
    State Communications 40, 563 (1981).
41. G. Kaindl, G. Wortmann, S. Roth, and K. Menke, Solid
    State Communications 41, 75 (1982).
42. W. Ross, A. Philipp, K. Seeger, K. Ehinger, K. Menke,
    and S. Roth, Solid State Communications 45, 933
    (1983).
43. M.S. Dresselhaus and G. Dresselhaus, Advances in
    Physics 30, 139 (1981).
44. J.E. Fischer, Physica 99B, 383 (1980).
45. J.E. Fischer, Intercalated Layered Materials, F. Levy
    Ed. (D. Reidel Publishing Co., Dordrecht, Holland),
    pp 481-532.
46. J.E. Fischer, Comments Solid State Phys. 9, 93 (1979).
47. D.C. Weber, J.P. Ferraris, P. Brant, W.B. Fox, A.W.
    Webb, and E.R. Carpenter, NRL Memorandum Report
    4335, R.B. Fox, Ed., Sept. 15, 1980.
48. C.R. Fincher, Jr., D.L. Peebles, A.J. Heeger, M.A.
    Druy, Y. Matsumura, A.G. MacDiarmid, H. Shirakawa,
    and S. Ikeda, Solid State Communications 27, 489
    (1978).
49. Y.W. Park, M.A. Druy, C.K. Chiang, A.G. MacDiarmid,
    A.J. Heeger, H. Shirakawa, and S. Ikeda, J. Polymer
    Sci.: Polymer. Lett. Ed. 17, 195 (1979); Y.W. Park,
    A.J. Heeger, M.A. Druy, and A.G. MacDiarmid, J.
    Chem. Phys. 73, 846 (1980).
50. C.R. Fincher, Jr., M. Ozaki, M. Tanaka, D. Peebles, L.
    Lauchlan, A.J. Heeger, and A.G. MacDiarmid, Phys.
    Rev. B 20, 1589 (1979).
51. F. Guinea, J. Phys. C: Solid State Phys. 15, 241
    (1982).
52. W. Spannring and H. Bassler, Chemical Phys. Lett. 84,
    54 (1981).
53. K.J. Donovan and E.G. Wilson, Phil. Mag. B44, 9 (1981).
54. E.J. Wilson, Chem. Phys. Lett. 90, 221 (1982).
55. L. Sebastian and G. Weiser, Phys. Rev. Lett. 46, 1156
    (1981).

56. R.H. Baughman and R.R. Chance, J. Polym. Sci: Polym. Phys. Ed. 14, 2037 (1976).
57. M. Kestesz, J. Koller, and A. Azman, Chem. Phys. 27, 273 (1978); E.G. Wilson, J. Phys. C: Solid State Phys. 15, 3733 (1982).
58. A.S. Siddiqui and E.G. Wilson, J. Phys. C: Solid State Phys. 12, 4237 (1979).
59. R.H. Baughman, N.S. Murthy, and G.G. Miller, J. Chem. Phys., 79, 515 (1983).
60. R.H. Baughman, N.S. Murthy, G.G. Miller, and L.W. Shacklette, J. Chem. Phys., in press.
61. J. Smook, M. Flinterman, and A.J. Pennings, Polym. Bulletin 2, 775 (1980).
62. T. Shimanouchi, M. Asahina, and S. Enomoto, J. Polym. Sci. 59, 93 (1962).
63. R.H. Baughman, H. Gleiter, and N. Sendfeld, J. Polym. Sci: Polym. Phys. Ed. 13, 1871 (1975); C. Galiotis and R.J. Young, Polymer, in press (June, 1983)
64. F.C. Frank, Proc. Royal Soc. A282, 9 (1964).
65. W.A. Little, J. of Less-Common Metals 62, 361 (1978).
66. K. Bechgaard, K. Carneiro, O. Eg, M. Olsen, F.B. Rasmussen, C. Jacobsen, and G. Rindorf, Mol. Cryst. Liq. Cryst. 79, 627 (1982).
67. P.M. Grant, Phys. Rev. B, in press.
68. A. Fournel, C. More, G. Roger, J.P. Sorbier, J.M. Delrieu, D. Jerome, M. Ribault, K. Bechgaard, J.M. Fabre, and L. Giral, Mol. Cryst. Liq. Cryst. 79, 261 (1982).
69. R.L. Green and G.B. Street, Chemistry and Physics of One-Dimensional Metals, ed. H.J. Keller, Plenum 1977, 167.
70. N.B. Hannay, T.H. Geballe, B.T. Matthias, K. Andres, P. Schmidt, and D. Macnair, Phys. Rev. Lett. 14, 225 (1965).
71. H.F. Mark, Die Naturwissenschaften 61, 423 (1974).
72. K. Soga, Y. Kobayashi, S. Ikeda, and S. Kawakami, J.C.S. Chem. Comm. 931 (1980).
73. G. Kossmehl and G. Chatzitheodorou, Makromol. Chem., Rapid Commun. 2, 551 (1981).
74. H. Nakanishi, K. Hasumi, F. Mizutani, M. Kato, K. Ishimura, and S. Fujishige, Senii Kobunshi Zairyo Kenkyusho Kenkyu Happyokai Shiryo 120, 92 (1981).
75. G. Thommes, Chemtech, pp 284-287 (May, 1981).
76. Mod. Plst. Int., pp 46-49 (Sept., 1982).

77. C.B. Duke and H.W. Gibson, Kirk-Othmer: Encyclopedia of Chemical Technology 18, pp 755–793, 1982 (John Wiley and Sons, Inc.).

78. G. Bahder, L.A. Bopp, G.S. Eager, C. Katz, R.G. Lukac, G.A. Schmidt, Development of Extruded Dielectric Underground Transmission Cables Rated 138KV, 230KV and 345KV, Electric Power Research Institute No. EPRI EL-428 (1977).

79. L. Teschler, Machine Design, 151–157 (Dec. 10, 1981).

80. J. Hocker, R. Merten, and B. Willenberg, European Patent Application EP 0062211A2.

81. J. Hocker, H.K. Muller, and B. Broich, European Patent Application EP 0062212A2.

82. J. Hocker and R. Dhein, European Patent Application EP 0062213A2.

83. D. MacInnes, Jr., M.A. Druy, P.J. Nigrey, D.P. Nairns, A.G. MacDiarmid, and A.J. Heeger, J.C.S. Chem. Comm., 1981, 317.

84. K. Kaneto, M. Maxfield, D.P. Nairns, A.G. MacDiarmid, and A.J. Heeger, J.C.S. Faraday Trans. 1, 78, 3417 (1982).

85. L.W. Shacklette, R.L. Elsenbaumer, R.R. Chance, J.M. Sowa, D.M. Ivory, G.G. Miller, and R.H. Baughman, J.C.S. Chem. Comm., 1982, 361.

86. L.W. Shacklette, R. L. Elsenbaumer and R.H. Baughman, J. de Physique, in press.

87. J.H. Kaufman, E.J. Mele, A.J. Heeger, R. Kaner, and A.G. MacDiarmid, J. Electrochem. Soc. 130, 571 (1983).

88. F. Beniere, D. Boils, H. Canepa, A. LeCorre, and J.P. Louboutin, J. de Physique, Colloq., in press.

89. M. Armand, J. de Physique, Colloq., in press.

90. A.F. Diaz and J. Cabrillo, unpublished.

91. A. Brokman, M. Weger, G. Marom, Polymer 21, 1114 (1980).

92. J.C. Dubois, M. Gazard, and M. Champagne, Presentation at the International Conference on the Physics and Chemistry of Conducting Polymers (Les Arc, France, Dec. 11–14, 1982).

93. A. Feldbaum, Y.W. Park, A.J. Heeger, and A.G. MacDiarmid, J. Polym. Sci: Polym. Phys. Ed. 19, 173 (1981).

94. T.C. Clarke, M.T. Kruunbi, V.Y. Lee, and G.B. Street, J.C.S. Chem. Comm. 384 (1981).

95. W.N. Allen, P. Brant, C.A. Carosella, J.J. DeCorpo, C.T. Ewing, F.E. Saalfeld, D.C. Weber, J. Syn. Metal 1(2), 151 (1980).

96. H. Mazurek, D.R. Day, E.W. Maby, J.S. Abel, S.D. Senturia, M.S. Dresselhaus, and G. Dresselhaus, J. Polym. Sci.: Polym. Phys. Ed. 21, 537 (1983).

97. M. Ozaki, D.L. Peebles, B.R. Weinberger, C.K. Chiang, S.C. Gau, A.J. Heeger, and A.G. MacDiarmid, Appl. Phys. Lett. 35, 83 (1979).

98. J. Kanicki, J. de Physique, Colloq., in press.

99. K. Misoo, S. Tasaha, S. Miyata, and H. Sasabe, private communication.

100. M.J. Cohen and J.S. Harris, Jr., Appl. Phys. Lett. 33, 812 (1978).

101. R.A. Scranton, J.B. Mooney, J.O. McCaldin, T.C. McGill, and G.A. Mead, Appl. Phys. Lett. 29, 47 (1976).

102. J. Tsukamoto, H. Ohigashi, K. Matsumura, and A. Takahashi, Metals 4, 177 (1982).

103. B.R. Weinberger, M. Akhtar, and S.C. Gau, Synthetic Metals 4, 187 (1982).

104. J. Kanicki, S. Bove and E. VanderDonekt, Mol. Cryst. Liq. Cryst. 83, 319 (1982).

105. T. Skotheim and I. Lundstrom, J. Electrochem. Soc. 129, 894 (1982).

106. T. Skotheim, L.-G. Petersson, O. Inganas, and I. Lundstrom, J. Electrochem. Soc. 129, 1737 (1982).

107. W.R. Salaneck, H.R. Thomas, C.B. Duke, A. Paton, E.W. Plummer, A.J. Heeger, and A.G. MacDiarmid, Synth. Met. 1, 21 (1979).

108. T. Skotheim, O. Inganas, J. Prejza, and I. Lundstrom, Mol. Cryst. Liq. Cryst. 83, 329 (1982).

109. R. Noufi, A.J. Frank, A.J. Nozik, J. Am. Chem. Soc. 103, 1849, (1981).

110. J.H. Lai, Macromolecules 10, 1253 (1977).

111. N.R. Bryrd and N.B. Sheratte, "Semiconducting Polymers for Gas Detection", NASA CR-134885 (1975).

112. S.D. Senturia, "Fabrication and Evaluation of Polymeric Early-Warning Fire-Alarm Devices", NASA CR-134764.

113. M.K. Malmros, US Patent No. 4,334,880 (issued June 15, 1982).
114. M.S. Wittingham and L.B. Ebert, in "Physics and Chemistry of Materials and Layered Structures" (F. Levy, ed.), Vol. 6, D. Reidel, Dordrecht, 1979.
115. M.A.M. Boersma, Cat. Rev. - Sci. Eng. 10, 243 (1974).
116. G. Wnek, unpublished.

HIGH TEMPERATURE POLYMERS

FOR ELECTRONIC DEVICES

J. Economy

IBM Research Laboratory
San Jose, California 95193

SYNOPSIS: In this paper, the potential advantages of polymers as dielectric insulating layers in advanced microelectronic devices are described along with the requirements for their use in chips. Some of the key results obtained from a detailed study of polyimides carried out within IBM are summarized along with efforts aimed at optimizing the polyamic acid precursor to achieve improved processibility. Finally, some preliminary results of work on a new type of high temperature polymer are described that appears to satisfy many of the shortcomings of existing systems.

INTRODUCTION

In recent years, the potential advantages offered by high temperature polymers as dielectric insulators for advanced microelectronic devices has generated considerable interest in such materials. In fact, development of the next generation of higher density memory and logic chips depends to a great extent on the availability of improved polymeric insulators. Because of these needs, much effort has been directed at trying to better characterize commercially available polyimides and determine to what degree they can be modified to meet these challenges. Alternatively, development of new types of polymers with properties tailored to the proposed use would be highly desirable, particularly if all the target properties could be achieved.

In this paper, the potential advantages of polymers as dielectric insulating layers in advanced microelectronic devices are described along with the requirements for their use in advanced chips. Some of the key results obtained from a detailed study of polyimides carried out within IBM are summarized along with efforts aimed at optimizing the polyamic acid precursor to achieve improved processibility. Finally, some preliminary results of work on a new type of high temperature polymer are described that appears to satisfy many of the shortcomings of existing systems.

PRESENT AND FUTURE REQUIREMENTS FOR POLYMERS AS CHIP DIELECTRICS

Shown in Fig. 1 is a schematic of the first commercial chip where a polyimide was used as part of a double dielectric insulating layer in conjunction with sputtered quartz.[1] This 64K bit memory chip known as SAMOS (Silicon Aluminum Metal Oxide Semiconductor) was introduced by IBM in late 1978. Advantages of the polyimide as a potential replacement for quartz include lower dielectric constant of 3.5 *versus* 4.5 and ease of processing into uniform coatings. In SAMOS, the polyimide/quartz double dielectric configuration was designed to reduce interlayer shorts, ionic contamination and metal migration. The polyimide was processed into a uniform film first by spin coating the polyamic acid, and then heating to imidize. Metal lines and interconnections were introduced by standard lithographic processes in which a photoresist was coated onto a partially imidized polyamic acid, and after UV exposure both resist and amic acid could be dissolved with base developer.

One feature that both polyimide and quartz have in common is their tendency to coat substrates in a conformal fashion, *i.e.,* they faithfully replicate the terrain upon which they are coated. For future generations of computers, it is anticipated that at least one and possibly two orders of magnitude increase in density of metal line circuitry on the chip will be required. This can only be achieved by increasing the number of levels of metal lines from 1–2 to 4–5 and by shrinking metal line widths from 3–5μ to 1μ and less. To meet these very aggressive design changes will require development of insulators which planarize during the coating step to reduce the problem of amplified topography that arises from applying several consecutive conformal coatings. With the drive toward finer line widths sloping profiles typical of the wet etching process of polyamic acid will not be acceptable. Dry etching processes, such as the use of oxygen plasmas through a mask, will be required to obtain more vertical profiles necessary for the tighter ground rules. Property requirements for the polymeric insulator will very likely be more stringent than those in SAMOS. Thus, outgassing of residual solvent or moisture from lower layers of insulation on the chip will have serious consequences for adhesion of upper insulating layers. In addition, since the chip may require processing temperatures of up to 400°C, one might expect significant amounts of volatiles associated with polymer degradation. The polymer must also have a glass transition higher than any of the processing temperatures to avoid possible movement of metal lines and vias associated with release of stresses above T_g.

In Table I are enumerated the various high temperature processes that a polymeric insulator may be exposed to. The processes for annealing, soldering or brazing may be carried out anywhere from a low of 300°C to a high of 425°C while the dry etching processes in an oxygen plasma may create local temperatures in excess of 200°C.

Before proceeding further, it is appropriate to consider the potential applicability of other high temperature polymers for this application. In Table II are enumerated the various classes of high temperature polymers along with potential shortcomings. From a cursory glance at this table, it is possible to eliminate from consideration inorganic polymers, fluoropolymers and phenolic type materials either because of their instability at high temperatures or insufficient rigidity. The rigid rod-like polymers are very difficult to process into thin coatings as are many of the heterocycles. By this process of elimination, one quickly arrives at the polyimides as being the most attractive not only because of their excellent thermal stability but also because of the commercial availability of a wide range of these materials. In fact, there are several very low molecular weight polyamic acid precursors which provide sufficient planarization prior to imidization to meet the requirements indicated earlier. Unfortunately, these oligomers typically contain polymerizable cycloaliphatics or ethynyl end groups all of which are thermally unstable at the higher processing temperatures indicated in Table 1. Another problem associated with processing at temperatures of 400°C are the stresses that develop due to the mismatch in thermal expansion coefficient between the silicon substrate and polyimide coating. These stresses are estimated to approach 4–5,000 psi and are undoubtedly much higher at points of stress concentration such as vias (interconnections between different layers of metal circuitry). Another concern would be the potential for degradation in electrical and mechanical properties with repeated cycling to temperatures of 400–425°C.

CHARACTERIZATION AND OPTIMIZATION OF POLYIMIDES

In the search for the appropriate polyimide composition, our work initially focussed on a very low molecular weight mixture of methylene dianiline and the diester of benzophenone diphthalic anhydride MDA/BDPA. This system gave acceptable planar coatings primarily as a result of the high solids loading of the coating solutions (~40–50%). However, after curing to 300°C and subsequent heating to 400°C, these coatings tended to crack particularly at points of stress concentration. The reason for this can be readily seen from the stress strain curves in Fig. 2 comparing the mechanicals of MDA/BDPA against a different composition derived from pyromellitic dianhydride and oxydianiline PMDA/ODA. The elongation at break of this MDA/BDPA was of the order of 2–3% while the PMDA/ODA system displayed outstanding mechanicals with an elongation at break in excess of 50%. MDA/BDPA also showed some tendency for outgassing at elevated temperatures which probably was associated with degradation reactions of the methylene bridge and from possible Schiff base formation between the amine and ketone during polymerization.

Because of the excellent thermal and mechanical properties of the PMDA/ODA system, a program was initiated by W. Volksen[2] to determine whether the M_w of the PMDA/ODA polyamic acid could be decreased sufficiently to permit increased solution loadings required for surface planarization without compromising the excellent mechanicals. A large number of samples of varying

M_w were prepared with selected data shown in Table 3. As can be seen from these results, the M_w of the polyamic acid solution must exceed 10,000 and preferably be close to 18,000 to achieve the desired mechanical properties. Unfortunately, at this M_w range, the solids content in the solution was too low to achieve the desired planarity. In this study, Volksen found that the agreement between calculated M_w *versus* observed M_w even at high molecular weights was very good when the starting materials were carefully purified.

A minor problem with the PMDA/ODA polymer was the tendency to produce small amounts of CO_2 on heating to above 365°C. Very likely this arises from the presence of a few percent isoimide which is formed during thermal polymerization. In fact, as shown in Table 4, depending on the method of converting the polyamic acid to imide one can obtain widely varying percentages of imide *versus* isoimide.[3] With respect to the thermally cured material, the problem of outgassing of CO_2 could be eliminated by simply heating the sample to 425°C[4] without any apparent loss in mechanical and electrical properties. In fact, even after heating under N_2 to 450°C for several hours, the PMDA/ODA polyimide retained a significant percentage of its strength and showed only slight deterioration in electrical properties.[5,6]

One area of concern in the program to optimize M_w for planarization was a report by Russian workers[7] that suggested that extensive branching accompanied imidization. This of course would shed considerable doubt on the conclusions drawn from the study on effect of M_w on mechanicals. P. T. Cotts[8] has clarified this question by determining for the first time the M_w of thermally cured polyimide by light scattering using H_2SO_4 as the solvent. As can be seen in Fig. 3, the M_w of thermally cured polyimide based on PMDA/ODA is within experimental error of the value for the polyamic acid. P. Cotts and W. Volksen were also able to elucidate the problems associated with changes in viscosity of the polyamic acid solution on standing. Thus, they found that the initial sharp drop in viscosity was due to a reequilibration of M_w/M_n to the most probable distribution, while the subsequent increase in viscosity with time was due to reorganization probably resulting in branching.

One unexpected problem of polyimide coatings is their tendency to swell anisotropically in polar solvents such as N-methyl pyrrolidone (NMP). Thus, films of Kapton® in NMP were shown by T. L. Smith[9] to swell up to 40% in the thickness direction and only several percent in the plane of the film. This anisotropic behavior has been further characterized by T. Russell and coworkers[10] by measuring the refractive index in the plane and perpendicular to the plane of the film using wave guide techniques. In this method, the refractive index is determined by guiding light from a He-Ne laser via a prism into the film which is supported on a substrate of lower refractive index. By this technique, one can follow changes in refractive index with curing temperature and at the same time quantitatively measure change in film thickness. As can be seen in Fig. 4, some orientation of the polyamic acid is already detectable at 80°C even though significant concentrations of solvent are still present. On heating to 215°C the film thickness decreases by 40% and for all intents the polyamic acid is completely imidized. Further heating above 215°C shows no further change in refractive index (orientation) or reduction in film thickness. It seems clear that the observed anisotropic swelling arises from the high degree of chain orientation that occurs during the spin coating and curing of thin films. It is interesting to note that the maximum observed swelling of 40% in thickness corresponds almost exactly to the change in thickness during imidization and solvent loss on heating from 80°C to 215°C.

The final question to be addressed in this paper concerning possible limitations in use of polyimides for microelectronic devices, is the issue of microporosity in polyimide coatings and its effect on moisture absorption and solvent pick up. Since the imidization of polyamic acid coatings is accompanied by both solvent and water evolution, it would seem reasonable to expect substantial concentrations of micropores. In some very recent studies in our laboratory, T. Russell[11] has examined the possible presence of micropores in polyimides through use of Small Angle

X-Ray Scattering (SAXS) techniques. This method is very sensitive to electron density gradients in a specimen and hence can be used to detect voids or a second phase of differing density. As can be seen from Fig. 5, there is essentially no detectable microporosity in the range of 20–300Å (<0.1% void content integrated over this range). Hence, one can conclude that the several percent moisture uptake by polyimide does not occur by diffusion through or accumulation in large pores but more likely has a chemical origin.

One other interesting feature of the SAXS pattern is the absence of any density gradient associated with the presence of ordered phases. This is especially surprising in light of the potential of the polymer backbone to assume a rod-like configuration over small distances. In fact, Yoon and Hupfer[12] have shown from an X-ray diffraction structure determination of a single crystal of the di-imide derived from 2/1 p-phenoxyaniline/PMDA that extensive electronic delocalization occurs at the imide-phenyl bond. As a consequence, PMDA/ODA polyimides can be approximated as a concatenation of 18Å long rods linked by a flexible ether swivel. Yoon, Takahashi, and Parrish[13] from WAXS of samples cured at 200°C concluded that even at this relatively low temperature some order can be detected as evidenced by two peaks, one typical of interchain distances found in nematic ordering (~4.6Å) and the other at ~15Å suggestive of some kind of ordering along the chain. The question of ordering in polyimide is now being examined in far greater depth over a much wider range of annealing temperatures by Yoon et al.[13] and hopefully they will be able to shed more light on this subject in the near future.

To summarize, the data generated in this study indicate that polyimide based on PMDA/ODA has outstanding stability up to 425°C (under inert conditions) and shows little loss in electrical and mechanical properties even after heating for several hours. There are subtle changes in composition at these temperatures associated with evolution of small amounts of CO_2; however, these changes do not appear critical. The exceptional mechanicals with an elongation at break in excess of 50% make this material unique compared to almost any rigid, high temperature polymer. In fact, the stress-strain curve resembles closely that of an elastomer. Problems such as conformabilty of the coating, moisture take-up and anisotropic swelling in strong solvents are serious and may be a deterrent to use of polyimide in advanced chip and module design.

DESIGNING A POLYMER FOR USE AS A CHIP DIELECTRIC

Up to now, the discussion has focussed on trying to adapt commercially available polymers for use in advanced chips. A far more satisfying and rewarding approach to chip dielectrics would be to tailor new polymers which would not compromise the excellent properties of the PMDA/ODA polyimide and yet solve the problems enumerated above. Since cost is only a modest consideration in chip applications, one could also explore far more exotic compositions. In addition, by designing materials which take into account possible opportunities and trade-offs in computer architecture and device design, material scientists would be in a position to influence the evolution of advanced computer hardware.

In the ensuing section, data are presented describing development of a polymer which was in fact designed to address many of the shortcomings of the polyimides indicated earlier. The design criteria for such a polymer are indicated in Table V. As noted previously, planarization depends strongly on the percent solids content in solution as well as the degree to which remaining solvent plasticizes the coating. Ideally, a prepolymer which melts and flows with a low viscosity prior to cure would permit the ultimate in surface planarization. If such a prepolymer is cured by an addition reaction, this would presumably minimize problems associated with microporosity as well as permitting fabrication of much thicker coatings. The key for achieving optimum thermal stability is to devise end groups in the prepolymer which react to form aromatic units. This issue has in the past been the major deterrent to development of processible high temperature polymers for the aerospace industries. Presumably for low moisture pick-up and low dielectric constant, the polymer should contain a minimal amount of polar groups and preferably consist of all aromatic units such as polyphenylene. It would also be desirable to devise an insulating layer that would function as a resist since this would

have the potential of greatly simplifying the lithographic
processing. Finally, achievement of the desired mechanicals
($E \sim 0.5 \times 10^6$ psi and e>10%), presents formidable problems
particularly for thermosetting resins; however, there are indications
that by optimizing the distance between cross-links or
entanglements there is hope for achieving significant improvement
in elongation characteristics. The system selected for our study
was one consisting of aromatic units with diacetylenes as reactive
end groups. Wegner[14] had reported in 1969 that certain
monomeric diacetylenes in the form of single crystals polymerized
to single crystal polymers on exposure to UV. In addition,
A. Hay[15] had reported that high molecular weight m-phenylene
diacetylene polymers decomposed on heating to 180°C but also
gave a carbon yield in excess of 90% on heating to 800°C. In
examining these earlier studies it appeared that the thermal
decomposition at 180°C might very well be a crosslinking reaction
and if so might give a direct approach to a high temperature
polymer by a simple addition reaction. The fact that only a
bimolecular reaction was necessary to form aromatic units and that
Hay indicated that the reaction was accompanied by a large
exotherm provided further encouragement to explore this avenue.
The radiation sensitivity of the diacetylene monomers observed by
Wegner suggested that such materials may act as negative resists.

Our initial approach involved preparation of readily accessible
oligomers such as that of m-phenylene diacetylene. The molecular
weight was controlled by oxidatively coupling 1,3-diethynyl
benzene in the presence of phenyl acetylene, which acted as a chain
terminator. These oligomers were found to melt and planarize well
below the temperature of decomposition (cross-linking), however,
the uncured coatings tended to crack probably as a result of the
tendency of this structure to crystallize. This problem was solved
by preparing the oligomer from oxidative coupling of
1,3,5-triethynyl benzene with phenylacetylene (PTEB).[16] The M_n
of the oligomer could be tailored to flow at temperatures ranging
from 80–135°C which is well below the cross-linking reaction at
180°C. This system had a relatively high concentration of
diacetylene units in order to enhance lithographic sensitivity;
however, this feature could also lead to a higher degree of

crosslinking and thus compromise the mechanical properties. In addition, the high diacetylene content would very likely result in an uncontrollable exotherm during crosslinking of thick films. For our purposes this system, because of its synthetic accessibility and potential ease of processing, would at least provide the crucial answers as to planarizability, aromatization of end groups and lithographic sensitivity. It should be noted that a controlled percent of ethynyl groups could be left in the prepolymer depending on the ratio of 1,3,5-triethynyl benzene to phenylacetylene and the reaction time. The intent here was to test the possibility of bimolecular reactions between ethynyl and diacetylene units to achieve aromatization.

A number of surfaces with differing topographies were coated with the PTEB oligomers to test planarizability. It was found that surface topographies with roughnesses ranging from as little as 750Å to as high as several microns could be effectively planarized by heating the oligomer to $120-130°C$ a temperature where the prepolymer flowed as a low viscosity liquid. In fact, this system provided the highest degree of surface planarization of any material tested in our program. Surface roughness measurement techniques such as Tallystep (sensitive to 250Å) and reflected light interference (sensitive to 125Å) indicated a surface planarity with a roughness less than 125Å. In these kinds of experiments, full planarization over a large surface with varying topographies is not possible since flow of the oligomers is limited to planarizing the immediate region.

The outstanding thermal and oxidative stability of the cured PTEB is shown in Fig. 6. Under helium, the polymer shows an incipient weight loss at $500°C$ increasing to a total of 5% at $700°$. In air, the polymer appears very stable to oxidative degradation with significant losses (>5%) beginning at $550°C$. Naturally, one must use these kinds of rising temperature thermograms with care since such data can be misleading. On the other hand, the unusually high thermooxidative stability suggests that polymerization of the oligomers may involve formation of aromatic units.

A DSC trace comparing the behavior of the PTEB oligomers containing 0, 15 and 45% ethynyl groups, respectively, is shown in Fig. 7. In all cases, the exotherm associated with curing appears as a very large spike between 150–180°C. The problem of runaway temperatures with curing of thick films is readily interpretable in light of these large exotherms. With care films of up to 10 microns could be cured with little difficulty. The appearance of the double peak with increasing concentration of ethynyl group presumably is related to two distinct processes associated with curing of the diacetylene and of the ethynyl.

To actually determine what reactions occur during cure is complicated by the insolubility of the crosslinked structure. In spite of our intuition suggesting that thermal aromatization could occur, there was no direct data to support this position. In fact Wegner's earlier work had shown a completely different cure mechanism of the single crystal diacetylenes. Furthermore, almost any mechanism one might propose for aromatization required some kind of forbidden electronic and or steric pathway. By resorting to solid state C^{13}NMR and using cross polarization techniques, it was possible to follow the curing process in an almost quantitative fashion.[17] The peaks associated with diacetylene and trisubstituted and monosubstituted phenyl rings could be assigned from solution NMR of the oligomeric PTEB. A sample was then heated and cured to 250°C in the presence of solvent to control the exotherm. In this manner, it was possible to isolate a sample in which 60% of the diacetylene was still present. Typically in a sample heated to 400°C, no diacetylene could be detected. From C^{13}NMR, it then was possible to show that the 40% diacetylene that had reacted had converted almost quantitatively to aromatic rings. In fact, from delayed decoupling experiments, it appears that approximately half of the newly formed aromatic rings contain no hydrogens. These results can be interpreted on the basis that a concerted Diels Alder polymerization of the diacetylenes takes place yielding condensed polynuclear rings. This is accompanied by a cycloaromatization reaction involving hydrogen of the adjacent phenyl rings. The first reaction is limited by the mobility of the oligomers and proceeds while the curing temperature still exceeds T_g. It should be noted that a similar NMR study of the curing of an oligomeric polyimide with ethynyl end groups by Sefcik *et al.*[18] shows that the primary reaction does not proceed by cyclotrimerization of ethynyls but rather by some sort of dimer coupling.

To summarize, the study on PTEB has demonstrated that one can tailor most of the desired properties into a polymer for use as a dielectric insulating layer. As shown in Table 6, summarizing the properties of PTEB, most of the goals were achieved even though this program was looked upon as a model study. Thus, it was possible to design a system that provided outstanding planarization and could be cured without evolution of volatiles to a thermally stable structure. By designing a composition consisting of only carbon and hydrogen, it was possible to achieve a material with very low moisture pick-up and low dielectric constant. The oligomeric PTEB shows strong absorption in the mid UV at 295, 315 and 335 and in fact preliminary studies indicated that the material could be imaged lithographically as a negative resist. The low elongation is in fact the only drawback of this material and is a direct result of the large number of diacetylene links which can participate in the curing process. Presumably, reducing the concentration of diacetylenes to end groups in oligomeric materials would not only act to reduce the brittleness but also eliminate the large exotherm observed with curing of diacetylenes.

Burlington Chip (SAMOS)

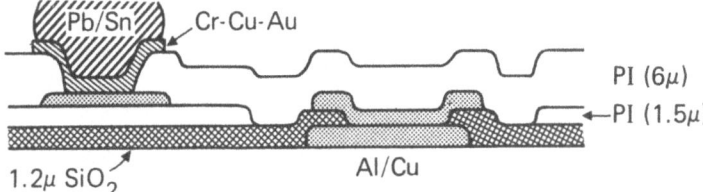

Advanced Multilevel Structure

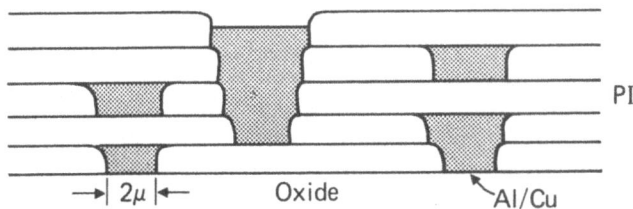

Figure 1. Cross-section of (a) a commercial FET chip with a polyimide coating as part of a double dielectric and (b) an advanced multilevel chip with four levels of metal circuitry and polyimide as insulator.

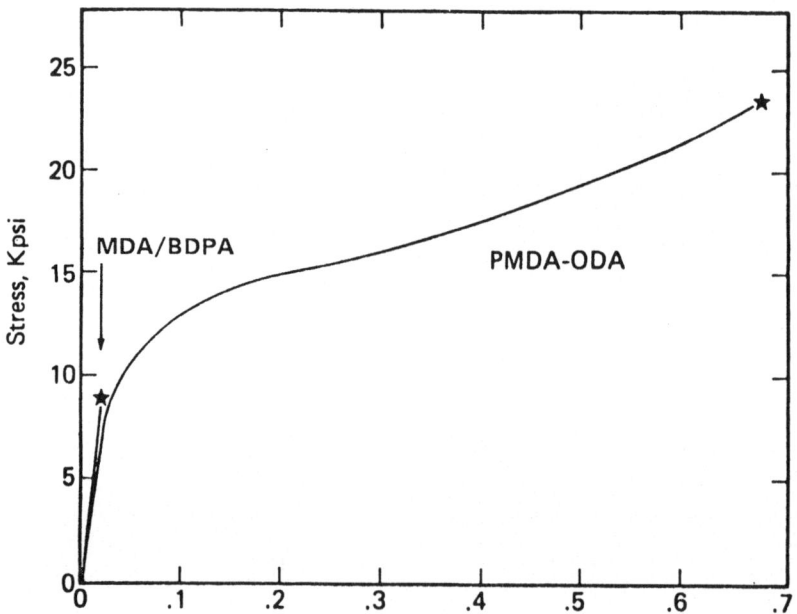

Figure 2. Comparison of stress-strain curve between MDA/BDPA and PMDA/ODA polyimide.

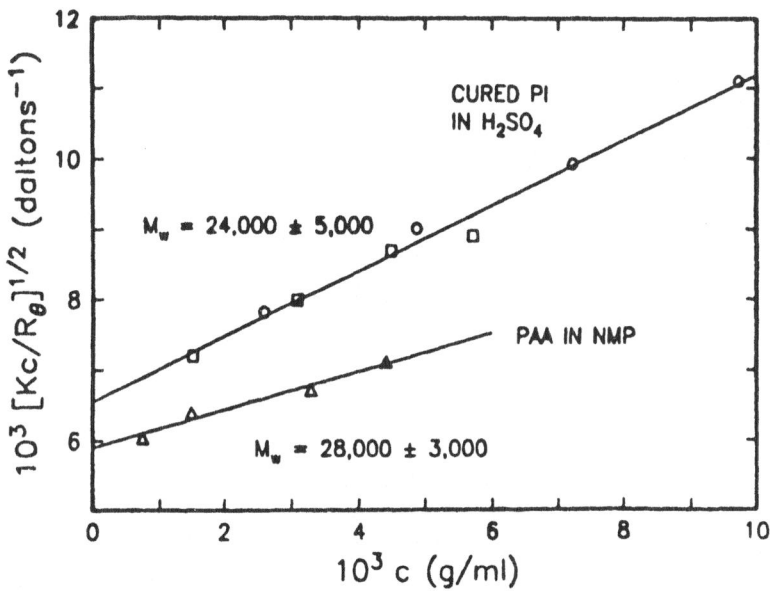

Figure 3. Molecular weight by light scattering of PMDA/ODA before and after cure.

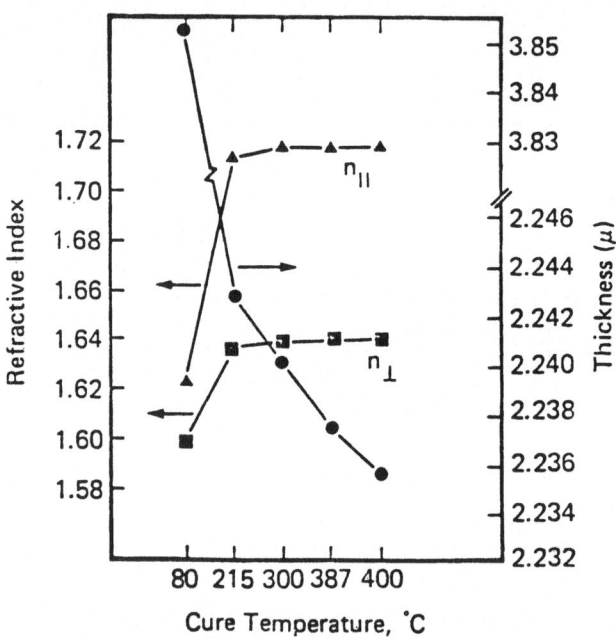

Figure 4. Change in refractive index and thickness during imidization of a polyamic acid coating.

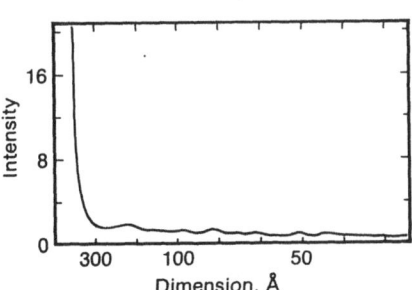

Figure 5. Small angle X-Ray scattering of cured PMDA-ODA.

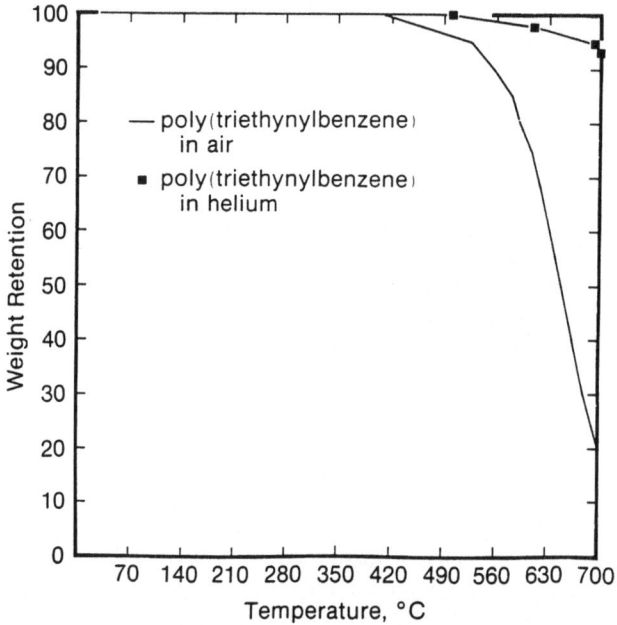

Figure 6. Thermal stability of PTEB in air and in helium.

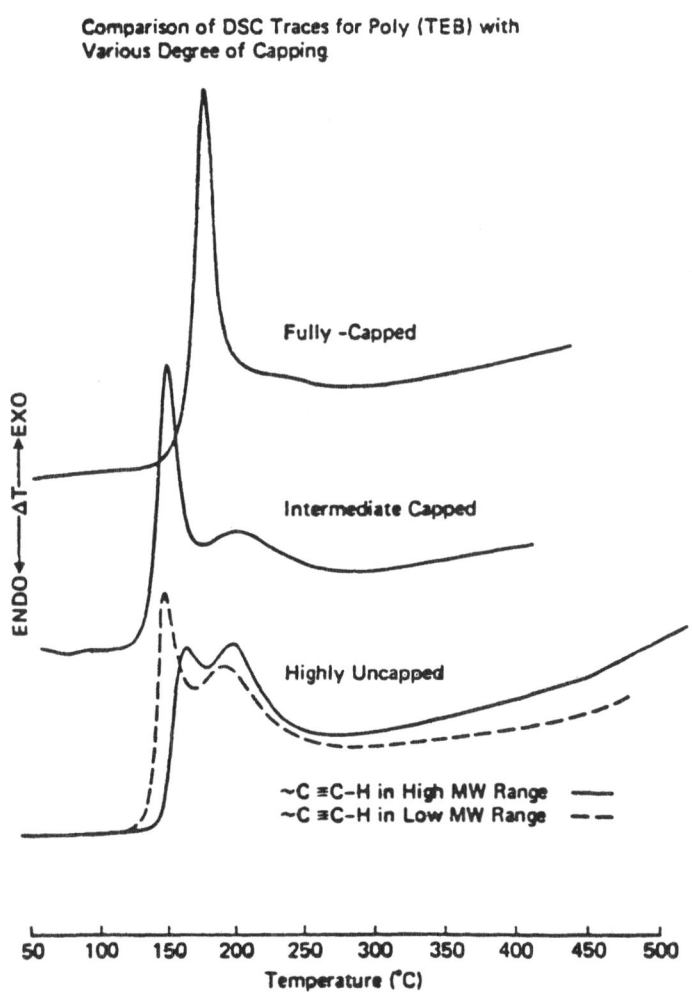

Figure 7. DSC traces for PTEB containing various percentages of ethynyl groups.

TABLE 1

Key Process Steps

	High Temperature Processing
• Al-Cu Metal Deposition	Annealing
• Chip/Module Interconnections	Pb Sn Soldering
• Module/Package Interconnections	Pin Brazing
• Lithography - lift off	Polar Solvents
- RIE	200°

TABLE 2

Alternate High Temperature Polymers

Candidates	Potential Problems
Inorganic Polymers	Reorganization Reactions 180–250°
• Silicones	Sensitivity to Hydrolysis
• Phosphazenes	Oxidative Degradation <200°
Fluoropolymers	Difficult to Process
• Teflon	Decompose at 350°
• KEL-F	
Heterocyclic Polymers	Coated from H_2SO_4 Solutions
• BBB Ladders	Imidazole Unit Unstable to Air ~220°
• Polybenzimidazoles	
Phenolic Based polymers	Thermal Stability (N_2) ~250–300°
• Phenylene Oxide	Oxidative Degradation <150°
• Phenol Formaldehyde	
Ethynyl Prepolymers	Uncontrolled Curing
• Thermid	Oxidatively Sensitive End Groups
• H Resin	
Rigid Rod-Like Polymers	Difficult to Process
• Aramids	
• Aromatic Esters	

TABLE 3

Mechanical Properties of PMDA/ODA VS M_w

M_w	Solids Content of Casting Solution (%)	Strain at Break (%)
5K	30	Brittle
10K	20	6
18K	15	40
37K	10	60
77K	8	60

TABLE 4

Isoimide Formation

	% Imide	% Isoimide
Chemical Cure		
(a) $(CF_3CO)_2O$	20	80
(b) $(CH_3CO)_2O$	80	20
Thermal Cure	95	5

TABLE 5

Design Criteria for a Dielectric Insulator

Desired Property	Structural Feature
Planarizability	• Low Prepolymer (melt before cure)
No Microporosity	• Cured by Addition Reaction
Thermal Stability (400°)	• Incorporate Aromatic Units
No Moisture Pick-up	• Eliminate H_2O Adsorbing Groups
Low Dielectric Constant	• Eliminate Polar Groups
Lithographic Sensitivity	• Photosensitive Prepolymer
High Mechanicals	• Optimize Contour Length (cross links)

TABLE 6

Summary of Properties for Polytriethynylbenzene

Processing Conditions

Solubility	Excellent
Melt Reflow	20–120°
Curing	150–200°

Properties

Planarizability	Excellent
Thermal Stability	>Polyimide
Moisture Pick-up	10^3X less than Polyimide
Dielectric Constant	2.7 (polymide ~3.5)
Lithographic Sensitivity	280–330 nm
Mechanicals	Elongation <2%

REFERENCES

1. R. A. Larson, IBM J. of Res. and Dev. 24(3), 268 (1980).

2. W. Volksen and P. M. Cotts, Extended Abstract of Polyimide Mtg., Mid-Hudson Section of SPE, 154 (November 10-12, 1982).

3. F. P. Gay and C. E. Berr, /J.Polym. Sci. A-1/ 6, 1935 (1968).

4. D. Y. Yoon and R. Herbold, (private communication).

5. R. Yang, M. J. Jackson, D. Sheider, and A. Bross, (private communication).

6. S. M. Zalar, Extended Abstract of Polyimide Mtg., Mid-Hudson Section of SPE, 143, (November 10-12, 1982).

7. P. N. Bribkova, V. V. Rode, and V. V. Korabok, Izvestiya Akad. Nowks SSR, Seriya Khimicheshaya 3, 568-575 (1970).

8. P. M. Cotts, Extended Abstract of Polyimide Mtg., Mid-Hudson Section of SPE, 143, (November 10-12, 1982).

9. T. L. Smith (private communication).

10. T. P. Russell, H. Gugger, and J. D. Swalen, J. Polym. Sci., Phys. Ed., accepted for publication, 1983.

11. T. P. Russell, Bull. Am. Phys. Soc. 28, 547 (1983).

12. D. Y. Yoon and B. Hupfer (private communication).

13. D. Y. Yoon, N. Takahashi, and W. Parrish, (private communication).

14. G. Wegner, Z. Naturforsch. 24b, 824 (1969).

15. A. S. Hay, J. Pol. Sci. R-1, 7, 1625 (1969).

16. J. Economy, M. A. Flandera, U.S. Patent Nos. 4,258,079, March 24, 1981; 4,273,906, June 16, 1983.

17. J. R. Lyerla (private communication).

18. M. D. Sefcik, E. J. Stejskal, R. A. McKay, and J. Schaefer, Macromolecules 12(3), 423 (1979).

AROMATIC IONOMER MEMBRANES: SYNTHESIS, STRUCTURE AND PROPERTIES

A. Eisenberg and E. Besso
Department of Chemistry
McGill University
Montreal, Quebec, Canada

H.L. Yeager and A. Steck
Department of Chemistry
University of Calgary
Calgary, Alberta, Canada

F.W. Harris and R.K. Gupta*
Department of Chemistry
Wright State University
Dayton, Ohio, U.S.A.

*Daychem Laboratories
Xenia, Ohio

SYNOPSIS

The syntheses of three new aromatic ionomers with equivalent weights of ca. 800, 600 and 400 have been carried out. Thus, 3,3'-(oxydi-p-phenylene)bis[2,4,5-triphenylcyclopentadienone] and 3,3'-(oxydi-p-phenylene)bis[2,5-diethoxycarbonyl-4-phenylcyclopentadienone] were copolymerized with 4,4'-diethynyldiphenyl ether in molar ratios of 70:30:100, 60:40:100 and 40:60:100 to afford the corresponding alkoxycarbonyl-substituted polyphenylene oxides, which were converted to the ionomers by treatment with potassium hydroxide.

The materials were studied in the form of thin films. Physical and mechanical properties were determined by DSC, torsion pendulum, stress strain, stress relaxation and water contact angle measurements. Transport properties were determined in alkaline solution. Evidence of the films' microporosity is seen from ion and water sorption, sodium ion self-diffusion and electrolytic conductance measurements. The latter results for lower equivalent weight samples in concentrated NaOH solutions demonstrate that these materials are promising candidates as separators in advanced alkaline water electrolysis cells.

375

INTRODUCTION

A joint research program among the above research
groups has been initiated to develop a high performance
polymer membrane-separator for alkaline water electrolyzers.
This research, sponsored by the National Research Council
of Canada, is part of Canada's contribution to research on
hydrogen production as a member of the International Energy
Agency (1). The IEA is composed of several countries which
pool the results of work in various fields of energy
research.

Hydrogen is now produced in very large quantities for
use as a chemical feedstock. The majority of this production is used for ammonia and methanol synthesis and for
petroleum refining, while significant amounts are also used
for oil and fat hydrogenation and as a metallurgical
reducing agent. In addition to these needs, hydrogen holds
great promise as an increasingly important energy form over
the next thirty years, as conventional fossil fuel reserves
are depleted (2). As the storage and transmission
difficulties of hydrogen are overcome, increasing uses of
this fuel, for heating, transportation, and energy storage
now seem likely.

Currently hydrogen is produced by the steam reforming
of natural gas, coal gasification, and water electrolysis.
While various thermochemical and photochemical means for
hydrogen synthesis have been proposed, full commercial
application will not be possible for many years. Thus, the
only viable means for hydrogen production from non-fossil
energy is the electrolysis of water, with the electricity
provided either by hydroelectric generation or nuclear power.

Present electrolyzers normally operate at 60°-90°C at
ambient pressures using concentrated KOH electrolytes (3)
(solid polymer electrolyte-acid electrolyzers are used for
specialized applications (4).) For both thermodynamic and
kinetic reasons, it would be desirable to increase the
temperature and pressure of operation to 150-200°C and 30
MPa. A major problem with these new conditions of operation
is the lack of availability of a separator to prevent the
mixing of product gases. Asbestos is universally used as a
separator-diaphragm for conventional alkaline water electrolyzers, but it dissolves above ca. 100°C. Several new

materials under consideration for use include the perfluorosulfonate polymers (Nafion, Du Pont and Co.), Teflon-bonded
potassium titanate, asbestos treated with binders, and a
fiber form of polysulfone (2). Each has certain advantages
and disadvantages with regard to the necessary properties
of chemical inertness, mechanical strength, low cost and
high ionic conductivity.

Excellent candidates for membrane materials in alkaline
electrolyzer applications are the aromatic ionomers (5).
These polymers, which were originally prepared for possible
use as windshield materials on supersonic aircraft, can be
cast into tough, flexible films that display outstanding
chemical and thermal stability (6). For example, they can
be immersed in a refluxing solution of potassium hydroxide
in ethylene glycol (196-198°C) without undergoing any
degradation. The thermal stability of these polymers in an
air atmosphere is also outstanding, as demonstrated by their
ability to withstand temperatures in excess of 200°C for
extended periods of time without any apparent degradation.
Although the ionomer's solubility is dependent upon its ion
content, samples with relatively low ion contents are only
soluble in dimethylsulfoxide (DMSO). The systems can also
be thermally crosslinked by heating them to 300°C under an
inert atmosphere.

Since it was likely that, for electrolyzer applications,
membranes would be needed with considerable ionic content,
the major objective of the synthetic portion of research
was the preparation of new aromatic ionomers with relatively
low equivalent weights. (The original materials had equivalent weights of 1200 or higher.) Specifically, the syntheses
of three new ionomers with equivalent weights of approximately 800, 600 and 400 were to be carried out. Thus, various
molar ratios of 3,3'-(oxydi-p-phenylene)bis[2,4,5-triphenyl-
cyclopentadienone](1a) and 3,3'-(oxydi-p-phenylene)bis[2,5-
diethoxycarbonyl-4-phenylcyclopentadienone](1b) were to be
polymerized with 4,4'-diethynyldiphenyl ether (2), and the
resulting polymers hydrolyzed to afford the corresponding
ionomers.

SYNTHESIS OF AROMATIC IONOMER PRECURSORS

Monomers

The syntheses of the biscyclopentadienones 1a and 1b

were carried out by the following routes (7):

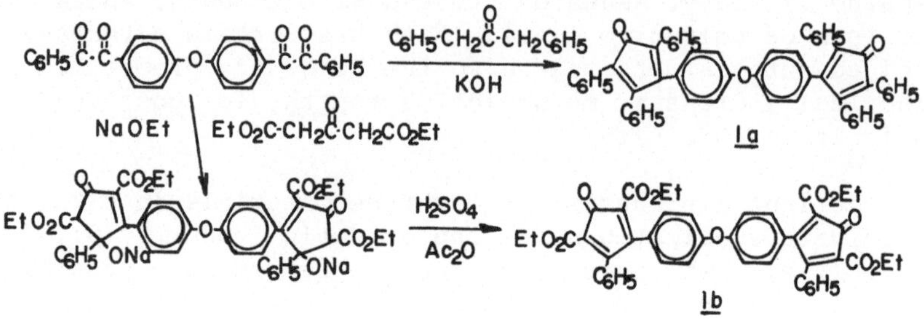

Chromatography of the dark purple 1a and the bright orange
1b on neutral alumina and silica gel, respectively, afforded
monomer-grade materials.

4,4'-Diethynyldiphenyl ether 2 was prepared from 4,4'-
diacetyldiphenyl ether via the following synthetic route (5):

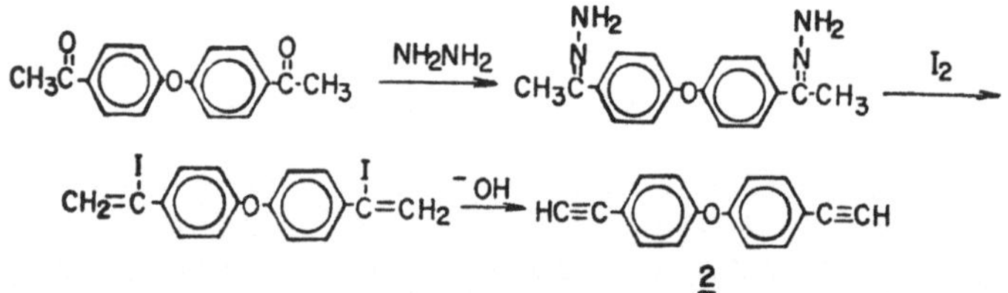

The white monomer was sublimed twice immediately prior to
its polymerization.

Polymers

Monomers 1a and 1b were copolymerized with 2 in molar
ratios of 70:30:100, 60:40:100, and 40:60:100 to produce
polymers 3a, 3b, and 3c, respectively. Although the Diels-
Alder reaction produces both meta and para catenation along
the backbone, only para catenation has been depicted. The
polymerizations were carried out in chlorobenzene (20%
solids w/w) contained in sealed polymerization tubes. The

$$\underline{1a} + \underline{1b} + \underline{2}$$

3a x = 0.7   y = 0.3   EW = 800

b x = 0.6   y = 0.4   EW = 600

c x = 0.4   y = 0.6   EW = 400

tubes were heated at 110°C for 18 hours and then at 210°C for 24 hours. Polymerizations were also conducted in a 300-ml, stirred, Parr pressure reactor using an identical heating cycle. In these runs, the carbon monoxide evolved during the polymerization was allowed to escape from the reactor. The off-white, fibrous polymers were isolated by precipitation in ethanol. The ionomer precursors are readily soluble in chlorinated-hydrocarbon solvents. Thin films can be cast from chloroform solutions that show infrared absorptions characteristic of alkoxycarbonyl-substituted polyphenylenes (7). Thermogravimetric analyses (TGA) of pellet samples in air and nitrogen atmospheres show initial weight losses between 365 and 550°C, which correspond closely to the calculated weight percentages of pendant ethoxycarbonyl groups contained in the copolymers. After the initial weight loss, the TGA curves closely resemble the TGA curves of phenylated polyphenylenes (8). The intrinsic viscosities ranged from 0.64 to 1.02 in chloroform at 30°C for these precursor polymers.

Nomenclature

In the subsequent discussion, mention of equivalent weight alone (e.g. 600EW) will be used to designate the ionomer (potassium salt) obtained by hydrolysis of the ester precursor. The precursor will be denoted by the equivalent weight followed by "ester" (e.g. 600EW ester).

STRUCTURE AND MECHANICAL PROPERTIES

Preparation

The polyphenylene ether ionomer membranes may be obtained by two different pathways. The first method entails a hydrolysis of the precursor (ester), in powder form, to the ionomer (salt) also in powder form. The purified ionic polymer is subsequently dissolved in DMSO, hot filtered, then cast on a glass plate at high temperature. The solvent is allowed to evaporate, yielding a dense, cohesive film which is further dried under reduced pressure until constant weight is achieved. The product of the method just described will be designated either as "homogeneous" film or by the equivalent weight followed by the suffix p (e.g. 600EW-p).

The second route entails the manufacture of a precursor film of the desired dimensions. This is done by casting a solution of the precursor onto a glass plate and allowing the solvent to evaporate at room temperature. The resultant film is then hydrolysed at high temperature and pressure. By varying hydrolysis conditions such as temperature, pressure or concentration, a wide range of results can be obtained. These films shall be designated as either "porous" or by the EW-f nomenclature.

Scanning electron micrographs of the porous ionomer film (9) indicate morphologies ranging from material with pore sizes of ca. 5 μm to a homogeneous material traversed by a network of channels of ca. 100 nm width. The latter structure has demonstrated excellent conductivity results, as will be discussed further, and will be the one meant when discussing the porous material.

Both homogeneous and porous films show a great affinity for water. The porous material of 600EW equilibrates at a water uptake of ca. 70% after boiling at 100°C for three days (9), whereas the homogeneous material retains ca. 15% water after the same treatment. This is further confirmation of the differing structures present. The water retained by the membrane falls into a minimum of two categories. The nature of the bonding has been investigated by differential scanning calorimetry at low temperature (9) and will be discussed further.

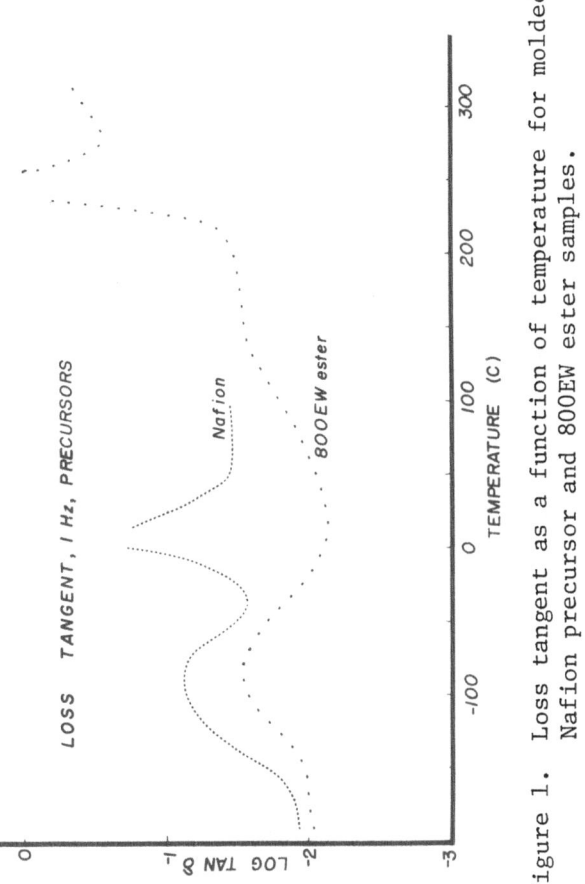

Figure 1. Loss tangent as a function of temperature for molded Nafion precursor and 800EW ester samples.

Characterization

Temperature stability tests were conducted on 600EW
ester and 600EW-f. The dry samples were allowed to equili-
brate at progressively higher temperatures, under reduced
pressure. Their weights were recorded as a function of
temperature. The results indicate no significant weight
loss (<2%) at temperatures below 280°C for the ester, whereas
the ionomer showed no significant weight loss below 340°C.
These figures confirm the excellent temperature stability of
the material. Interestingly, the temperature stability
testing provided the first clear evidence of a high degree
of hydrophilicity in the ionomer membrane. If the weighing
was not performed extremely rapidly, or in a hermetic
environment, the sample weight of 600EW-f was observed to
increase by as much as 10% over a period of two minutes.

Dynamic Mechanical Studies

Torsion pendulum experiments (ca. 1 Hz) were performed
on molded bulk samples of the ester precursors and the
corresponding ionomers where possible. A molded bulk sample
of 600EW-p could not be obtained since the material decomposes
extensively at molding temperatures. The glass transition
temperatures obtained are shown in Table I and do not vary
significantly in the precursors. As the ionic content
increases, a trend towards higher glass transition temperatures is observed, of the order of 3-5°/mole % ions. The
curves obtained for the 800EW ester and 800EW-p are shown
in Figs. 1 and 2 respectively, along with those for Nafion-
Na of 1200EW for comparison. The difference in glass
transition temperatures obtained by the two methods is
probably due to a small amount of degradation which in-

Table I Transition temperature (°C) for the dry polymers

| SAMPLE | TORSION PENDULUM | | D S C | |
	Precursor	Ionomer	Precursor	Ionomer
1200EW[10]	257	280	–	–
800EW-p	250	330	292	399
600EW-p	255	–	287	421
600EW-f	–	–	290	417
400EW-f	–	–	285	470
Nafion 1200[12]	20	220		

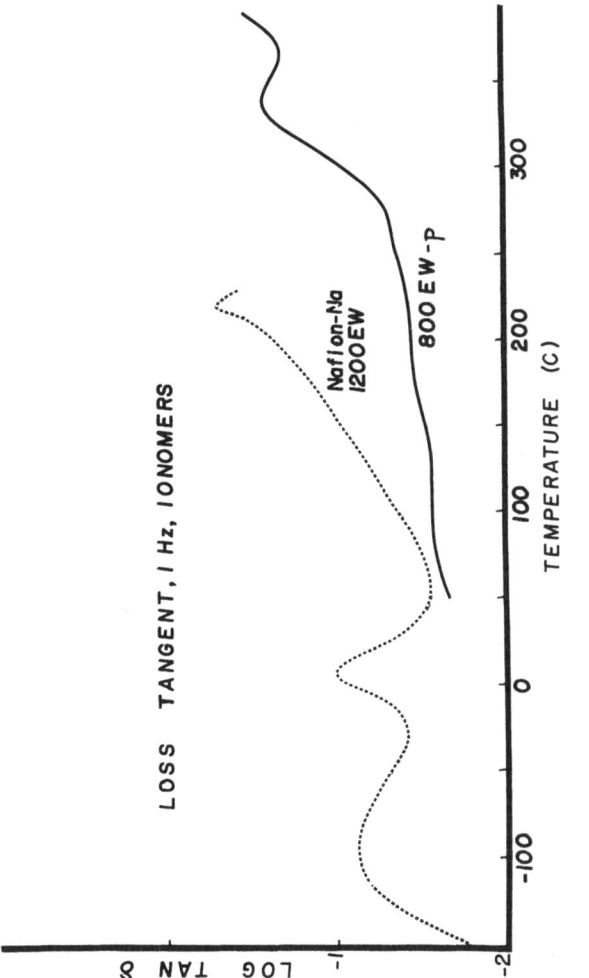

Figure 2. Loss tangent as a function of temperature for molded Nafion and 800EW ionomers.

evitably occurs at molding conditions of temperature and
pressure. The samples analyzed by DSC were prepared from
solution and are thus undegraded.

Vibrating reed experiments (ca. 500 Hz) were performed
on the molded, bulk ester samples. The loss tangent showed
a maximum at ca. -60°C in both 600EW ester and 800EW ester.
This is probably due to pendant phenyl group vibrations as
assigned in earlier work on 1200EW material (10).

Stress-Strain and Crystallinity

Stress-strain measurements were performed on the quasi-
dry membranes (i.e. containing a water content in equilibrium
with ambient humidity conditions), on an Instron model 1122
instrument. Tests were conducted on 1200EW-p and 800EW-p
films of ca. 100 μm thickness at varying temperatures. The
higher temperature was chosen to provide a distance below
the glass transition (i.e. an undercooling) comparable to
that of Nafion 1200 at room temperature. The results obtained
for modulus, break strain and tensile strength are shown in
Table II. The modulus of the aromatic ionomers is one order
of magnitude greater than that obtained for Nafion, reflecting the much greater stiffness of these systems. However,
Nafion exhibits a very much larger strain at break,
reflecting the fact that it is above its brittle-to-ductile
transition temperature (13) at 25°C. The 1200EW-p and 800EW-
p are below that point, the brittle-to-ductile transition
occurring at ca. 200° C for the 800EW-p. The tensile
strength of the three materials is similar.

Testing of the 600EW ester film at 25°C repeatedly
showed very large strains to break (>150%). This is illustrated in Fig. 3, with two ionomers and the non-ionic poly-

	T_{exp}	(T_g-T_{exp})	Modulus (MPa)	Tensile Strength (MPa)	Break Strain (%)
1200EW-p	100	180	1180	36	7
800EW-p	150	180	1170	25	13
Nafion 1200	25	200	200	36	185

Table II   Stress-strain data on dry polymers

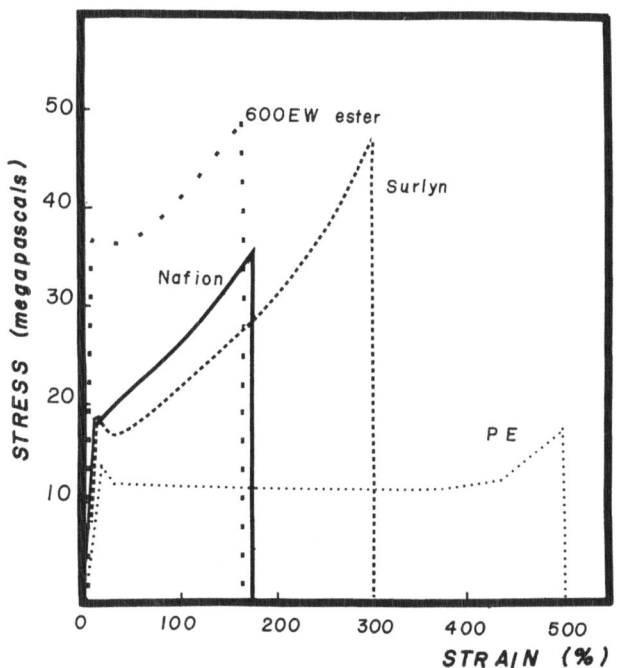

Figure 3. Stress strain curves comparing 600EW
ester, Surlyn-Na, Nafion-Na and
polyethylene at 25°C.

ethylene shown for comparison. A definite yield point,
followed by significant plastic elongation, can be observed.
The large elongation to break suggested the presence of
crystallinity in the sample. This was investigated further
by wide angle X-ray diffraction of 600EW ester, strained and
unstrained, as well as of 600EW-f. Microdensitometer traces
of the resultant exposures are shown in Fig. 4 and clearly
indicate crystallinity, enhanced in the strained ester sample.

Stress Relaxation

Stress relaxation experiments performed in the dry state
on the aromatic ionomer films were complicated by the hydrophilic nature of the materials. The curves obtained to date
are characteristic of the water plasticized system rather
than the dry material. More reliably, experiments were

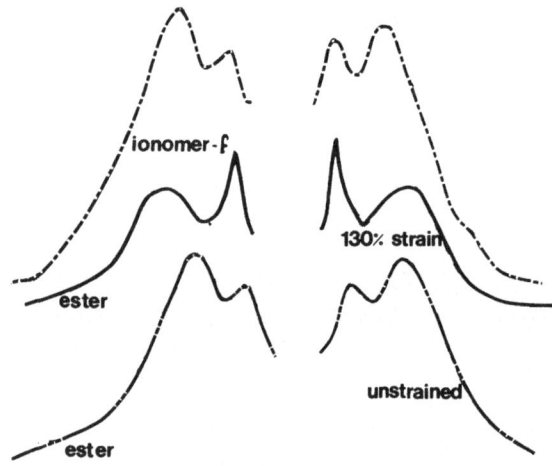

Figure 4. X-ray diffraction curves for strained and
           unstrained 600EW ester samples and for
           600 EW ionomer.

conducted under water (12,13) at temperatures ranging between 20 and 80°C. The results obtained for Nafion (12),
1200EW-p and 600EW-f (13) are shown in Fig. 5 as isochronal
modulus-temperature plots. Again, the modulus of Nafion
appears lower than that of the aromatic ionomers. This is
explained by the fact that, in addition to the plasticizing
effect of ca. 10% water absorption, Nafion is much closer to
its glass transition temperature than either of the other
two systems. The aromatic 1200EW-p was shown to absorb much
less water than 600EW-f, this being the reason for the lower
modulus of the latter.

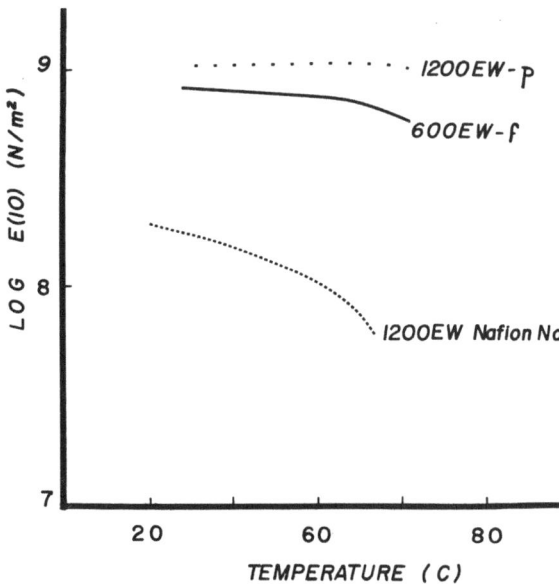

Figure 5.   Isochronal underwater stress relaxation
            behaviour of 1200EW-p, 600EW-f and Nafion
            1200 ionomers.

Differential Scanning Calorimetry

The types of bonding for the water absorbed by the membranes was studied more extensively by DSC (9). As would be expected from their different structures, porous. films absorb much more water than homogeneous films. In the porous film, the water appears in well-defined distributions when examined in the freezing temperature range (-50 to 20°C). The freezing of the water held in the membrane pores occurs in distinct distributions, suggesting varying sizes of pores or of surfaces. The remaining water, held by the ion pairs, does not freeze and is calculated by difference. In homogeneous film, all the water appears to be associated with the ionic groups.

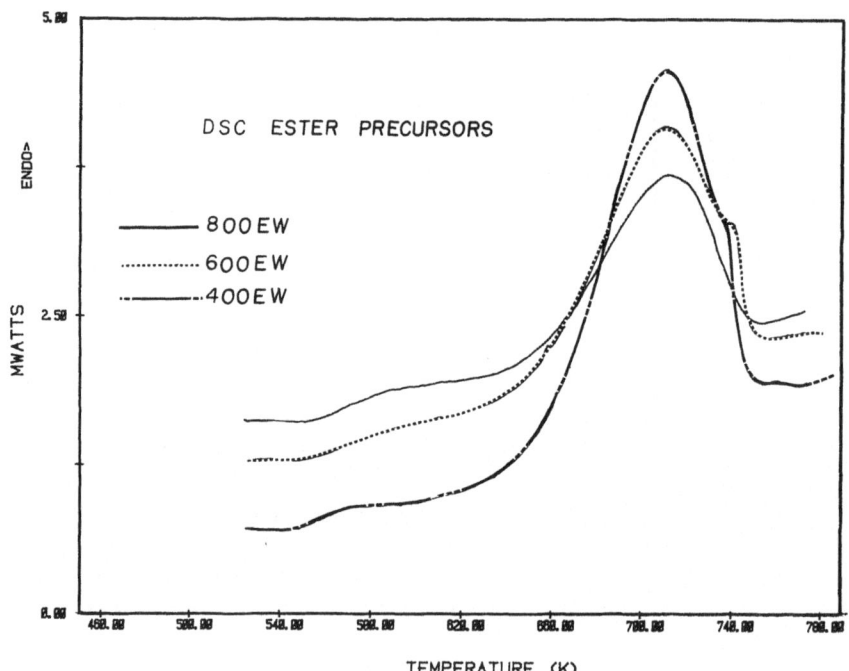

Figure 6.   Differential scanning calorimetry data for the ester precursors.

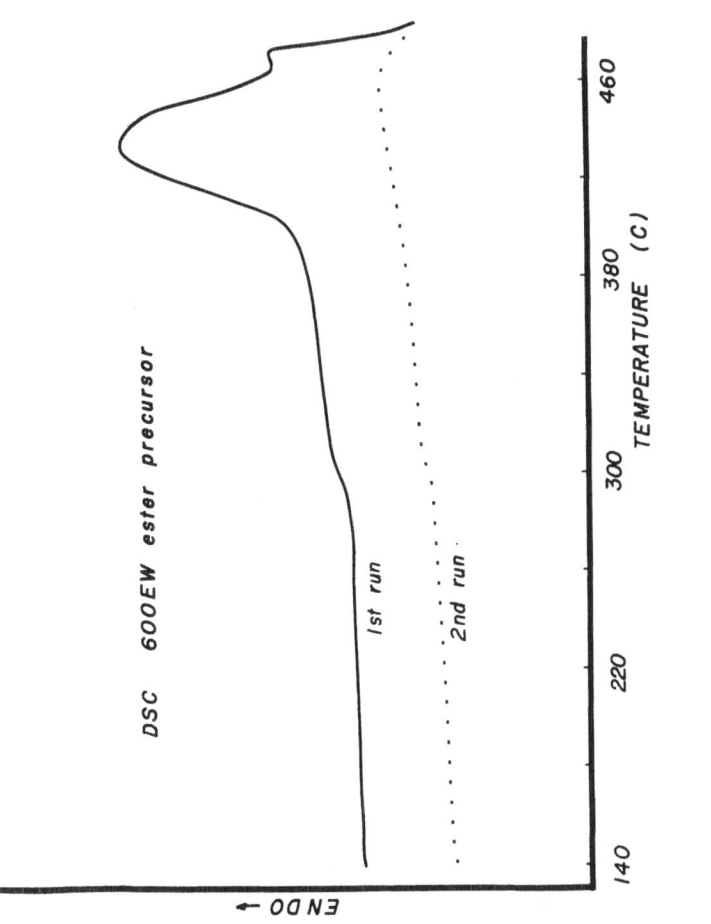

Figure 7. DSC data obtained for solution-cast 600EW ester (1st run) and for quenched sample (2nd run).

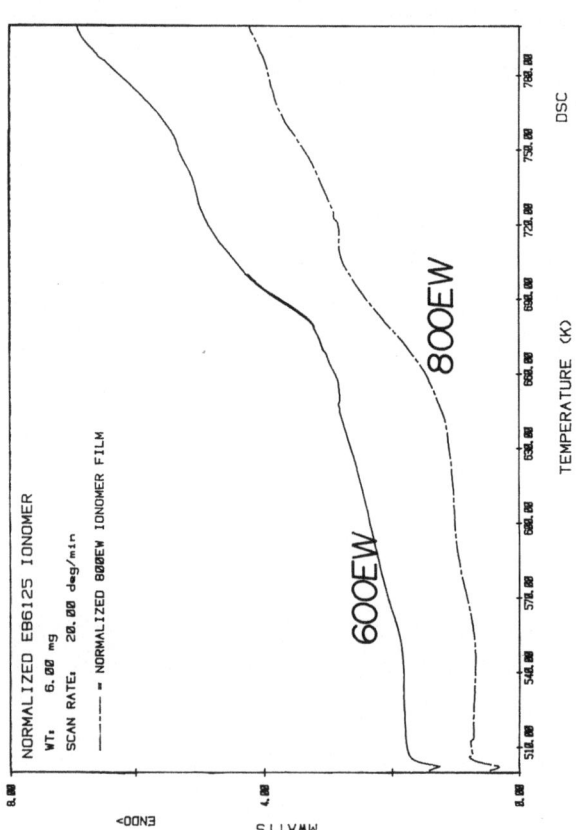

Figure 8.   DSC data obtained for 800EW and 600EW ionomers.

High temperature DSC studies were used to determine the glass transition temperature. The ester precursors display almost identical T_g's at ca. 290°C, as shown in Fig. 6. The endotherm peaking at 440°C is due to decarboxylation of the material. This is corroborated by thermogravimetric analysis and by the fact that the enthalpy of transition increases with increasing ionic content. A second run shows the disappearance of this high temperature peak (Fig. 7). The ionomers show a glass transition at temperatures which increase with increasing ionic content (Fig. 8). These are summarized in Table I. A minor transition occurring at ca. 290°C is probably due to residual ester groups.

Surface Wettability

Both the Nafion and aromatic ionomer systems show an affinity for water. However, they differ significantly in one respect. An important feature of the aromatic ionomers, particularly in view of their possible use in aqueous environments, is their excellent surface wettability. Although highly speculative, one possible interpretation is the following. On comparing the chemical structures of Nafion and these systems, one sees a large difference in the distance from backbone to ionic groups. In Nafion, the ions are attached to long hydrophobic sidechains, making the migration of these sidechains towards the interior of the membrane a reasonable pathway for the reduction of surface free energy. In the aromatic ionomers, the ionic groups are attached directly to both sides of the backbone phenylene groups, making their presence at the membrane surface much more likely and thereby increasing wettability. Although the membranes compared here vary in ionic content, Nafion 1200 is approximately twice as dense as the aromatic ionomer (2.1 vs. ~1.3), indicating comparable ionic density.

SOLUTION PROPERTIES OF THE HYDROLYZED IONOMER FILMS

Three types of experiments have been employed to evaluate these ionomer films in solution environments; sorption of water and electrolyte, ionic self-diffusion, and electrical conductance measurements.

Sorption Studies

The ion exchange capacities and water sorption properties of 400EW and 600EW polymer films are listed in Table

Table III Carboxylate Polymer, Na^+ form, 0.1 $\underline{M}$ NaOH, 25°C

Nominal Equivalent wt.	$\dfrac{mmol\ -CO_2^-}{g\ dry\ polymer}$	Weight % H_2O	$\dfrac{mmol\ -CO_2^-}{mL\ wet\ polymer}$
400	2.44	66	1.10
600	1.57	32	1.25

III. The ion exchange capacities (mmol exchange site per g dry polymer), were determined from the Na^+ content of the polymer using ^{22}Na radio-tracer in dilute solution. These values correspond well to those calculated from the nominal equivalent weights. It is seen that these polymer films sorb very large amounts of water, which suggests possible microporosity. Because the 400EW film sorbs far more water than the 600EW film, the volume concentration of exchange sites is actually smaller for the lower equivalent weight material.

Sorption characteristics for these films in concentrated NaOH solution at 80°C are listed in Table IV. The concentration of Na^+ is seen to rise abruptly in these environments, due to the sorption of electrolyte solution. This electrolyte sorption is also reflected in the mole ratio of water to sodium ion in the polymers. This ratio is quite generally found to be smaller in the polymer phase than in solution, even in cases where a large fraction of sodium ion content is due to electrolyte sorption (14). This can be ascribed to the lower dielectric constant and less aqueous character of the polymer phase. However, in Table IV it is seen that the ratios closely approach those of the solution phase for these materials. This also suggests that microporosity is a prominent feature of these aromatic carboxylate films, as reflected in the highly solution-like environment which these ratios indicate.

Sodium Ion Self-Diffusion Measurements

The measurement of cation self-diffusion coefficients in the polymer phase can provide significant information

Table IV   Carboxylate Polymer, Na^+ form, NaOH Solution, 80°C

Equiv. Weight	$\underline{M}$ NaOH	$\dfrac{\text{mmol } Na^+}{\text{mL polymer}}$	mmol H_2O/mmol Na^+	
			polymer	solution
400	5.0	3.08	9.2	10.8
	11.0	5.88	4.5	4.6
600	5.0	2.14	7.6	10.8
	11.0	4.10	4.2	4.6

about the nature of the ionic transport mechanism. Sodium ion self-diffusion coefficients were measured here as a function of equivalent weight, temperature, and solution concentration using radiotracer techniques described previously (15). Results are shown in Figures 9 and 10, in which Arrhenius plots of the logarithm of D_{Na}^+ is plotted vs the reciprocal of absolute temperature.

The dilute solution, 0-40°C values in Figure 9 show relatively large diffusion coefficients for the 400 and 600 equivalent weight polymer films. Also shown are values for sodium ion diffusion in water and in a perfluorinated carboxylate polymer membrane (16). The latter material is an ion clustered polymer with excellent ionic diffusion properties, although it is not microporous. These aromatic carboxylates are not capable of the ion clustering morphologies which generate large ionic diffusion coefficients, and thus microporosity is again indicated as an underlying reason. Activation energies of diffusion, calculated from the slopes of the lines in Figure 9, are 26.5, 19.6, and 22.4 kJ mol^{-1} for the 400EW, 600EW, and perfluorocarboxylate polymers respectively. All of these values are only slightly larger than that for sodium ion diffusion in pure water in this temperature range, 19.1 kJ mol^{-1}. Thus a solution-like diffusion mechanism is inferred.
This behavior is repeated for concentrated NaOH solution environments in the 70-90°C temperature range as well, as shown in Figure 10. Again, large diffusion coefficients result compared to the nonporous membrane and activation energies remain in the 20 ± 5 kJ mol^{-1} region. The large drop in diffusion coefficients which is seen between 5 M

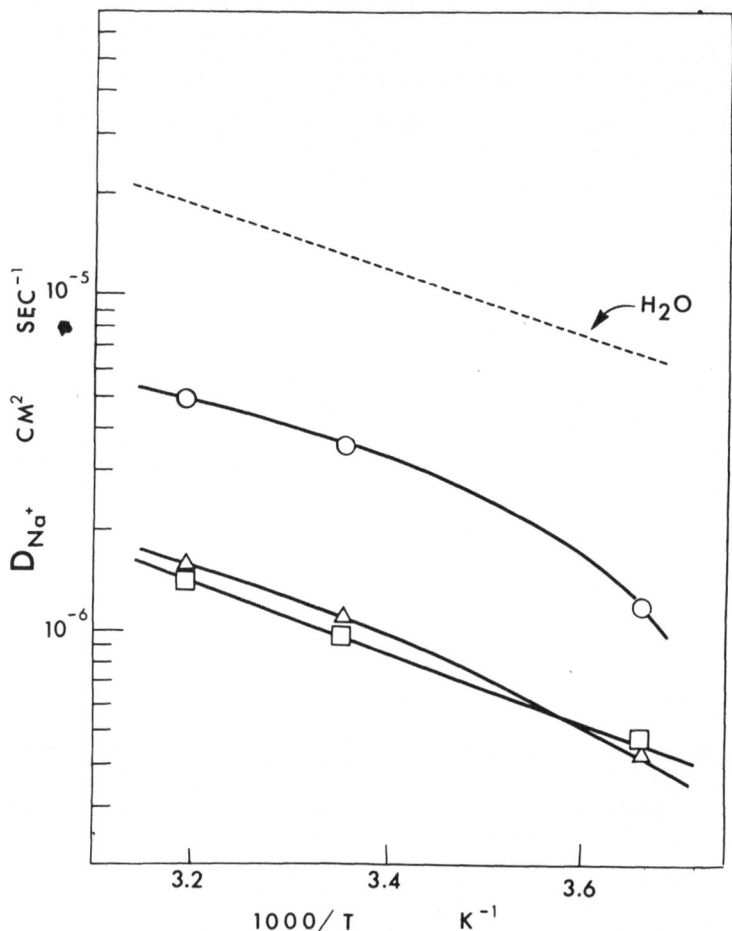

Figure 9.   Arrhenius plots of sodium ion diffusion
in aromatic carboxylate polymer films,
0.1 $\underline{M}$ NaOH solution. ◯- 400EW, ☐-
600E$\overline{W}$, △- perfluorinated carboxylate
membrane.

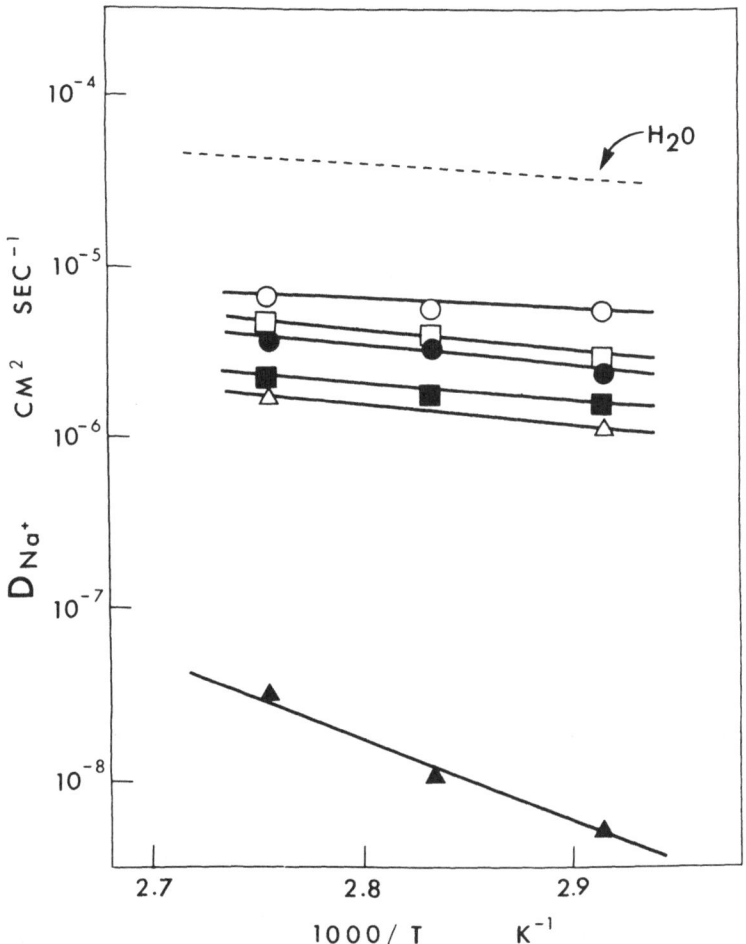

Figure 10. Arrhenius plots of sodium ion diffusion in
aromatic carboxylate polymer films; open
symbols, 5 M NaOH, closed symbols 11 M NaOH
solution. O,● - 400EW; □,■ - 600EW: Δ,▲-
perfluorinated carboxylate membrane.

and 11 M NaOH for the perfluorocarboxylate polymers can be attributed to the effect of dehydration on the diffusion mechanism in this homogeneous film (14). This effect is much less pronounced for the microporous aromatic carboxylate films.

Electrical Conductance Measurements

The results of conductance measurements in 10 M NaOH at a membrane current density of 5 kA m^{-2} are shown in Figure 11. Values are also included for the solution and for typical asbestos diaphragms at 100°C, and for a homogeneous perfluorosulfonate Nafion film. These results show that the specific conductances of the aromatic carboxylate films are more like a porous separator such as asbestos than a nonporous membrane film such as Nafion. In addition, conductance generally increases with decreasing equivalent weight. This may be due to an increase in generated microporosity with increasing ester content when these materials are hydrolyzed into the ionomer form.

The decrease in specific conductance with increasing temperature for the 400EW film is rather unusual. These measurements were repeated with a second film of identical history. Results are shown in Figure 12. The closed symbols represent original measurements which were made in order of increasing temperature. The open symbols refer to a second series of experiments which were performed immediately after the first series. It appears that some sort of annealing effect has taken place at the highest temperature, so that values at lower temperatures are subsequently lower. The result is that a cycled 400EW film performs in a very similar fashion to 600EW films.

The dependence of specific conductance on current density at 160°C is shown in Figure 13. These films show ohmic behavior, even at a current density as high as 10 kA m^{-2}. In contrast, the nonporous Nafion film shows an increase in specific conductance, perhaps due to a morphological change caused by the effects of large amounts of electroosmotically transported water. Such effects would be expected to be minimal for a microporous material.

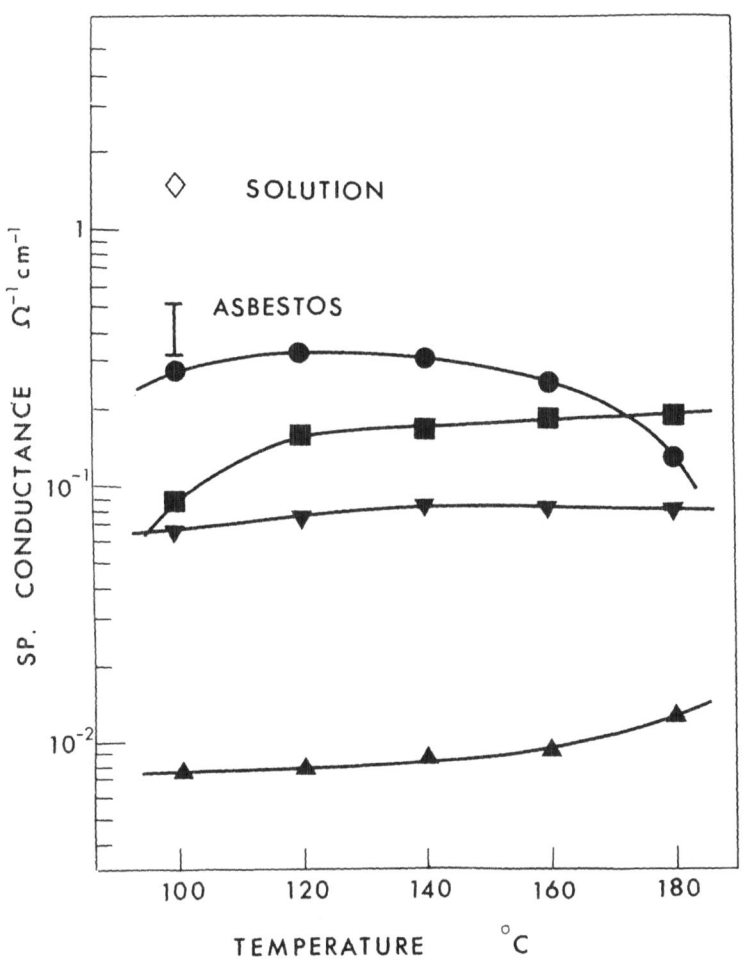

Figure 11. Specific conductance of polymer films in 10 M
NaOH, 5.0 kA m^{-2}. ● - 400EW, ■-600EW, and ▼-
800EW aromatic carboxylate; ▲1150EW Nafion
perfluorinated sulfonate membrane.

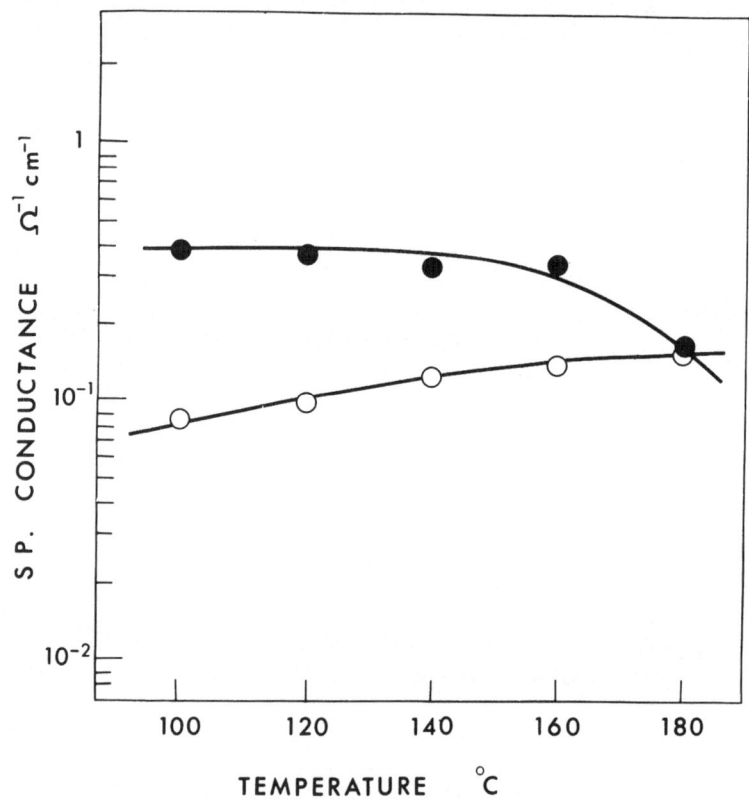

Figure 12.   Specific conductance of 400EW aromatic carboxylate film in 10 M NaOH, 5.0 kA m². ● , first
series of experiments; ○, second series of
experiments.

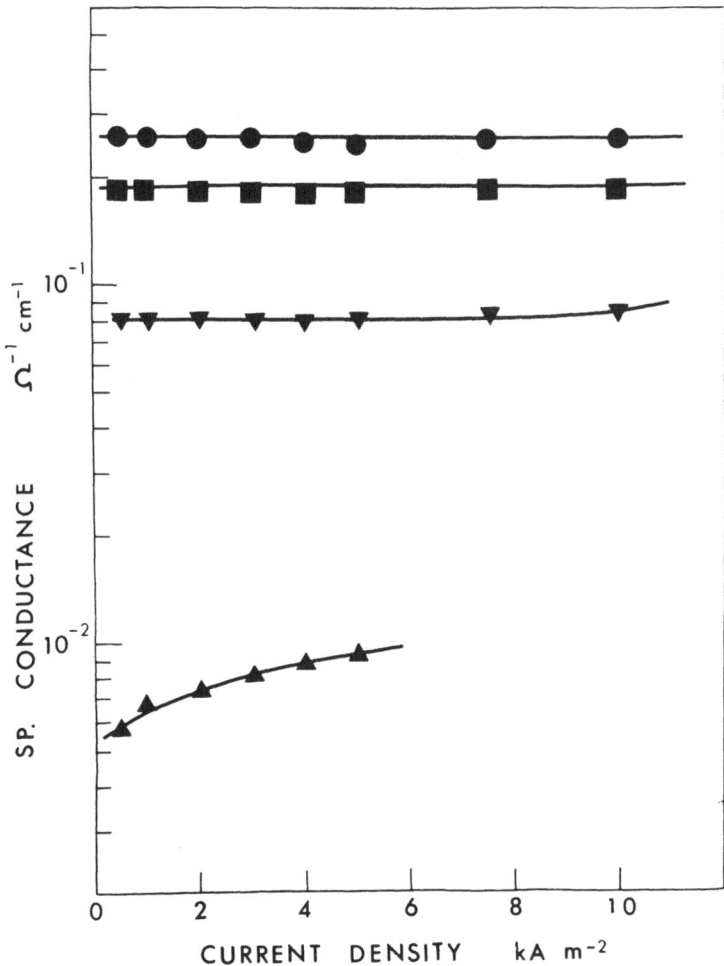

Figure 13. Specific conductance of polymer films in 10 M
NaOH at 160°C. Symbols are the same as in
Figure 11.

References

1. J.B. Taylor, Chemistry in Canada, February, 1980.
2. M. Hammerli, "The Potential Role of Electrolytic Hydrogen in Canada", Atomic Energy of Canada, Ltd., Publication No. AECL-7676, Chalk River, Ontario, 1982.
3. B.V. Tilak, P.W.T. Lu, J.E. Colman, and S. Srinivasan, in "Comprehensive Treatise of Electrochemistry" Vol. V, J. O'M. Bockris, B.E. Conway, E. Yeager and R.E. White, Eds. Plenum Press, New York, 1982, Chapter 1.
4. L.J. Nuttall, in "Hydrogen: Production and Marketing", A.C.S. Symposium Series 116, M.W. Smith and J.G. Santagelo, Eds., American Chemical Society, Washington, D.C. (1980), Chapter 10.
5. F.W. Harris, B.A. Reinhardt, and R.D. Case, Polymer Preprints, 19(2), 394 (1978).
6. F.W. Harris, B.A. Reinhardt, R.D. Case, S.M. Padaki, V. Sudarsanian, and W.A. Feld, "Synthesis and Modification of Carboxylated Polyphenylenes and Phenylated Polyimides", Technical Report AFML-TR-76-9, Air Force Materials Laboratory, WPAFB, March 1976.
7. F.W. Harris and B.A. Reinhardt, Polymer Preprints, 15(1) 691 (1974).
8. J.K. Stille, H. Mukamal, and F.W. Harris, J. Polym. Sci., Part A1, 5, 2721 (1967).
9. R. Legras, A. Eisenberg et al., to be published.
10. M. Rigdahl, A.B. Reinhardt, F.W. Harris and A. Eisenberg, Macromolecules, 14, 851 (1980).
11. E.H. Andrews, Fracture in Polymers, Oliver & Boyd, London (1968).
12. T. Kyu and A. Eisenberg, J. Polym. Sci., in press.
13. K. Yamada and A. Eisenberg, to be published.
14. Z. Twardowski, H.L. Yeager and B. O'Dell, J. Electrochem. Soc. 129, 328 (1982).
15. H.L. Yeager, B. Kipling and R.L. Dotson, J. Electrochem. Soc. 127, 303 (1980).
16. H.L. Yeager, Z. Twardowski and L.M. Clarke, J. Electrochem. Soc. 129, 324 (1982).

SOME RECENT ASPECTS OF POLYMER FLAMMABILITY

Eli M. Pearce

Polymer Research Institute and Chemistry Dept.
Polytechnic Institute of New York
333 Jay Street
Brooklyn, N.Y. 11201

SYNOPSIS

There are various stages in the pyrolytic degradation and combustion of volatile products leading to the flammability behavior of polymers. Mechanisms to decrease flammability involve modifying condensed phase or volatile phase reactions. The reported studies relate our recent progress in the understanding and controll of these mechanisms.

Some general known relationship between flammability and polymer structure are reviewed. Our studies on relating polymer end groups and molecular weights to flammability indicated that the known thermal degradation mechanisms for nylon 6 and polyethylene terephthalate (PET) are, in part, related to their flammability behavior.

The question of the flammability behavior of similar flame retarding structures when used as additives or as comonomers in PET is discussed. For the case of structures related to tetrabromo-bisphenol-A, there was little difference, but for those containing triphenylphosphine oxide related structures a switch from volatile phase to condensed phase mechanisms was possible.

Efforts in regard to understanding char formation mechanisms and improvements in the amount of char have been studied for systems such as polystyrene, cardopolymers, phenol-formaldehyde resins, aromatic polyamides and styrylpyridine based polymers. Increased char formation occurred when thermally stable thermally induced crosslinking structures occurred and/or thermally stable aromatic rings were produced.

Crosslinking could be induced in polystyrene structures containing vinylbenzyl chloride as a comonomer which enhanced char formation. For the cardopolymers, the phenolphthalein based polymers were shown to have increased char formation because of the rearrangement of the lactide group to a thermally stable ester crosslink. Although other cardopolymers showed/improvements in flammability, the mechanisms were not explored. Vapor phase and condensed phase mechanisms were applicable for substituted phenolic resins but mechanisms for the latter were not elucidated.

Specific halogen substituted aromatic polyamides have given substantial increases in char formation due to the formation, in part, of thermally stable benzoxazole units. Another system in which ring formation accounts for increased char formation were those which contained styrylpyridine units. In this case a Diels Alder addition reaction could account for these results.

INTRODUCTION

Factors associated with polymer flammability have been previously reviewed.[1,2,3,4] Stages in the polymer flammability process can be simply described as a) preheating, b) decomposition to give volatile compounds and residuals, 3) ignition of the volatiles in the presence of oxygen, and d) combustion. This may be considered as a sustained cyclic process since the combustion process itself is exothermic. Mechanisms to decrease the flammability of polymer have involved the interruption of the cyclic process at any stage.

Recent research by us has elucidated a number of basic mechanistic concepts and related chemistry important for the understanding of approaches to improved flammability. This will be the subject of this paper.

SOME FLAMMABILITY-STRUCTURE RELATIONSHIPS

The Oxygen Index Test can be used as a relative laboratory method for studying one aspect of polymer flammability -- a parameter related to extinguishment. This method developed by Fenimore and Martin[5] is widely used. In this method, the minimum volume concentration of oxygen in an oxygen/nitrogen mixture capable of sustaining candle-like burning of the polymer is determined. Relatively rankings of polymers can be made as long as certain limitations are recognized. The test method has measured only one aspect of the complex phenomena associated with polymer flammability; the absolute values were dependent on the shape and dimensions of the test object (e.g. fibers and films give lower values than molded plastics) although relative to each

other the results have significance; certain thermoplastic formu-
lations were capable of showing melt flow away from the flame
front and thus could give somewhat higher and erroneous oxygen
index values.[6]

Some General Relationships

        Polymers which can be largely degraded at relatively low
temperature (eg. less than 400°) to give monomer, dimer or other
low molecular weight combustible volatiles gave low oxygen index
values. Thus many of the important commodity polymers such as
polyethylene, polypropylene, polyformaldehyde, polystyrene and
polymethyl methacrylate will have oxygen indices of less than 21.

        Polymers containing aromaticity in the backbone usually had
improved oxygen indices, -- for example meta-phenylene isophthalamide (OI=32) vs. nylon 6,6 (OI=23).

        Polymers containing halogen or phosphorus also showed
increased oxygen indices, -- for example polyvinyl chloride
(OI=45) vs. polyethylene (OI=18). Of course, the principle of
intentionally adding halogen or phosphorus compounds to polymers
is one of the main approaches to flame retarding polymers.

        If the polymer degraded at elevated temperature to monomer
and the monomer is "non-flammable", a high oxygen index was
obtained, -- for example, polytetrafluoroethylene (OI=95).

Relationships to Degradation Mechanisms

        Few studies have been attempted to systematically relate
oxygen index values to known degradation mechanisms. We have
attempted to study two cases, nylon 6 and poly (ethylene
terephalate) (PET).

Nylon 6[6]

        We have shown that the oxygen index of nylon 6 can be related
to its amine and carboxylic acid end group concentrations and to
its molecular weight by the empirical relationship-

$$(OI)_{N6} = 13.5(1 + \frac{1}{\eta}) + \frac{2050}{13.5a+ac}$$

where (a) and (c) were a function of the amine and carboxylic acid
end groups, respectively, and η was the reduced viscosity in
m-cresol. It has been well known that the thermal degradation
of nylon 6 occurred primarily by an amide "back-biting" reaction
to generate caprolactam and that this reaction was carboxylic
acid catalyzed. Thus, according to the primary degradation

mechanism, higher amine and/or carboxylic acid end groups should give lower oxygen index values, and this was quantitatively obtained by use of the empirical equation. A correction must be made due to the melt flow which occurred away from the flame front thereby removing a potential source of fuel. The melt flow should be related to the molecular weight of the polymer (in our case, to the reduced viscosity) and thus, experimentally, the higher the molecular weight, the lower the oxygen index.

Poly(ethylene terephalate) (PET)[7]

Similar studies to that of nylon 6 were attempted for PET. We attempted to correlate variations in the PET structure with end-group concentration, molecular weight and diethylene glycol content. The only relationship that could be found was that the log of the oxygen index was proportional to the log of 1/[COOH]. Since it was known that a major step in PET thermal degradation was hydrolytic and that this was carboxylic acid catalyzed, this relationship seemed plausible.

There also were many studies reported on the flame retardation of PET by incorporation of halogen or phosphorus[8]. Similarly, we have studied a combination of flame retardants, hexabromobenzene and triphenyl phosphate, in PET. The luster and color of the spun PET fiber was preserved and had a fabric oxygen index of 28 vs. 21 for unmodified PET[9]. The retention of luster was unusual since most flame retardants were incompatible and thus opacify the system.

For the case of PET flame retardation, we were interested in determining whether certain flame retardant structures functioned differently when the structure was incorporated as an additive or when it was incorporated as a comonomer into the backbone of the chain.

Tetrabromobisphenol-A was incorporated as an additive and its bishydroxyethyl ether was incorporated as a comonomer at the 4-8 mole % bromine basis[7]. The oxygen indices were somewhat higher for the additive approach (e.g. at 6 mole % bromide, OI of 17 vs. 22 on pellet samples) probably due the additive flame retardants ability to diffuse more rapidly from the polymer matrix and thus volatilize, thereby moderating the gaseous combustion process.

Triphenylphosphine oxide as an additive to PET was compared to an analogous structure, phenylphosphonyldi (p-methyl benzoate), used as an additive and as a comonomer at the 5 mole% phosphorus level. The PET control had an OI (pellet) of 17 vs. 25 for $\phi_3P=0$ additive, and 29 for phenylphosphonyldi (p-methyl benzoate) used either as an additive or as a comonomer. Thermogravimetric analysis indicated that $\Phi_3 P=O$ volatilized completely and served as a vapor phase flame retardant whereas the pheynylphosphonyldi (p-methyl benzoate) as an additive or comonomer showed an increased residue at 500° and thus indicated condensed phase activity. The similarities between this additive and comonomer results suggested that ester interchange probably occurred during elevated temperature studies and that the additive was probably incorporated as a comonomer.

The results on both the bromine and phosphorus examples indicated that differences in flame retardation could occur for similar flame retardant structures and were dependent on whether they were used as additives or comonomers.

RELATIONSHIPS BETWEEN POLYMER STRUCTURE AND CHAR FORMATION

Among the mechanisms available for polymers with improved flammability behavior are those systems which showed an increased propensity for char formation which could in many cases, be related to polymers with improved thermal stability. The structural features leading to improved thermal stability have been summarized by Wright[10], and include structures having strong chemical bonds and maximized resonance stabilization -- characteristics associated with polymers containing aromatic functionality as a major component in the polymer chain backbone. Laboratory scale evaluation of char propensity has been previously reviewed[2]. Parker, Fohlen and Sawko[11] have related char formation to polymer aromaticity. Char formation is usually determined from the weight residue at 700°C by dynamic thermogravimetric analysis under N_2 since this measured the results from pyrolytic degradation rather than thermoxidative degradation. Van Krevelan[12] has shown a linear correlation with TGA-char formation for a large number of char forming polymers.

Polystyrene

Polystyrene was a polymer which underwent facile chain scission at relatively low temperatures ($\sim$350°C) and gave predominently monomer and dimer.

It occurred to us that improved flame retardation should be obtainable if thermally stable crosslinks between the aromatic rings could be obtained at high temperature. Model systems were designed using copolymers of |styrene and vinylbenzyl chloride in the presence of either Sb_2O_3 or ZnO.[13] The mechanism proposed was that Lewis Acids were formed from combination of chlorine with the inorganic oxide and this then promoted Friedel-Crafts alkylation reactions. The crosslinking occurred by reaction of the benzyl carbenium ion with another benzene ring. In practice, the results showed major improvement in both oxygen index and char formation (e.g. polystyrene - OI=19, char (500°C)=9% vs. copolymer containing 13.5% vinylbenzyl chloride - OI=49, char (500°C)=47%.) These results illustrated the advantage of forming high temperature stable crosslinks as a route for reducing flammability.

Cardopolymers

We have recently investigated the char propensity and correlation with oxygen index for a number of polymers containing cardo (looped functionality) units. Polymers with looped functional groups have been reviewed[14] and several of these have been mentioned by Korshak[15] and by Morgan[16].

Phenolphthalein Based Polymers

A number of phenolphthalein or phenolphthalein based polymers were investigated. These have included diglycidyl ethers[17,18] polycarbonates[17,19,20,21] and polyesters.[17,20]

The studies on the diglycidyl ethers of phenolphthalein (DGEPP) used a trimethoxyboroxine (TMB) cure at elevated temperature and were compared to the bisphenol-A based systems (DGEBA)[17,18]. The char residue at 700° for the cured DGEPP system was 47% vs. 26% for the DGEBA system and the OI's were 40 and 20, respectively. A linear relationship with % char or OI and copolymer (DGEPP + DGEBA) composition were obtained. Improvements in char formation using triethanolamine borate as the curing agent were also found.[22]

The phenolphthalein polycarbonate (char yield (800°)=54%, OI=38.0) also showed flammability improvement when compared to the bisphenol-A polycarbonate (char yield (800°)=21%, OI=26.5). Linear relationships were also obtained as a function of % char or OI vs. copolymer composition and could be used for predictions.[17,19]

Monomers containing various substituents on the phthalide ring of phenolphthalein could be easily prepared by condensation of the appropriate phthalic anhydride with two moles of phenol. The effect of these substituents in polycarbonates have been previously discussed.[20,21] A most interesting derivative was the polycarbonate based on 4,5,6,7-tetrabromophenolphthalein which had an oxygen index of about 75 and a char yield of 32%.[20,21] Study of its copolymers with bisphenol-A or phenolphthalein polycarbonate showed non-linear relationships with interesting enhancement of char and oxygen index at intermediate compositions.[17,21]

Similar results are obtained from phenolphthalein or its phthalide ring substituted monomers when converted to polyesters of isophthalic or terephthalic acid.[20]

These monomers could be readily converted to phthalimidines by reaction with NH_3 or aniline where the oxygen of the phthalide ring became either an -NH or -Nϕ, respectively.[20] No appreciable enhancement of char or oxygen index was noted when compared to the phthalide derivatives.

We studied the complex degradation of phenolphthalein polycarbonate at high temperature using various techniques and have concluded that a prime factor in increased char formation was related to the opening of the internal ester phthalide ring to thermally stable ester crosslinks between chains.[23]

Bisphenol - fluoreneone (BPF)

Fluoreneone could be condensed with two moles of phenol and gave BPF. BPF was converted to diglycidyl ethers[17,18], poly-carbonates[17,24] and polyisophthalates[17,24]

The diglycidyl ether of BPF was cured with TMP and the char yield studied for the homopolymer and copolymers with the diglycidyl ethers of DGEBA. The homopolymer gave a char yield at 700° of 42%[17,18]. A synergistic effect was noted in the copolymer systems with a maximum of close to 40% char obtained at about a 20 mole % composition of the fluoreneone based diglycidyl ether. This produced a useful reduced flammability system containing no halogen or phosphorus.

The diglycidyl ether of BPF was also studied using tri-ethoxyboroxine, triisopropoxyboroxine, and triphenoxyboroxine as curing agents and these also showed high char formation.[25]

Bisphenol-fluoreneone based polyisophthalates and polycarbonates were also prepared and showed expected high char formation, and these were compared to the tetramethyl substituted bisphenol-fluoreneone polymers.[17,24] The unsubstituted polyisophthalates gave higher char yields than the polycarbonates, (59% vs 49%) and the tetramethyl methyl substituted polymers gave much reduced char formations (22-34%).

Other Cardopolymer Systems

The following additional cardopolymer systems were evaluated as di- or tetraglycidyl ethers using trimethoxyboroxine (TMB) and diaminodiphenylsulfone (DDA) as curing agents[26] -

| | % char curing agent | |
epoxy based on –	TMB	DDS
3,6-dihydroxyspiro-[fluorene-9,9-xanthane]	50	41
10,10-bis (4-hydroxyphenyl)-anthrone	48	34
9,9,10,10-tetrakis –(4 hydroxyphenyl) anthracene	52	41

Phenolic – Derived Polymers

A number of phenol-formaldehyde resins were synthesized and studied in relation to their flammability characteristics.[26] The resins were cured by various processes - formaldehyde or s-trioxane under acidic conditions, formaldehyde under basic conditions, or by reaction with terephthaloyl chloride to give ester crosslinks. The oxygen index increased from 33 for unsubstituted phenolics to about 75 for meta-halogen substituted phenolics. The formaldehyde cured phenolic gave an OI of 75 compared to 50 for the trioxane cured phenolic and 40 for the terephthaloyl chloride cured phenolic. Copolymers with different weight percentages of halogen substituted phenols as resols showed large increases in oxygen index with only small changes in char. This indicated the presence of presently unexplained condensed phase reactions.

Aromatic Polyamides

A number of wholly aromatic polyamides based on substituted m and p- phenylene diamines and isophthaloyl and terephthaloyl chloride were prepared and evaluated for OI and char.[28] The effect of having chlorine substituents ortho to the amine function was unusual in that large increases in OI were obtained for the aramids with retention of high levels of char comparable to that of the unsubstituted aramids. Since all the chlorine was removed upon heating, a decrease in char yield was expected. An increased char yield indicated unexpected condensed phase reactions at high temperature.

In order to understand these systems, studies were initiated on the degradation and degradation products obtained from poly (1,3-phenylene isophthalamide), poly (chloro-2,4-phenylene isophthalamide) and their model compounds.[29,30,31] Among the key degradation reactions were the hydrolytic scission of the amide function for the unsubstituted and chlorosubstituted polymers, homolytic scission of the carbon chlorine bond followed by radical coupling, and the formation of benzoxazole structure from the chlorosubstituted polyamide (Equation 1)

(1)

HCl +

The proposed mechanism requires a tautomeric shift which would be prevented if the -NH group was converted to the -NCH₃. We have prepared such polyamides and it appeared that the benzoxazole did not form.[32] Studies have continued in comparing F, Cl and Br substituents and appropriate dihalo polymers. Additional studies have been done on the effects of electron withdrawing substituents on the thermal degradation of aromatic polyamides.[33]

Styrylpyridine-Based Polymers[34]

The influence of a pyridine group on the reactivity of the methyl group in 2-, 4-, 2,4-, and 2,6-pyridine has been evaluated by condensing the methyl derivatives of pyridine with aromatic aldehydes in the presence of acetic anhydride and acedic acid. The reactivity of these methyl derivatives has been explained on the basis of the resonance effect of the N-H bond pushing electrons into the ring. The ring would thus acquire a slight positive charge on the 2 and 4 positions, and readily facilitate the formation of a resonance carbanion from the methyl group

which, in turn, is responsibile for the condensation with aromatic aldehydes. The reactivity of methyl groups in different positions on the pyridine ring with an aromatic aldehyde can be related to the yield of the final product (styrylpyridine). Therefore, the order of reactivity for this reaction is 4-picoline>>2,6-lutidine>2,4-lutidine. The activation energy of the condensation reaction between different methyl pryidines and excess benzyldehyde, measured by ultraviolet spectrophotometry, is 34.14 K/J mole for 2-picoline and 20.38 kJ/mole for 4-picoline. The yield of 2-styrylpyridine increased with increasing reaction time and temperature. When the styryl substitution is on various positions of the pyridine ring, shift in the maximum absorption wavelength was not observed as compared to 2-styrylpyridine. However, the position of maximum absorption wavelength of the hydroxystyryl on pryidine was dramatically shifted as compared to styrylpyridine.

Glycidyl ethers of epoxy resins were prepared from 2,4-di
(p-hydroxystyryl) pyridine (2,4-DGESP), 2,6-di(p-hydroxystyryl)
pyridine (2,6-DGESP), and 2,4,6-tri(p-hydorxystyryl)-pyridine
(2,4,6-TGESP), in order to study the relationships between structure
and polymer degradation routes. To prepare a highly crosslinked
material with good thermal stability, trimethoxyboroxine (TMB) was
used as the curing agent. The resins changed color from light
yellow to orange after mixing with TMP, and to dark brown after
the curing cycle. The relative char yield of the three different
resins, as measured by TGA under N_2, were 2,4-DGESP = 2,6-DGESP
> 2,4,6-TGESP. The char yield of the cured 2,6-DGESP varied
slightly with different amounts of the TMB curing agent, and was
higher than that of the uncured 2,6-DGESP. The oxygen index
increased as a function of thermal curing time for the 2,6-DGESP
epoxy resin. An intermolecular Diels-Alder reaction with
2,6-DGESP was proposed as a primary reaction route during thermal
curing.

The styrylpyridine based polyarylates and their model compounds
were synthesized. The order of the relative char yield and oxygen
indices were polyarylates > p,p'-bis(β-2-vinylpyridyl)diphenyl
terephthalate (p,p'-BVPDPT) = p,p'-2,6-(β-2-vinylpyridyl)diphenyl
dibenzoate >(p,p'-2,6'-2,6-(β-2-vinylpyridyl)phenyl benzoate
(p-VPPB). This is also probably due to the Diels-Alder reaction
of the C=C double bonds present in the main chain with the
pryidine group. The result is a highly crosslinked C-C bond char.
The styrylpyridine based model compounds of the ester and
carbonate underwent the Photo-Fries arrangement when irradiated
with UV light to give o-hydroxybenzophenone related structures.
In addition to undergoing the Photo Fries rearrangement, it was
observed tht the less sterically hindered ester model compounds
also underwent dimerization and isomerization reactions.

s-Triazine polymers of di-(p-cyanostyryl)pyridine were
prepared from (2,4-D(p-CN)SP) and 2,6-di(p-cyanostyryl)-pyridine
(2,6-D(p-CN)SP) to study the relationships between structure and
polymer heat resistance. The relative char yields of the polymers
and model compounds as measured by TGA, were: polymer of 2,4-di(p-
cyanostyryl)pyridine (2,4-SPT)≅ polymer of 2,6-di-(p-cyanostyryl)
pyridine (2,6-SPT)>2,4-D(p-CN)SP-2,6-D(p-CN)SP > 2(p-CN)SP. For
the three styrylpyridine prepolymers (PSP), there was no particular
difference in char yield and oxygen index, because of their similarity in chemical composition, structure, and an intermolecular
Diels-Alder reaction occurring during thermal curing.

REFERENCES

1. E. M. Pearce and R. Liepins, Environ. Health Perspect., 11, 59 (1975).

2. E. M. Pearce, Y. P. Khanna, and R. Raucher, Thermal Analysis in Polymer Flammability in Thermal Characterization of Polymeric Materials, edited by E. Turi, Academic Press, New York, 1981 pps. 793-843.

3. S. W. Shalaby and E. M. Pearce, Intern. J. Polymeric Mater., 3, 81 (1974).

4. E. M. Pearce, S. W. Shalaby, and R. H. Barker, in Flame Retardant Polymeric Materials, Volume I, M. Lewin, S. Atlas and E.M. Pearce, Eds., Plenum, New York (1975), pps. 239-90.

5. C. P. Fenimore and F. J. Martin, in Flame Retardant Polymeric Materials, Volume I, op. cit., p. 371.

6. H. K. Reimschusessel, S. W. Shalaby and E. M. Pearce, J. Fire and Flammability, 4, 299 (1973).

7. A. B. Desphande, E. M. Pearce, H. S. Yoon, and R. Liepins, J. Appl. Poly. Sci. Appl. Poly. Symp., 31, 257 (1977).

8. E. L. Lawton and C. J. Setzer, in Flame Retardant Polymeric Materials, Vol. I, op. cit., p. 193.

9. P. J. Koch, E. M. Pearce, J. A. Lapham, and S. W. Shalaby, J. Appl. Poly. Sci., 19, 227 (1975).

10. W. W. Wright, in Degradation and Stabilization of Polymers, G. Gueskens, ed., Halsted Press-Wiley, New York (1975), p. 43.

11. J. A. Parker, G. M. Fohlen and P.M. Sawko, "Development of Transparent Composites and Their Thermal Responses" paper presented at Conference on Transparent Aircraft Enclosures, Las Vegas, Nevada, Feb. 5-8, 1973.

12. D. W. van Krevelan, Chimia, 28, 504 (1974); Polymer, 16, 615 (1975).

13. Y. P. Khanna and E. M. Pearce, in Flame Retardant Polymeric Materials, Vol. II, op. cit.

14. S. V. Vinagradova and Y. S. Vygodskii, Russian Chem. Revs., 43 (7), 551 (1973).

15. V. V. Korshak, <u>The Chemical Structure and Thermal Characteristics of Polymers</u>, translated by J. Schmorak, Israel Program for Scientific Trnaslations, Jerusalem, 1971. pps. 10,14,16,75, 76,238,350,394.

16. P. W. Morgan, <u>Macromol.</u>, <u>3</u>, 536 (1970).

17. E. M. Pearce, S. C. Lin, M.S. Lin and S. M. Lee, in <u>Thermal Methods in Polymer Analysis</u>, W. W. Shalaby, ed., Franklin Institute Press, 1978, pp. 187-198.

18. S. C. Lin and E. M. Pearce, <u>J. Poly. Sci., Chem.</u>, <u>17</u>, 3095 (1979).

19. M. S. Lin and E. M. Pearce, <u>J. Poly. Sci., Chem.</u>, <u>19</u>, 2151 (1981).

20. M. S. Lin and E. M. Pearce, <u>J. Poly. Sci., Chem.</u>, <u>19</u>, 2659 (1981).

21. M. S. Lin, Exploratory Synthesis of New Fire Safe Polymeric Materials, Doctoral Dissertation, Polytechnic Institute of N.Y., June 1980 (Adviser: E. M. Pearce).

22. B. Arada, S. C. Lin, and E. M. Pearce, <u>Intern. J. Polymeric Mater.</u>, <u>7</u>, 167 (1979). Epoxy Resins IV.

23. M. S. Lin, B. J. Bulkin, and E. M. Pearce, <u>J. Poly. Sci., Chem.</u>, <u>19</u>, 2773 (1981).

24. J. Lo, Flammability and Photo-stability of Selected Polymer Systems, Doctoral Dissertation, Polytechnic Institute of N.Y., June, 1981 (Adviser: E. M. Pearce).

25. C. S. Chen, B. J. Bulkin, and E.M. Pearce, <u>J. Appl. Poly. Sci.</u>, <u>27</u>, 1177 (1982).

26. C. S. Chen, B. J. Bulkin and E. M. Pearce, <u>J. Appl. Poly. Sci.</u>, <u>27</u>, 3289 (1982).

27. Y. Zaks, J. Lo, D. Raucher and E. M. Pearce, <u>J. Appl. Poly. Sci.</u>, <u>27</u>, 913 (1982).

28. A. K. Chadhuri, B. Y. Min, and E. M. Pearce, <u>J. Poly. Sci., Chem.</u>, <u>18</u>, 2949(1980).

29. Y. P. Khanna, E. M. Pearce, B. D. Forman and D. A. Bini, <u>J. Poly. Sci., Chem.</u>, <u>19</u>, 2799 (1981).

30. Y. P. Khanna, E. M. Pearce, J. S. Smith, D. T. Burkitt, H. Njuguna, D. M. Hindenlang and B. D. Forman, J. Poly. Sci. Chem., 2817 (1981).

31. Y. P. Khanna and E. M. Pearce, J. Poly. Sci., Chem., 19, 2835 (1981).

32. A. C. Karydas and E. M. Pearce, unpublished results (to be published in 1983).

33. Y. P. Khanna and E. M. Pearce, J. Appl. Poly. Sci., 27, 2053 (1982).

34. H. Yan and E. M. Pearce, unpublished results (to be published in 1983).

ACKNOWLEDGEMENT

We acknowledge the support through the years from various organizations, the National Bureau of Standards, the Army Research Office, and the NASA-Ames Research Center.

AUTHOR INDEX

SUBJECT INDEX

417